THE EVOLUTIONARY SYNTHESIS

The Evolutionary Synthesis

Perspectives on the Unification of Biology

EDITED BY

Ernst Mayr and William B. Provine

Harvard University Press
Cambridge, Massachusetts, and London, England 1980

Library of Congress Cataloging in Publication Data
Main entry under title:

The Evolutionary synthesis.

　　Bibliography:　p.
　　Includes index.
　　1.　Evolution.　2.　Evolution—History.　I.　Mayr,
Ernst, 1904-　　II.　Provine, William B.
QH366.2.E87　　575　　80-13973
ISBN 0-674-27225-0

Contents

Preface

The theory that the diversity of life (plants and animals) is the product of evolution by common descent was almost universally accepted by biologists soon after the publication of *On the Origin of Species* (1859). Although Darwin's specific explanation—gradual evolution by natural selection—was immediately adopted by Wallace, Hooker, Gray, Bates, Poulton, and Weismann, it was rejected by most biologists and bitterly attacked by many others. It seemed to make some temporary headway in the 1870s, but lost ground again in the 1880s and 1890s and almost received a fatal setback through the mutationist theories of the early Mendelians. In the first decades of the twentieth century the schools of saltationism, orthogenesis, and neo-Lamarckism had decidedly more followers than selectionism. Indeed, only a handful of authors between 1900 and 1920 could be designated as pure selectionists. On the whole—and admittedly this is an oversimplification—two camps were recognizable, the geneticists and the naturalists-systematists. They spoke different languages; their attempts in joint meetings to come to an agreement were unsuccessful. In the early 1930s, despite all that had been learned in the preceding seventy years, the level of disagreement among the different camps of biology seemed almost as great as in Darwin's days. And yet, within the short span of twelve years (1936-1947), the disagreements were almost suddenly cleared away and a seemingly new theory of evolution was synthesized from the valid components of the previously feuding theories. Huxley (1942) referred to this episode in the history of biology as the "evolutionary synthesis."

The historian would like to know what happened during these crucial years. What factors were responsible for the breaking down of the barriers between the separate camps? What important insights were contributed by the specialists of the various biological disciplines that constitute the field of evolutionary biology? What misunderstandings had to be removed? Why were the thirties and forties so favorable to the synthesis?

The Committee on the Recent History of Science and Technology of the American Academy of Arts and Sciences conceived the idea of organizing a conference composed of two workshops charged with the task of attempting to answer these questions. Happily, some of the architects of the synthesis were still alive and they were invited. Also invited were some of the leading evolutionists of the next generation, as well as a number of historians of biology and philosophers of science. J. Huxley, B. Rensch, and G. G. Simpson, unfortunately, were prevented from participation by illness or conflicting engagements. That the conference was none too early is sadly demonstrated by the subsequent death of three participants (Th. Dobzhansky, I. M. Lerner, and E. Boesiger) and two correspondents (J. Huxley and B. L. Astaurov).

The major objective of the conference was to elicit as much information as possible about any factor, scientific or otherwise, that had had a positive or negative influence on the occurrence of the synthesis. Some participants had prepared formal papers; others presented their views informally. All of them made major additional contributions in response to questions in the ensuing discussion periods. Alexander Weinstein, who had been invited as a former student of T. H. Morgan, provided particularly valuable information.

The entire conference was recorded on tape and transcribed. Most unfortunately, the machine went on strike during E. B. Ford's presentation, and it was impossible to restore the missing part. All participants were asked to edit the transcripts of their own discussions, while the overall supervision and coordination were in the hands of Ernst Mayr and William Provine. During this editorial process most of the discussions were consolidated with the major presentations, and material was eliminated that did not relate directly to the synthesis. Biographical essays were gathered in a separate section.

Following Ernst Mayr's prologue, we have divided the book into two major sections with a smaller concluding section. The first section contains analyses of the contributions of the various biological disciplines to the evolutionary synthesis. The evolutionary synthesis was, however, genuinely different in different countries. Because a historical understanding of the progress of the synthesis must encompass this diversity, the second major section is devoted to analyses of the evolutionary synthesis in different countries.

The two sections are closely interrelated. Nearly every essay in the second section deals with the contributions to the synthesis of several of the fields examined in the first section.

The third section contains a discussion of general interpretive issues in the evolutionary synthesis, Will Provine's epilogue, and the biographical essays.

We should note here that because of the variety of acceptable systems of transliteration from the Russian, we have made no attempt to impose consistency in this respect from one chapter to another; rather, we have permitted the individual contributors to follow the method of their choice.

ACKNOWLEDGMENTS

The workshops could not have functioned so successfully, without the efficiency and never failing helpfulness of John Voss, Patricia Flaherty, and Shirley Hazen, of the staff of the American Academy. Our special thanks are due Alexandra Oleson, conference coordinator.

We are deeply grateful to William Kimler for his preparation of the detailed index to this book.

The conference was made possible by National Science Foundation grant GS-42934 to the American Academy of Arts and Sciences.

E.M.
W.B.P.

THE EVOLUTIONARY SYNTHESIS

Prologue:
Some Thoughts on the History
of the Evolutionary Synthesis*

Ernst Mayr

The term "evolutionary synthesis" was introduced by Julian Huxley in *Evolution: The Modern Synthesis* (1942) to designate the general acceptance of two conclusions: gradual evolution can be explained in terms of small genetic changes ("mutations") and recombination, and the ordering of this genetic variation by natural selection; and the observed evolutionary phenomena, particularly macroevolutionary processes and speciation, can be explained in a manner that is consistent with the known genetic mechanisms. The objective of this conference is to examine the rapid changes in evolutionary biology that occurred in the period of the synthesis (from approximately 1936 to 1947), to reconstruct the sequence of events leading to the synthesis, and to identify the factors responsible for the preceding disagreements.

To meet these objectives, the conference faces a formidable task. It is extremely difficult to explain and reconstruct something that happened forty years ago. Memories of past events have become dim; the situation looked then, and probably still looks today, different to representatives of different specialized fields; and finally, the situation was different in each country and sometimes even in different scientific centers in the same country. To resolve these problems, we have invited some speakers to report on the relation between the synthesis and various disciplines of biology (such as systematics, genetics, paleontology, and botany), others to speak on its occurrence (or not) in various countries (USSR, Germany, France). We hope that this approach will illuminate the synthesis from many different angles.

Historiography of science must avoid two great dangers. Chauvinism exaggerates the importance of whatever field or country a given scientist represents and tends to belittle the contribution of others. Butterfield

---

*Dedicated to Bernhard Rensch, one of the architects of the evolutionary synthesis, on his eightieth birthday (January 21, 1980).

(1957) has called the second danger the "whiggishness" of science writing —that is, the application of the hindsight of modern understanding in the evaluation of past events, combined with a suppression of all inconvenient phenomena.

No one can entirely avoid either of these shortcomings; sometimes they even provoke illuminating controversy. Yet all of us must keep a careful watch for manifestations of both pitfalls so that we can correct misleading statements before still other inaccuracies are added to the all too rich repertory of myths in science.

The conference has five specific objectives:

(1) To define the concepts that were dominant in various fields of biology and in various countries in the period preceding the synthesis

(2) To identify misunderstandings and other factors that delayed a reaching of consensus

(3) To identify the respective contributions made by various individuals and various biological disciplines, such as genetics, cytology, systematics, and paleontology

(4) To determine the factors that induced some authors to resist the synthesis

(5) To determine how the actual synthesis was achieved.

My own task in opening the conference is to attempt to specify the principal problems posed by the synthesis. I shall try to describe the state of evolutionary biology as it existed in the 1920s and early 1930s. I shall attempt to specify the objections raised by the opponents of the Darwinian theory and to identify the source of these objections. Finally, I shall try to focus attention on the relative importance of the contributions made by various branches of biology and the specific role played by certain key figures.

Most of what has so far been written about developments in evolutionary biology in the 1920s and 1930s has been written from the viewpoint of genetics. My own treatment is clearly affected by the fact that I am a systematist by background. Inevitably my interpretation of many developments differs from that of a geneticist. Future historical research must determine which of opposing interpretations seems to represent the situation more accurately. My major purpose here is to tickle the memory of the participants and to give them an opportunity to elaborate on my comments or to correct them if my recollections or interpretations are faulty. Nothing would be worse for our purposes than to gloss over existing difficulties, discrepancies, and contradictions.

The Opposition to the Darwinian Theory

The Darwinian (selectionist)[1] interpretation of evolution is now so nearly universally accepted among biologists that the present generation of evolutionists can hardly comprehend the opposition that the theory of natural selection still encountered in the 1920s and 1930s. During that period, nearly all the major books on evolution, including those by Berg, Bertalanffy, Beurlen, Boeker, Goldschmidt, Robson, Robson and Richards, Schindewolf, and those of all the French evolutionists (such as Cuénot, Caullery, Vandel, Guyénot, and Rostand) were more or less antiselectionist. The general texts on zoology and botany, even when adopting Darwinism, usually gave a good deal of space to Lamarckism as a legitimate possible alternative (see chapter 9). Darwinism was even less popular among nonbiologists. The philosophers, in particular, were almost unanimously opposed until relatively recent years (Cassirer, 1950; Grene, 1959; Popper, 1972).

The Darwinians were fully aware of the continuing popularity of non-Darwinian evolutionary theories. As recently as 1932, T. H. Morgan found it advisable to use an entire chapter to refute the hypothesis of an inheritance of acquired characters. An extraordinary amount of space is likewise devoted to the refutation of anti-Darwinian arguments by Haldane (1932), in the various books of the synthesis (Dobzhansky, 1937; Huxley, 1942; Rensch, 1947; Simpson, 1944) and in the postsynthesis literature (see, for example, Fisher, 1954).

The very few books on evolution written by authors who were firm adherents to neo-Darwinism (such as Haldane, 1932) had various shortcomings. None of these was greater than their attempt to explain evolution in terms of changes in gene frequencies. This explanation left most nongeneticists thoroughly dissatisfied because events at the level of the gene did not at all explain the organismic phenomena studied by paleontologists, systematists, ecologists, and students of behavior.

A peculiar myth popular among the geneticists at that time illustrates their failure to understand the real meaning of the Darwinian theory. Many held that the acceptance of evolution by natural selection depended on the maturation of genetics. The facts do not support the validity of this claim. First, Darwinism together with all of its consequences

---

1. The term "Darwinism" in the following discussions refers to the theory that selection is the only direction-giving factor in evolution.

was accepted not only by Darwin, Hooker, Wallace, and Weismann, but also by Poulton, K. Jordan, and many naturalists in the nineteenth century before the birth of genetics. More important, Darwinism was rejected by three of the founders of Mendelism—Bateson, de Vries, and Johannsen—the first evolutionists who truly understood Mendelian inheritance. Nevertheless, genetics subsequently did make a decisive contribution to the synthesis, but it is only one of the multiple sources of the synthesis.

Different Schools of Evolutionism

The number of competing theories of evolution in vogue before the synthesis is quite bewildering. The frequently used dichotomy, Darwinism versus Lamarckism, is not very satisfactory because both labels usually lumped rather different theories. Evolutionary theories can be classified by numerous criteria; my own arrangement is only one of many possible ones. I have chosen two classifying criteria: whether the author was an essentialist or believed in the uniqueness of individuals (population thinking), and whether the author allowed only for hard inheritance or admitted also "soft" inheritance. By "soft" I mean whether the author believed that the genetic basis of characters could be modified either by direct induction by the environment, or by use and disuse, or by an intrinsic failure of constancy, and that this modified genotype was then transmitted to the next generation. Soft inheritance is usually referred to as a belief in an inheritance of acquired characters, but soft inheritance includes a broader range of phenomena. It is also sometimes called Lamarckism or neo-Lamarckism, even though Lamarck's own theory was only one subdivision in this group of theories. Using these two criteria, I have designed a discrimination grid (table 1) that separates the best-

Table 1 Criteria for classifying evolutionary theories

Based on—	Also allowing for soft inheritance	Exclusively hard inheritance
Essentialism	Orthogenesis Geoffroyism	Saltationism
Populationism	Darwinism	Neo-Darwinism Synthetic theory

known theories of evolution. I have avoided the term Lamarckism because it refers to several heterogeneous theories and most often to ones not supported by Lamarck himself.

These various theories may be characterized as follows:

(1) Geoffroyism, which ascribes evolutionary change to the direct influence of the environment. The product of this induction is transmitted to future generations, by means of an inheritance of acquired characters. Geoffroyism was adhered to by most neo-Lamarckians.

(2) Orthogenesis, a rather heterogeneous assortment of theories, all of which ascribe evolution to a built-in tendency or drive toward progress and ever greater perfection.[2] Lamarck's original thesis, widely adopted not only by biologists but particularly by sociologists, anthropologists, and philosophers, H. F. Osborn's aristogenesis, and Teilhard de Chardin's ideas belong to this tradition.

Population thinking (see chapter 4) is absent in both Geoffroyism and orthogenesis. However, it is also not strictly correct to list them under essentialism, as I have done, because both schools believe in a change through time. Their representatives, however, never make it quite clear whether they believe in genuine change or merely the unfolding of an immanent potentiality.

(3) The saltationism (macrogenesis) school, which postulates the origin of new types by discontinuous variation. Belief in such a process goes as far back as the Greeks and was particularly strong in the eighteenth and nineteenth centuries. The mutation theory of de Vries and other Mendelians is in this tradition as well as the later theories of R. Goldschmidt ("systemic mutations") and of certain paleontologists, particularly Schindewolf. T. H. Morgan had a strong saltationist tendency in his earlier writings that had not yet entirely disappeared from his 1932 book.

(4) The original Darwinism, which postulates slow, gradual evolution through natural selection utilizing abundantly available genetic variation. However, it also admits a certain amount of effectiveness of use and disuse and other manifestations of soft inheritance.

(5) Neo-Darwinism, which differs from the original Darwinism primarily by excluding all possibility of an inheritance of acquired characters. Neo-Darwinism was established by Weismann (1883) and adopted by Wallace (1889) and other Darwinians.

This classification into five schools of evolutionism is an oversimplification. Most authors between 1860 and 1940 adopted a mixture of these

2. Adherents of this theory might be referred to as orthogenesists. It would be as inappropriate to call them orthogeneticists as it would be to call believers in Genesis geneticists.

theories. For instance, a simultaneous belief in "mutations" for discontinuously varying characters and in Geoffroyism for continuously varying characters was particularly widespread. It was often not realized that certain elements in some of the mixed theories were incompatible. As time went on, the existence of a certain amount of natural selection was more and more widely admitted, but this was considered by many evolutionists to be quite compatible with an inheritance of acquired characters. Table 1 is only a first guide, but many of the authors of mixed evolutionary theories do not fit readily in any one of the four categories. The only thing these various non-Darwinian theories of evolution had in common was that they denied natural selection as the exclusive mechanism of evolution. This opposition to selection was based on a great variety of objections.

The Widening Split in Biology

The rise of evolutionism after 1859 coincided with an increasing separation of zoology and botany into new special fields, such as embryology, cytology, genetics, behavioral biology, and ecology. Simultaneously, the gap widened between the experimental biologists and those anatomists, zoologists, botanists, and paleontologists who had been raised as naturalists and who worked with whole organisms. Each group not only dealt with different subject matter but also asked basically different questions. When it came to the interpretation of evolutionary phenomena, their conclusions often were diametrically opposed. As early as 1894 Bateson suggested that "the Discontinuity of Species results from the Discontinuity of Variation" (p. 568), whereas most zoologists more than ever believed in Darwin's gradual evolution.

The rift became aggravated by the rediscovery of Mendel's rules in 1900, which resulted in the birth of genetics and eventually led to an ever increasing understanding of the principles of inheritance and of the nature of the genetic material. As essential as a knowledge of genetics is for a full understanding of the process of evolution, it spread only slowly through biology. Most naturalist-evolutionists learned about genetics from the writings of those geneticists who wrote books about evolution —that is, from Bateson, de Vries, Johannsen, and T. H. Morgan. Unfortunately none of these four geneticists understood evolution. All four tended to think as essentialists and failed to appreciate the nature of species as biological populations. All four downgraded or denied altogether the importance of natural selection, but instead considered mutation pressure as the major directive force in evolution.

These authors wrote more extensively on evolutionary questions than other early Mendelians and their views were believed by the nongeneticists to be the official evolutionary dogma of Mendelism. When I speak of "Mendelians," I have in mind Bateson, de Vries, and Johannsen. The heyday of this Mendelian period was from 1900 to about 1908. When Castle, East, Baur, Jennings, and other geneticists made findings that fitted far better into the Darwinian picture of evolution by gradual change, they did not base comprehensive evolutionary theories on their new knowledge. As a result, these advances in genetic understanding were largely unknown to the nongeneticists. Again and again de Vries, Bateson, and Johannsen, as late as the 1920s and 1930s, were attacked by the zoologists, botanists, and paleontologists (see chapters 4 and 9, which describe the situation in systematics and in Germany). Awareness of this friction is fundamental to understanding why the synthesis was delayed until the 1930s.

The particular evolutionary problem in which the naturalists were most interested was speciation (for example, Poulton and K. Jordan in England, and D. S. Jordan and Sumner in the United States). They continued the Darwinian tradition of the Galápagos finches, Wallace (1855), and Darwin's *Origin*. And everything they found seemed to confirm Darwin's insistence on gradualism and to refute the saltationism of the Mendelians. While Bateson claimed that "the transformation of masses of populations by imperceptible steps guided by selection is, as most of us now see, so inapplicable to the fact, that we can only marvel both at the want of penetration displayed by the advocates of such a proposition, and at the forensic skill by which it was made to appear acceptable even for a time" (1913, p. 248), Rothschild and Jordan (1903, p. 492) came to almost exactly the opposite conclusion: "Whoever studies the distinction of geographic varieties closely and extensively, will smile at the conception of the origin of species *per saltum.*" Poulton (1903) and D. S. Jordan (1905) rejected the saltationist claims of the Mendelians with equal emphasis (see also chapter 4).

The Mendelians not only ignored the comprehensive findings on speciation made by these systematists, but they also took a dim view of natural selection. Johannsen (1915) for instance, asserted that selection cannot produce a deviation from the mean in self-fertilizing species, "and even the most careful selection experiments with cross-fertilizing plants and animals confirm most convincingly our interpretation of an inability of selection to achieve more than a mere isolation or separation of previously existing constitutionally different organisms: Selection of differing individuals creates nothing new; a shift of the 'biological type' in the direction of selection has never been substantiated" (p. 609). He repeats his claim with almost the same words on page 613.

Although Johannsen admits that some authors had produced results with selection, he says he is forced to ignore these findings because they were "carried out with material that was not homogeneous genotypically," as if Darwin had ever claimed that selection would be successful when a group of genetically identical individuals are exposed to it. Johannsen, like other early Mendelians, had no understanding whatsoever of natural populations. He finally concluded that it is "completely evident that genetics has deprived the Darwinian theory of selection entirely of its foundations." For this reason "the problem of evolution is still an entirely open question" (p. 659).

Similar statements were made by de Vries and several of the other early Mendelians. The reaction of the naturalists to the claims that speciation is caused by mutations and that selection is irrelevant is recorded in numerous statements in the literature. One by Osborn (1927) is best known: "Speciation is a normal and continuous process; it governs the greater part of the origin of species; it is apparently always adaptive. [De Vriesian] mutation is an abnormal and irregular mode of origin . . . it is . . . a disturbance of the regular course of speciation" (again, see chapter 4, Systematics). When Dobzhansky (1937, p. 40) referred to this statement as "a profound misunderstanding of the genetic conception of the mechanism of evolution," he overlooked the fact that Osborn talked about speciation and was evidently unaware of Morgan's transfer of the term mutation. The evolutionary literature of the 1910s and 1920s shows convincingly that there was an extraordinary communication gap between the various disciplines of biology. As late as the 1930s and 1940s systematists referred to "de Vriesian mutations," many of them quite unaware of the fact that the term mutation had since been transferred by Morgan to a different class of phenomena.

The result of the thoroughly erroneous claims about evolution made by the leading Mendelians resulted in an almost complete alienation between experimental geneticists and naturalists. As Sewall Wright (1967) stated correctly, all sorts of non-Darwinian theories "continued to flourish at least to the 1930s," even though by then they were "discredited theories."

Three major factors were responsible for this lag. First, the naturalists were not sufficiently familiar with the advances in genetics made after 1910 and did not understand that their objections to the claims of Bateson, de Vries, and Johannsen were no longer valid against the new genetics represented in the writings of Castle, Jennings, Morgan, Sturtevant, Muller, East, Baur, Fisher, and Haldane. Second, little effort was made in the genetic literature to refute the theories of the naturalists, which these nongeneticists thought would explain evolutionary phenomena far better than Mendelian genetics. Furthermore, the geneticists ignored the rich literature on geographic variation and speciation, as D. S. Jordan (1905) rightly complained, and as was still evident from the writings of Morgan, Muller, and many other geneticists in the 1920s and 1930s. Finally, the Mendelians and mathematical-population geneticists failed to build a useful bridge to evolutionary biology. Actually, nothing in the supposedly evolutionary writings of T. H. Morgan, R. A. Fisher, and J. B. S. Haldane explained the multiplication of species, the origin of higher taxa, and the origin of evolutionary novelties. Their interpretations concerned the gene level in a single gene pool.

Conceptual Differences between Geneticists and Naturalists

The opponents in the evolutionary controversies failed to recognize that they belonged to the two different biologies that I characterize as the biology of proximate causes and that of ultimate causes (Mayr, 1961). The functional biologist is interested in the phenotype and its development resulting from the translation of the genetic program within the framework of the environment of the respective individual. It is this interaction between the translation of the genetic program and the environment that we refer to as proximate causes. The evolutionist is interested in the origin of the genotype, in the historical reasons of antecedent adaptation and speciation responsible for the particular genetic program that now exists. This analysis deals with ultimate causes. There are few biological problems for which we cannot study either the proximate or the ultimate causation, as I showed in 1961 for bird migration.

In the first third of this century, no one seems to have realized the existence of these two markedly different kinds of causations. As a result, even when dealing with the same phenomenon, the opponents talked past each other. For example, in Morgan's caustic discussion of the theory of sexual selection, he remarks that Darwin's speculative approach would never lead anywhere. How much superior, he says, is the modern experimental study of sexual dimorphism which has produced such interesting results: "This evidence has directed attention away from a discussion that seems rather barren, to the field where specific and sometimes quantitative data may be obtained. I refer to the role that certain internal secretions or hormones play in the development of the secondary sexual characters" (1932, p. 156). Morgan ends with the statement: "The sharp contrast between the older speculative procedure, that tried to explain secondary characters, and a new experimental way of studying the same problem is evident from what has been said." Morgan quite clearly did not understand that the clarification of the biochemical mechanism by which the genetic program is translated into the pheno-type tells us absolutely nothing about the steps by which natural selec-tion had built up the particular genetic program. It does not tell us why one species has sexual dimorphism and another species does not, nor does it tell us anything about the function of the sexual dimorphism in the life history of the species.

It is often overlooked how fundamental a contribution to biological theory Darwin made by bringing the study of ultimate causes into the realm of science, causes that until then had been considered to belong to the realm of the supernatural. Any question concerning the "Why?" of an adaptation or of any other biological phenomenon was answered prior to Darwin with "It is due to design" or "It is the result of a natural law laid down by the Creator." Both answers excluded the given phenomenon, for all practical purposes, from scientific analysis. Darwin's theory of natural selection provided the first rational approach to the study of ulti-mate causes. Arguments about causal explanations based *either* on proxi-mate *or* on ultimate causes began almost immediately after the publica-tion of *On the Origin of Species*. There was a well-known controversy between Haeckel, who insisted that morphological phenomena could never be understood fully except in terms of descent, and the botanist Goebel, who insisted that the metamorphosis in plants was a purely func-tional phenomenon that could be studied only experimentally. The zool-ogist Koelliker advanced similar arguments against Haeckel, stressing that a given morphological type was the result of the inner workings of physical and chemical laws rather than of its evolutionary history. In-deed, Koelliker even claimed that many of the taxa of the systematists might be polyphyletic, consisting of species or genera that are similar, not because they descended from a common ancestor, but because they are produced by the same constellation of physico-chemical laws.

There was no doubt in the minds of the experimental biologists that their methods were more objective, more scientific, and therefore supe-rior to the "speculative" approach of the evolutionary naturalist. This attitude can be documented by numerous quotations. Morgan (1932), for example, states that it was only the experimental method introduced in 1900 that permitted "an objective discussion of the theory of evolution in striking contrast to the older speculative method of treating evolution as a problem of history" (p. 13). In this statement, Darwin and all natural-ists are clearly classified as practitioners of "the older speculative method of treating evolution." Biology will make progress only by "an appeal to experiment . . . the application of the same kind of procedure that has long been recognized in the physical sciences as the most dependable one in formulating an interpretation of the outer world" (p. 14). Speculation permits only generalizations; only the experiment permits the development of the theory. As a result, the evolutionary "theory was more of the order of a broad generalization than of a scientific theory based on controlled experimental data" (p. 15). Proclaiming loudly that the evolutionists did not base their interpretations on the approved experimental method of science and that therefore their conclusions were speculative and unreliable was of course not the best way to bridge the wide gap between camps.

The complete preoccupation of most experimentalists with the study of proximate causes led them simply to ignore the existence of evolution and the problems it posed. For example, Hartmann's 869-page *General Biology* (1947) devotes only 23 pages to a discussion of evolutionary problems. Strasburger, in the fourteenth edition (1919) of the most famous botany textbook, assigned only about 6 of 643 pages to the presentation of evolutionary biology. Most of this space is simply a listing of the evidence for evolution with only the skimpiest of discussions of the causes under the headings of Lamarckism (p. 181) and Darwinism (p. 182). There were some exceptions, particularly in England, at least at Oxford, where a substantial Darwinian tradition among laboratory and experimental biologists was handed down from Lankester through Goodrich to J. Huxley and de Beer and to E. B. Ford and Haldane. However, almost everywhere else the gap between the two camps in biology seemed almost unbridgeable.

In addition, two other basic reasons led to the sharp separation between the geneticists and the naturalists. Each camp dealt with different levels in the hierarchy of evolutionary phenomena and with very different dimensions.

Hierarchical Level

If we look at the hierarchy of evolutionary phenomena, beginning with the phenomena of macroevolution (such as higher taxa, origin of evolutionary novelties, adaptive radiations), through the lower hierarchical levels of species and populations, down to the hierarchical level of genes and macromolecules, we perceive at once that the naturalists and the experimental geneticists were concerned with different levels. Naturalists repeatedly objected that they could not see any connection between changes in gene frequencies, described by the geneticists, and the evolutionary events at the hierarchical levels of species and higher taxa. As recently as 1973 this is perhaps the most cogent objection against population genetics raised by Grassé (p. 286). Pattee (1973) has correctly emphasized the difficulties of transferring the findings and conclusions of one hierarchical level to another one, particularly to a higher one. This difficulty is well illustrated by the communication gap of the presynthesis period.

The thorough study of the gene level, carried out by the experimental geneticists, was a necessary but not a sufficient condition for the synthesis. It was necessary to gain the new understanding of genetics, but unfortunately it led to a rather misleading conception of evolution reflected in the almost universally adopted definition of evolution as a change of gene frequencies or, as Dobzhansky phrased it, "A change in the genetic composition of populations" (1937, p. 11). There was an unspoken conviction among geneticists that evolution was understood as soon as the behavior of genes was understood.

This view had various consequences in the interpretation of evolutionary phenomena. It encouraged Morgan, Muller, and other geneticists of this period to consider evolution essentially as the result of mutation pressure. When selection was considered, it was perceived as pertaining to single genes. The literature of that period was very little concerned with the effects of any kind of gene interaction. Fitness tended to be considered as the sum total of the fitnesses of individual genes. The calculations of the mathematical-population geneticists, when based on this single gene approach, seemed as futile to the average naturalist as the elegant calculations of the biometricians demonstrating that Mendel was wrong. The evolutionary change of gene frequencies in populations was sometimes empirically demonstrated by the replacing of individual beans in a bean bag, which induced me to dub this approach as "bean-bag genetics" (Mayr, 1959). This approach seemed to be a most unrealistic representation of what actually goes on in highly variable natural populations.

In retrospect it is quite clear that—contrary claims notwithstanding—the new genetics by itself did not constitute a synthesis. It is simply not true that evolution can be explained as a change in gene frequencies. Falling victim to their own definition of evolution, the new geneticists failed to explain the very phenomena that occupied the attention of the most active students of evolution, such as the multiplication of species, the origins of evolutionary novelties and higher taxa, and the occupation of new adaptive zones. Changes in gene frequency are a by-product of adaptation and of the origin of evolutionary diversity (induced by natural selection) and not the other way around. As necessary as the advances in the understanding of genetics were, they were not sufficient for the synthesis.

The Conflict of Dimensions
Nearly all the classical work in population genetics before the synthesis dealt with the behavior of individual genes (and their alleles) within a

single gene pool. Even when considered a population, it was a closed population (Mayr, 1959). It was a nondimensional system, except for the very minor involvement of time in a sequence of generations. Where a dimension was introduced, as in Sewall Wright's stimulating model of a genic landscape, it was not reflected in the mathematics or graphs that, with very few exceptions, were straight bean-bag genetics.

By contrast the naturalists worked with the two-dimensional system of geographic variation and speciation, with the multidimensional resource space of ecology, and with the millions or billions of years and generations of the time dimension encountered by the paleontologist. To be sure, there is no real discontinuity between the nondimensional gene pool of the geneticist and the dimensional systems of the naturalist. The system of the geneticist is, so to speak, only a cross-section or instantaneous snapshot of a multidimensional system. Nevertheless, these differences in dimensionality contributed to the differences in thinking that separated geneticists and naturalists.

Faulty Concepts and Misunderstandings

The conceptual worlds of the experimentalists and the naturalists were very different for the three reasons just stated. Their paradigms (as defined by Kuhn) were markedly different. On the whole, they also worked in different institutions and read different journals. That there should be a considerable chasm between these two worlds—at least in most places—is not surprising. Even when there was a dialogue between them, as at the Tübingen conference of 1929, the clash of the different paradigms prevented a meeting of the minds (Weidenreich, 1929) and led to mutual misunderstandings and various misconceptions. The conceptual advances made by either group were not perceived by their opponents; in fact, they were usually unknown to them.

As a result, the construction of a unified and comprehensive theory of evolution during the first three decades of the century was impossible. The naturalists had wrong ideas on the nature of inheritance and variation; the experimental geneticists were dominated by typological thinking that resulted in their distrust of natural selection and a belief in the importance of pure lines and mutation pressure. Like the naturalists, the geneticists had many misconceptions about the nature of variation. They had at least as great an ignorance of the excellent taxonomic literature on species and speciation as the naturalists had of the genetic literature. A detailed discussion of the various subjects that were controversial prior to the synthesis will shed light on the nature of the prevailing misconceptions.

The Genotype-Phenotype Confusion

Few young geneticists of today, who have learned from the beginning the fundamental difference between the DNA blueprint and the translated proteins of the phenotype, realize how slowly the insight of the total difference between genotype and phenotype spread and found acceptance. Even Johannsen, who had in 1909 provided the terminological handle to discriminate between these two phenomena, continued to confuse them occasionally in his own subsequent discussions. De Vries never made a clear distinction between mutations of the phenotype and those of the genotype. Most naturalists described phenomena concerning the phenotype; the geneticists, those concerning the genotype. The use of the same terms, such as mutation, to describe aspects of the phenotype by naturalists and to describe aspects of the genotype by geneticists led to endless confusion.

To quote another example, to the geneticist, "blending" means the complete fusion of the parental germ plasms resulting in an intermediate genotype. The students of whole organisms, however, applied "blending" to the appearance of intermediate phenotypes in crosses (particularly between full species) when there was a failure of segregation in the F_2 and F_3 generations. In 1940 the leading American ichthyologist, Carl Hubbs, who was fully familiar with Mendelian genetics, used the naturalists' definition of blending: "Through a study of hundreds of specimens of natural fish hybrids, representing dozens of interspecific and often intergeneric combinations in several families, I have become convinced that as a very general rule the systematic characters of fishes show blending inheritance. The same sort of inheritance seems to operate when subspecies and races of fishes have crossed. A large body of information indicates that simple Mendelian segregation very seldom results when crosses have been made between natural forms of any vertebrate group" (p. 205).

Such a lack of clear segregation in species crosses was encountered by nearly every author who had crossed individuals belonging to different populations or taxa. This is what Darwin referred to in most of the statements that were subsequently interpreted as indicating his belief in blending inheritance (Ghiselin, 1969, p. 168). This is the blending which M. Wagner and other naturalists had in mind when they insisted on the importance of geographic isolation. The claim, I believe first made by R. A. Fisher and subsequently widely adopted, that blending inheritance was almost universally believed in before 1900, has only partial validity. Blending simply meant different things to different authors. Although I long ago called attention to this difference in interpretation (1942, p. 68), it has been largely ignored by the historians of genetics, resulting in a misinterpretation of the actual beliefs of a number of authors.

Soft Inheritance

The most far-reaching consequence of the failure to distinguish between genotype and phenotype was that it fostered a belief in soft inheritance. I use this term to designate the belief in a gradual change of the genetic material itself, either by use and disuse, or by some internal progressive tendencies, or through the direct effect of the environment. As the genetic material itself is affected, these changes are (by definition) inherited. The concept of soft inheritance includes the inheritance of acquired characters, but it is more comprehensive because it also includes various theories of a direct influence on the germ plasm (Geoffroyism). All theories of soft inheritance deny the complete constancy of the genetic material that we now know to exist. Darwin himself believed in a certain amount of soft inheritance. Along with Wagner and others, he believed isolation to be necessary to make genetic changes permanent, that is, to harden them. These authors believed in a very slow and gradual effect, sometimes extending over thousands of generations.

The almost universal belief in soft inheritance was a major stumbling block in the path of a neo-Darwinian interpretation. As firmly as most naturalists believed in natural selection, they also believed in one or another kind of soft inheritance—either in the effectiveness of use and disuse (as with Darwin) or in direct induction by the environment. Most supporters of soft inheritance thought that it would greatly help natural selection if the genetic material itself was somewhat plastic, particularly with respect to fine adjustments. They thought natural selection and the phenomena, often combined under the heading "inheritance of acquired characters," were not opposing but synergistic forces. Several recent histories of genetics have expressed the opinion that a belief in natural selection and one in soft inheritance were mutually exclusive. This view is wrong. Darwin, for instance, fully believed that genetic changes induced by use and disuse (soft inheritance) were just as much subject to selection as spontaneous genetic changes.

No modern evolutionist can have an idea of the enormous quantity of literature on the inheritance of acquired characters published between the 1870s and the 1920s. Some of this literature is reviewed by Semon (1912), Plate (1913, pp. 436-496), Kammerer (1924), and Detlefsen (1925). An experimental substantiation of the inheritance of acquired characters was repeatedly claimed by Fischer, Kammerer, MacDougal, Tower, McBride, Heslop-Harrison, Sr., and several others. Hindsight helps us interpret the positive results of some experimenters as fraud (Tower, Kammerer, and others), experimental error, or a failure to distinguish between phenotype and genotype. The naturalists and taxonomists who believed in soft inheritance took the results of these studies at face value because they had no means of testing the validity of the experimental research. As soon as one set of experiments had been either refuted or found to be nonrepeatable, another set would be published. The last perhaps were the *Dauermodifikationen* of Jollos (1934), refuted by Plough and Ives (1935).

The belief in soft inheritance was greatly prolonged by the dichotomy of variation into continuous and discontinuous variation. Again and again it was asserted that discontinuous variation, the result of mutations, was the product of chromosomal genes, but that these genes had to do only with superficial characters. The more fundamental distinctions between species and higher categories, however, were postulated to be cytoplasmic, as Loeb theorized in 1916, and as accepted by Plate (1913, 1925), Bather (1927), and many others. It was a popular belief in my student days (1925-1926). Federley (1929) devoted many pages to a refutation of this hypothesis, and as late as 1938 even Haldane thought such a refutation was necessary.

Soft inheritance was so widely accepted in the early Mendelian period because it seemed to explain observed variation in nature far better than mutationism. Also, the hypothesis of direct adaptation was a very fruitful heuristic principle and led to numerous beautiful papers on adaptive variation, demonstrating a close correlation between environmental factors and adaptive characters. Later it was possible to adopt these results with a minimum of change by reinterpreting them in terms of micromutations and selection, for instance, in Heslop-Harrison's study on the variation of the winter moth (1920).

The naturalists, who still saw mutation (in the de Vriesian sense) as a drastic and disturbing event and who had not yet accepted the importance of very small selective differences, listed a number of phenomena they thought were consistent with the theory of soft inheritance but inexplicable under the theory of neo-Darwinism. Rensch (1929, pp. 166-171) cited four such phenomena. Most of the older generation retained their belief in soft inheritance until the 1930s and 1940s. In discussions with two such distinguished biologists as F. Weidenreich (in 1948) and P. Buchner in 1954, I found that both were still firm believers in soft inheritance, as was Smirnov with whom I discussed these problems in Moscow in 1972. S. Freud reaffirmed his Lamarckian position as late as 1938.

Fortunately, my own generation not only was not as firmly imprinted with this belief, but also kept up better with the advances of genetics.

Even though we had been raised believing in soft inheritance, we were able to give up this belief as soon as the terminological difficulties (such as mutation) had been eliminated and as soon as those facts had been reinterpreted in neo-Darwinian terms which had previously seemed irreconcilable in view of the mutationist interpretation of de Vries, Bateson, and Johannsen.

It was perhaps the greatest contribution of the young science of genetics to show that soft inheritance does not exist. Mendel had taken this completely for granted, at least in his experiments with the pea (*Pisum*). None of his factors ever changed in the slightest. Weismann (1883) was an uncompromising advocate of hard inheritance and Goette (1875) and His (1874, p. 158) rejected soft inheritance even earlier. The history of the belief in soft inheritance and its eventual refutation has been described by Bliakher but would benefit by a more comprehensive treatment.

The leading Mendelians unanimously supported hard inheritance, which fitted perfectly with their typological thinking. Unfortunately, they primarily cited large discontinuous mutations as evidence for hard inheritance, evidence that did not at all explain any of the phenomena quoted by the naturalists in favor of soft inheritance, such as gradual geographic variation and climatic adaptation. As a result, a belief in soft inheritance was still universal among naturalists around 1910 and continued strong for several decades. Several essays in *Fifty Years of Darwinism* (1909), particularly E. B. Wilson's refutation, demonstrate the strength of the belief in an inheritance of acquired characters at that time (see also chapter 8). Advances on several fronts were necessary before the apparent discrepancy between the geneticists' and the naturalists' evidence could be resolved.

Variation

Darwin's explanatory theory had two major components, variation and selection. The never-ending and often confused arguments on the nature of variation, from the 1890s (Bateson, 1894) to the 1930s (Robson and Richards, 1936) and 1940s (Goldschmidt, 1940), show how long it took until the nature of genetic variation was fully understood and to what extent the interpretation was affected by philosophical commitments.

No other concept was as detrimental as the prevalence of typological thinking (essentialism). Species and populations were not seen as highly variable aggregates consisting of genetically unique individuals, but rather as uniform types. T. H. Morgan always acted as if a mutation occurred in a homozygous wild type. Goldschmidt (1940) did not speak of populations but of "taxonomic units." A species, for him, could not evolve into another one by variation and selection, but new species could only originate saltationally as new types (systemic mutations). Most orthodox taxonomists also spoke of species as if they were well-defined types.

This kind of thinking affected the interpretations of mutation, selection, and speciation. As long as a mutation was thought of in terms of affecting *the* wild type, there was little hope of understanding the role of mutation in evolution. The classical view that the wild type gene was almost always superior to all other alleles (except for some highly improbable mutations) led to a complete misunderstanding of the role of selection. Gradual geographic variation, as described in Rensch's climatic rules, conflicted with typological thinking and was often explained, after Johannsen's bean experiments, as being the result of nongenetic modification. As Goldschmidt (1940, p. 79) stated correctly, this view was particularly true for the botanical literature. Bateson's argument (1909) that intermediates between species do not exist, that speciation by selection for "impalpable" differences is therefore refuted, and that the only conceivable process of speciation is by sudden saltation, was based on typological thinking.

A true understanding of natural selection, speciation, and adaptation was not possible until population thinking had displaced typological thinking. Population thinking apparently had two major sources (see chapter 4, Systematics). It was widespread among the naturalists, particularly those who dealt with large samples of natural populations. The population thinking of the naturalists had a particularly important impact on the Russian school. The second source, as in Darwin's day, was the animal and plant breeders. Those geneticists, like Castle, East, R. A. Emerson, and Sewall Wright, who had the closest contact with breeders, also stayed most successfully away from typological thinking. Mathematical geneticists like Fisher and Haldane applied statistical methods largely to the frequency of genes in statistical rather than natural populations, although both occasionally applied their methods to material from natural populations (for example, see Fisher, 1931).

Two Kinds of Variation

There was a widespread belief in the nineteenth century, frequently referred to by Darwin, in the existence of two kinds of variation: discontinuous variation producing sports or mutations, and continuous variation represented by universal individual variation (see also chapter 4). Darwin always thought that individual variation was more important and the major material for natural selection and he almost entirely ignored sports after the Jenkin review (1867). T. H. Huxley and others disagreed; Bateson (1894) pleaded the importance of discontinuous variation in an entire book. He did not make much headway, however, until the redis-covery of the Mendelian laws. The Mendelians stressed the evolutionary importance of discontinuous variation, represented by striking muta-tions, and minimized the importance of individual variation. Indeed, after Johannsen's report on his bean experiments there was even a ten-dency to make the overt or unspoken assumption that most individual variation consisted of nongenetic modifications and was thus of no evo-lutionary interest.

The believers in soft inheritance were forced by the fact of Mendelian inheritance to concede hard inheritance for saltational mutations, but they compensated for this retreat by claiming that mutations, or discon-tinuous characters segregating according to Mendelian laws, had nothing to do with species differences and evolution (see chapter 4). Anderson (1939) invoked a chromosomal "ground substance" and Plate (1913) postulated that the seat of continuous variation was in the cytoplasm. Osborn (1927) stressed the importance of individual variation for evolu-tion and the irrelevance of mutations. Many similar references depre-cated the importance of mutation and emphasized the importance of a somewhat mysterious individual variation. Dobzhansky (1937) reported quite correctly a widespread belief "that continuous variability was dif-ferent in principle from the discontinuous one . . . It was said that only the latter was clearly genic while the former was alleged to be non-Men-delian and to be due to some vague principle which assiduously escapes all attempts to define it more clearly" (p. 57). As late as 1938 Haldane took considerable pains to refute the thesis that the cytoplasm controls species differences (p. 87).

To dislodge the proponents of soft inheritance from their last toehold, it was necessary to establish two important facts: (1) that there is no dif-ference between large and small mutations—that is, between de Vriesian mutations and Darwin's individual variation; and (2) that the compo-nents into which Darwin's individual variation can be dissected show the same hard inheritance as sports or conspicuous mutations.

The evidence making these two statements quite irrefutable was soon supplied by geneticists, beginning with the work of Nilsson-Ehle (1909), East (1910, 1916), Baur (1925), Emerson and East (1913), and Castle (1916). By the 1920s it was reasonably clear to those who kept up with the genetic literature that there was only one kind of variation. Yet many geneticists and nearly all taxonomists continued to act for several de-cades more as if this evidence did not exist. As late as 1932 Morgan re-ferred to "Darwin's postulate that the individual variations, everywhere present, furnish the raw materials for evolution. This the mutationists would deny" (p. 110). The controversy between Osborn and Morgan and much of the continuing argument during the 1930s indicate that confusion about the meaning of mutation was a major remaining stumbling block.

Mutation

The young biologist of today who knows that gene mutations are the only raw material of evolution is usually surprised to learn that mutations had a low repute among naturalists in the 1920s. De Vries, when reintroducing the term mutation into the biological literature, had defined it as a drastic change that could lead in a single leap from one species to another (for earlier usages see Mayr, 1963). Any malformation that bred true was by definition a mutation. When I was a student in the 1920s we used to make fun of the "new species" of *Drosophila* that arose from *D. melanogaster* by the mutations of white eye, yellow, and crumbled wing. Osborn (1927) well expressed the feelings of the naturalists.

Geneticists, in retrospect, have not been too happy with this phase in the history of evolutionary genetics. Sewall Wright (1967) stated: "De Vries' speciation by major mutation was soon abandoned as the primary cause of evolutionary transformation" (p. 246). But all the evidence contradicts this chronology. To be sure, Wright's statement may be valid as far as the working geneticists are concerned, as is evident, for instance, from the writings of East, Johannsen (1923), and the Morgan school. However, owing to the usual communications lag, the naturalists (see chapter 4) continued to interpret mutations in the de Vriesian sense as conspicuous changes of the phenotype and either denied completely any evolutionary significance of such mutations (Weidenreich, 1929; Osborn, 1927) or restricted the term to cases of conspicuous polymorphism and discontinuous geographic variation, as was done by F. M. Chapman (1923) and E. Stresemann (1926). Even later, mutations were occasionally invoked to explain physiological or structural adaptations, as was the position of G. K. Noble, an exceptionally well-informed biologist at the American Museum of Natural History, who expressed such ideas to me as late as 1939 and made similar statements to G. G. Simpson (personal communication). In a 1928 paper on the origin of arboreal frogs, Noble wrote, "The conclusion seems obvious that the tree climbing apparatus developed before the frogs began to climb trees." In other words, among the many mutations that had occurred in frogs, one gave it a "tree-climbing apparatus" that induced its possessors to start climbing trees. As late as 1949 B. Patterson stated in the explanation of a new higher taxon of mammals: "A single orthodox mutation resulting in the development of claws is believed to have been the starting point of the new adaptive shift represented by the Stylinodontinae" (1949, p. 277). I am not concerned here with everyone's foolishness but rather with the importance of realizing the communications lag. Mutation to the naturalists of the twenties and thirties still meant de Vriesian mutation. One almost has to go so far as saying that the generation which had become imprinted with the de Vriesian mutation concept had to die out before the new concept of mutation, introduced by Morgan, could take hold.

The difficulty was aggravated by the fact that the naturalists were interested in ultimate and the experimentalists in proximate causes. A drastic change in the phenotype was the characteristic earmark of a mutation for the naturalists. Whether they were studying diversity or adaptation, it was changes in the phenotype that were important for them. The chief criterion of the mutation for the student of proximate causation was the behavior in genetic crosses, which is why Morgan showed no hesitation in transferring the term from the drastic phenomena described by de Vries to any kind of genic change, no matter how slight (Allen, 1968, p. 130). Chetverikov (1926) fully appreciated the danger of confusion and attempted to introduce the term "genovariation" for Morgan's new concept of mutation, alas unsuccessfully. He rightly stated, "The term mutation has several different meanings even in genetics, the 'mutation' of de Vries by no means corresponding to the 'mutation' of Morgan" (p. 170).

De Vries had compounded the confusion by never making it clear whether mutation referred to a change in the phenotype or in the genotype. He spoke of mutation as if it were a unitary phenomenon, which is true only for the drastic change in the phenotype of de Vriesian mutations. We now know that, genetically speaking, de Vries's *Oenothera* mutations were a horrible mixture of segregates of balanced chromosomal complexes, cases of polyploidy, and a few cases of true gene mutations. It took the evolutionists some forty or fifty years to free themselves of the albatross that de Vries had hung around the neck of the term mutation.

Mutation Pressure
Even after the waning of the de Vriesian concept of mutation, mutation pressure remained a dominant evolutionary concept for the geneticists. They supported the so-called classical view of species structure—that is, most species are essentially pure lines that have to wait for the occurrence of a mutation to be able to evolve. According to Haldane (1932, p. 57), "The fundamental importance of mutation for any account of evolution is clear. It enables us to escape from the impasse of the pure line. Selection within a pure line will only be ineffective until a mutation arises." Morgan (1932, p. 111) was even more specific: "Since Darwin's time two interpretations of natural selection have been expressed." Morgan rejects

one: "Selection of the innumerable differences always present will lead to progressive steps of evolution." According to the other interpretation, "These small differences are not everywhere present at all times but appear relatively infrequently as mutations." In other words, there is an essentially homozygous wild type occasionally affected by a mutation. Any feeling of a population or of a gene pool was still absent. Muller's evolutionary writings of the 1920s also feature mutation pressure with natural selection hardly mentioned. Precisely those geneticists who had the greatest interest in evolution were the ones who rejected the existence of a vast supply of genetic variation most emphatically. "The concept of evolution as proceeding through the gradual transformation of masses of individuals by the accumulation of impalpable changes is one that the study of genetics shows immediately to be false" (Bateson, 1909, p. 289). With "impalpable changes" Bateson presumably had soft inheritance in mind, but the champions of individual variation among the naturalists conceived this assertion as an affront against their views. How strongly mutation pressure was believed in at that period is further documented in the statement of a well-known botanist: "We know at present that mutation pressure and the scattering of variability in small populations can produce new varieties and that chromosome vagaries and chance hybridization [that is, allopolyploidy] can give rise to new species, not only without the aid of natural selection, but actually in opposition to it" (Zirkle, 1941, p. 74). Natural selection for him was merely a process of elimination of the unfit. Even such a champion of natural selection as Dobzhansky implied that the rate of evolution might be determined by the rate of mutation (1937, p. 37). Morgan (1932) is impressed by the power of mutation pressure to such an extent that he argues: "Under the circumstances, it is a debatable question whether still to make use of the term natural selection as a part of the mutation theory, or to drop it because it does not have today the same meaning that Darwin's followers attached to his theory." Morgan's attention was entirely directed to the proximate causation of genetic variation, and because the Darwinians did not reject soft inheritance "we can no longer say that the individual variants, everywhere present, sufficed to supply the material for natural selection." For him the process of variation caused evolution and not the action of selection on the individuals thus produced.

Recombination
The most astonishing aspect of the evolutionary analyses of this period is the scant mention of recombination. Even though Morgan sometimes refers to it in other publications, it nowhere enters his 1932 arguments

where he poses an either/or choice between mutability and individual variability: "The kind of variability on which Darwin based his theory of natural selection can no longer be used in support of that theory because . . . selection of the differences between individuals, due to the then existing genetic variants, while changing the number of individuals of a given kind, will not introduce anything new. The essential of the evolutionary process is the occurrence of new characteristics" (p. 149). He ignored the possibility that recombination could produce new characters of selective significance.

It would be decidedly whiggish to suppress the fact that even some of the members of our conference slighted recombination in the 1930s. Dobzhansky, who later did much to establish the evolutionary importance of recombination, hardly referred to it in the first edition of his book (1937). Almost his only reference to recombination is in connection with hybridization. Genotypes, he says, consist of harmoniously co-adapted genes. "Inter-breeding of species results in the breakdown of the existing systems, and the emergence of a mass of recombinations" (p. 229). Without quoting Lotsy's paper (1916), Dobzhansky seems to adopt Lotsy's idea that such hybridization might be beneficial because it could produce new recombinations that "might be in fact better than the old ones, that is to say, new and higher adaptive peaks may be discovered." Geographic isolation "debars the organism from exploring greater and greater portions of the field of gene combinations, and hence decreases the chance of the discovery of new and higher adaptive peaks. Isolation is a conservative factor that slows down the evolutionary process" (p. 229). Earlier in the book for didactic reasons, he asks these questions: "Is evolution caused by mutations or is evolution 'a result of recombination and permutation of the gene elements?' " (p. 39). Dobzhansky emphatically rejects the second alternative, which, unfortunately, he presents in the form of Lotsy's hybridization theory. Dobzhansky's treatment in this chapter almost reads as if one had to make a choice between mutation or recombination. This impression was corrected in the later editions.

Recombination was, of course, well known since 1900 as one of the basic Mendelian processes and described in every genetics textbook. Remarkably, only a few evolutionary geneticists used it as a source of material for selection—among others, the plant geneticists East and Baur, and even more so, Chetverikov. No one, however, made a greater contribution to the understanding of recombination and its evolutionary importance than C. D. Darlington. He showed how many chromosomal mechanisms and "genetic systems" regulate the amount of inbreeding or outbreeding—that is, the amount of recombination. Through his 1940 paper in Huxley's *New Systematics*, his work became widely known among the zoologists who had not read his earlier more strictly cytological or genetical publications.

Natural Selection

The second of the two principal components of Darwin's theory of evolution was natural selection. The understanding and general acceptance of this causal factor encountered even more difficulties than variation (see also chapter 4). Well-known zoologists and botanists only a few decades ago doubted its efficacy. As late as 1938, Buchner in his textbook of zoology quotes, with evident approval, Johannsen's claim that selection cannot gain anything new. Robson and Richards (1936) referred to selection in a manner that was representative of a widespread feeling: "We do not believe that natural selection can be disregarded as a possible factor in evolution. Nevertheless, there is so little positive evidence in its favor . . . that we have no right to assign to it the main causative role in evolution." No doubt various contributors to our conference will cite statements made between 1920 and 1950 that also express a distrust in the importance of natural selection.

Darwin and Wallace had inferred the important role of natural selection on the basis of their experience as naturalists. Not surprisingly, in the first two or three decades after 1859, natural selection had far more adherents among the naturalists than among other biologists. Hooker, Gray, Bates, F. Müller, Weismann, K. Jordan, and Poulton were all either systematists or, like Weismann, biologists who worked with natural populations.

Darwin's contemporaries among the experimental biologists were almost without exception rather cool, if not hostile, toward natural selection. They were essentialists and automatically equated natural selection with elimination. They, particularly the anatomists, thought in terms of types.

There were numerous reasons for this resistance against the acceptance of natural selection:

(1) The essentialists saw selection as a purely negative process that could eliminate the unfit deviations from the type but could not play a constructive role in evolutionary advance. Bateson and Zirkle referred to selection in this way; so did for the most part Goldschmidt and Morgan. Other geneticists, like Castle, East, Jennings, Baur, and Wright, saw things more correctly.

(2) Morgan, and perhaps others, saw in natural selection a potential proximate cause. Does the term the survival of the fittest "mean, for in-

stance, that natural selection is an active agent in evolution, which in itself brings about progressive changes; or do such terms mean only that it acts as a sieve for the materials that present themselves as variation?" (1932, p. 110). He adopts the second alternative and concludes, "The argument shows that natural selection does not play the role of a creative principle in evolution" (p. 131).

(3) As stated by Haldane, "I believe that the opposition to Darwinism is largely due to a failure to appreciate the extraordinary subtlety of the principle of natural selection" (1932, p. 111). He admits "there is no doubt that innumerable characters [of animal and plant species] show no sign of possessing selective value, and, moreover, these are exactly the characters that enable a taxonomist to distinguish one species from another" (p. 113). Actually, evolutionists during the ensuing forty years spent much time in discovering previously unsuspected selective advantages of characters, erroneously considered neutral.

(4) Darwin's method of drawing inferences from observations was considered pure speculation: proof, it was claimed, could be supplied only by experimental procedures.

In addition to such general reasons, numerous very specific objections were raised against the validity of selectionism. No one has yet tabulated and analyzed them all, but the more important ones can be found in Kellogg (1907), Delage and Goldsmith (1909), Tschulok (1922), Plate (1925), Hertwig (1927), and various French authors, such as Caullery, Cuénot, and Vandel. Without endeavoring to rank them in order of their cogency, the following objections were raised most frequently:

The species argument: Existing species are separated by "bridgeless" gaps. Intermediates between species are not observed. Because only the nonessential (accidental) characters of species vary (never the essential ones), the sterility barrier that separates species could not possibly have evolved gradually.

The blending argument: New variants cannot be considered incipient species because they will cross with the parental type as soon as they arise, and thus revert back to it by blending.

The origin of evolutionary novelties: Selection cannot explain the origin of new structures, because incipient new organs, like rudimentary wings, can have no selective value unless they are large enough to be fully functional.

The origin of higher categories: Even if it could be demonstrated that related species are occasionally derived from each other, higher taxa are far too distinct and too different to permit their origin to be explained on the basis of gradual evolution by natural selection.

The integrated nature of evolutionary change: The theory of natural selection is based on the variation of individual features. Yet in most phyletic lines there is a simultaneous change of very many morphological-physiological characters, indicating a change of the genotype as a whole. This observation conflicts with the assumption of single mutations as the material of natural selection.

These claims were answered by counterclaims; still, during the first three decades of the century, these controversies did not lead to clarification or to a meeting of the minds, but actually widened the gap between various schools of evolutionists and particularly between naturalists and experimental biologists.

Chance and Accident

The role that chance plays in evolution has been controversial ever since 1859. Yet this argument has not played a very important role either in delaying or in furthering the synthesis. It was, of course, important to refute the claim that Darwinism ascribed adaptation to pure chance. No Darwinian had ever claimed this, however. Darwinian evolution is a two-step process and the second of these steps, natural selection, is not a chance phenomenon. To be sure, stochastic processes are all-important in the production of variability, being involved in mutation, cross-over, the separation of chromosomes at meiosis, and in other prezygotic processes. Yet the crucial factor in Darwinian evolution, the contribution to the gene pool of the next generation, is largely determined by physiological and other competitive properties of individuals—that is, by non-chance factors. I say "largely" because there are almost invariably multiple possible answers to selection pressures and chance may indeed play a role in which of these answers is chosen. Ever since Gulick (1873) stressed the fortuitous nature of the geographic variation of Hawaiian snails, there has been a controversy in the evolutionary literature over the importance of such a stochastic contribution. The change of evaluation of this factor is clearly represented in succeeding editions of Dobzhansky's book (1937-1970). At the moment, of course, it is again in the focus of attention under the label of "non-Darwinian evolution." That there is a strong random component in evolutionary change is obvious from the fact that many populations are established by an infinitesimal fraction of the parental population. This principle, designated by me as the Founder Principle (1942), but clearly previously recognized by Rensch (1939, pp. 183, 191), is so important because the pivotal role of small populations in evolution is increasingly being recognized.

Factors Contributing to the Delay in the Synthesis

The coming of the synthesis was delayed not only by the communication gap between experimentalists and naturalists and the misunderstandings that resulted from it, but also by several special factors. In addition to the slow demise of the belief in the existence of soft inheritance, the physical scientists almost unanimously had no use for Darwin's theory of natural selection (Hull, 1973). They insisted that the solution to the problems of evolution would have to be found by asking entirely different questions and by using entirely different methods. They insisted that the hypothetico-deductive approach was pure speculation and that the looked-for laws could be found only by induction from experiments. This attitude is well reflected in a discussion of the problems of evolution by Caullery (1931, pp. 400-401), but also in some of the discussions of de Vries, Johannsen, and Morgan. The feeling of the physical scientists that the evolutionists are still very much in the dark was quite strongly expressed at a conference a few years ago (Moorhead and Kaplan, 1967). Unfortunately, the physical scientists may well have prevented experimental geneticists from looking in the literature of the naturalists for part of the answers to still-open questions.

The pendulum always swings too far after the termination of a controversy. Certainly this was true for Bateson's fight with the biometricians. The failure of the biometricians to explain Mendelian inheritance led Bateson to state that "with the discovery of segregation it became obvious that methods dispensing with individual analysis of the material are useless" (1909, p. 107). In spite of promising beginnings made by the biometricians and also by Nilsson-Ehle (1909) and East (1910), the analysis of quantitative characters by statistical methods was delayed for decades by the premature and ill-conceived endeavor of the biometricians (see also Wright, 1967). In turn, this delayed the resolution of the seeming conflict between Darwinian gradualism and mutational discontinuity.

In an Australian symposium, R. A. Fisher (1959) points out how wrong the early Mendelians were. In 1909 defenders of selection like Weismann and Poulton

> were clearly in a minority whose views were regarded with much scepticism, for strange as it must seem to the present generation, many were persuaded that the element of discontinuity inherent in a particulate theory of inheritance implied a corresponding discontinuity in the evolution of one specific type from another.
>
> The early Mendelians could scarcely have misapprehended more thoroughly the bearings of Mendel's discovery, and of their own ad-

vances, on the process of evolution . . . They thought of Mendelism as having dealt a death blow to selection theory.

More attention to the History of Science is needed, as much by scientists as by historians, and especially by biologists, and this should mean a deliberate attempt to understand the thoughts of the great masters of the past, to see in what circumstances or intellectual milieu their ideas were formed, where they took the wrong turning or stopped short on the right track.

The Foundations of the Synthesis

At the end of the 1920s the situation in biology did not seem at all promising for a meeting of the minds in the near future. Many general college texts still presented five or six equivalent evolutionary theories. The taxonomists were split into a number of different camps, some still upholding the most extreme typological tradition of the Linnaeans, others fully adopting very sophisticated populational approaches. Most paleontologists were either saltationists or orthogenesists, while those whom we believe to have been neo-Darwinians failed to write general papers or books. The geneticists had emerged from de Vriesian mutationism but many still upheld typological concepts and certain genetic interpretations (such as fitness of genes, mutation pressure) that made communication with the naturalists difficult. The naturalists, for their part, were hopelessly behind in their understanding of the mechanisms of heredity and of the enormous advances in genetic theory. There seemed to be an unbridgeable gap between the camps, the students of proximate causes and the students of ultimate causes. As early as 1922, Bateson made an impassioned plea for a closer collaboration between geneticists and systematists. Both groups are to be blamed for the division, he said, but "if we cannot persuade the systematists to come to us, at least we can go to them. They too have built up a vast edifice of knowledge which they are willing to share with us, and which we greatly need. They too have never lost that longing for the truth about evolution which to a man of my date is the salt of biology, the impulse which makes us biologists." Yet another fifteen years passed before Bateson's wish was fulfilled. The members of the two camps continued to talk different languages, to ask different questions, to adhere to different conceptions. In retrospect, it is quite clear that in the first three decades of the twentieth century the foundations were laid for the eventual synthesis. There was a growing understanding of the evolutionary process, even though the two camps concentrated on different hierarchical levels and even though the knowledge of the advances made by one group percolated only very incompletely to the other.

The creation of a new evolutionary terminology greatly contributed to the eventual synthesis. At least some of the misunderstandings resulted from the lack of an appropriate and precise terminology for certain evolutionary phenomena. Nearly all the architects of the new synthesis contributed terminological innovations. Dobzhansky (1937b) coined the term *isolating mechanisms* for the totality of the barriers maintaining the reproductive isolation of species (and previously described in great detail by Du Rietz, 1930); Rensch, the term *Rassenkreis*, subsequently designated by Huxley *polytypic species*; Serebrovsky coined *genofond*, subsequently referred to as *gene pool*; I revived Poulton's term *sympatric* and provided the antonym *allopatric*. *Balanced polymorphism, climatic rules, founder principle, gene flow, introgression, isolate, sibling species, stabilizing selection*, and *taxon* are some other helpful terms coined or adopted during this period (see Mayr, 1978).

Perhaps the most important advance was a change in attitudes toward selection. I have described the negative attitude of the early Mendelians toward selection. A misinterpretation of Johannsen's experiments on beans persuaded a remarkably large number of biologists that the power of selection was of negligible importance. It induced Morgan to say, as late as 1932, "the implication in the theory of natural selection that by selecting the more extreme individuals of the population, the next generation will be moved further in the same direction, is now known to be wrong" (1932). Provided there is sufficient mutation "even without natural selection evolution might have taken place." Goldschmidt belittled selection in a different way, when admitting that "selection may wipe out one type or isolate a new type" but saying nothing about the possibility of selection leading to the construction of a new type (1940, p. 4).

Morgan, and to a lesser extent Goldschmidt, were basically essentialists and their difficulty with natural selection is not surprising. Unless one adopts population thinking and considers every individual as representing a uniquely different genotype, natural selection does not make much sense.[3] The field naturalists who had stressed the uniqueness of individuals even before 1859 had far less trouble with natural selection than the laboratory workers. This is the thinking that makes the work of Chetverikov and his school so influential. It was this heritage which gave Dobzhansky's *Genetics and the Origin of Species* (1937) its tremendous impact. Dobzhansky devoted the entire sixth chapter to natural selec-

---

3. Goldschmidt thought of the origin of new types in terms of proximate causation. In 1952 I asked Goldschmidt how the population in which a new hopeful monster occurred would react to it. He answered, after a considerable pause, "I have never thought of it that way."

tion. His treatment clearly reflects how strongly natural selection still had to fight for general recognition. Dobzhansky's presentation was particularly effective because he treated selection not merely as a theory but as a process that can be substantiated experimentally. Among other things, he showed that there is no conflict between adaptive geographic variation (as, for instance, reflected in Rensch's climatic rules) and selection. As a result, naturalists no longer needed to take refuge in Lamarckian explanations as they had been forced to through the arguments of the mutationists.

By the time Morgan and Goldschmidt made their antiselectionist statements, they were decidedly in the minority. The belief in the power of selection was spreading steadily. Support for selectionism came primarily from two directions. First, artificial selection was demonstrated as a highly effective method through the brilliant selection experiments of Castle (Castle and Phillips, 1914), MacDowell (1917), Sturtevant (1918), Payne (1920), and the work of numerous animal and plant breeders. The results of the Illinois maize selection experiments for high protein and oil content were particularly impressive. Second, mathematical geneticists (Norton, Fisher, Haldane, and others) showed that even very small selective advantages can have a major evolutionary impact if selection is continued over a sufficiently long period (Provine, 1971).

The Contributions of Post-Mendelian Genetics

The dispute over the power of selection was, to a large extent, carried out within genetics. The three Mendelians (de Vries, Bateson, and Johannsen) who were most interested in evolution were also the ones most strongly antiselectionist. Because they were among the most distinguished geneticists in the first decade of the century, they contributed through the prestige of their writings to the opinion that Darwin's theory of natural selection was refuted. As Nordenskioeld evaluated Darwinism in his otherwise authoritative *History of Biology* (1920-1924): "The theory has long ago been rejected in its most vital points . . . the objections made against the theory on its first appearance very largely agree with those which far later brought about its fall." The contemporary European literature indicates that this statement reflected a widespread, if not prevailing, opinion.

Even though the Mendelians were not the only ones to attack Darwinism, they were certainly the most influential. It is particularly interesting for the historians that it was a reaction to mutationism within the camp of the geneticists themselves which contributed to the rejection of the de Vries-Bateson concept of evolution. New findings made by geneticists, primarily between 1908 and 1927, showed that the findings of genetics do not conflict with natural selection and population thinking.

In addition to confirming the importance of selection, genetic research during this period led to several important findings. The most important, perhaps, was the refutation of soft inheritance. It became possible to explain in terms of micromutations what had previously seemingly necessitated the postulation of soft inheritance. Castle's willingness (1919) to reinterpret the results of his selection experiments with hooded rats signaled the final victory of hard inheritance within the camp of genetics (Provine, 1971). This left no alternative in the explanation of gradual evolution but selection.

Various geneticists, from Nilsson-Ehle, East, and Baur to Morgan and Chetverikov, also demonstrated that seemingly continuous variation is caused by discontinuous genetic factors that obey the Mendelian rules in their mode of inheritance exactly like de Vries's mutations. In other words, there is only one kind of genetic factor. There is a complete gradation from the most cataclysmic mutations to those the effects of which are virtually unnoticeable. Furthermore, some mutations are beneficial, thus controverting the concept of mutation as a strictly deleterious process.

The discovery of all sorts of interactions among genes, such as pleiotropy (the effect of a single gene on several different components of the phenotype) and polygeny (multiple factors that determine a character by several independent genes) further helped to explain gradual evolution. Allelic as well as epistatic interactions provide a much greater plasticity to the genotype than envisioned by the early Mendelians.

A new branch of genetics emerged that, as Dobzhansky correctly emphasized in 1937, "has as its province the processes taking place in groups of individuals—in populations—and therefore is called the genetics of populations . . . The rules governing the genetic structure of a population are distinct from those governing the genetics of individuals" (p. 11). Because the major aspect of nature that affects the genetic composition of populations is the environment, population genetics—when related to natural populations—has also been designated ecological genetics (Ford, 1964).

The naturalists might claim credit with equal justification as the geneticists for having initiated population genetics. All the pioneers (Schmidt, Goldschmidt, Sumner, Chetverikov, Ford) were naturalists or, at least (like Goldschmidt), were stimulated by naturalists to undertake work with natural populations. It led to the particularly convincing work of Sumner (summarized in 1932) who conclusively proved the genetic basis of adaptive geographic variation (Provine, 1979) and of gradual trends in geographic variation (for the work of Chetverikov, see chapter 8).

One of the particularly important results of these analyses was the discovery of high genetic variability of wild populations, consistent with population thinking, but refuting the previously prevailing concept of essentially homozygous "wild-type" species. The evidence for this new concept of populations was well presented by Dobzhansky (1937).

Unfortunately, most of the evidence leading to the new interpretation of evolutionary genetics was published in the specialized genetic literature and was not used as the basis of evolutionary conclusions. Curiously, neither Morgan (1932) nor Goldschmidt (1940), perhaps the leading geneticists to write books on evolution except for Haldane (1932), based their conclusions on the newer findings. As a result and as a consequence of the far more general phenomenon of a communications lag between different disciplines, these findings did not penetrate the evolutionary literature of the naturalists (broadly defined). In the United States, England, and Germany, the naturalists in the 1920s and early 1930s still argued against the obsolete interpretation of evolution by the mutationists even though the new insights of the younger geneticists had already refuted the mutationists' arguments.

Advances Made by the Naturalists

Systematics, contrary to widespread misconceptions of biological historiography, was not at all in a backward and static condition during the first third of the twentieth century. Indeed, this period witnessed a revitalization of the field, manifested in evolutionary studies of all kinds. Population thinking was widely adopted and, as a consequence, variation within and between populations was actively studied, which led to the development of the biological species concept, to the widespread adoption of polytypic species taxa, and to the study of species in space and time as adapted systems. It led to the development of a new school of systematics variously called the *new systematics* or *population systematics* (Mayr, 1942).

These changes were particularly well reflected in the breeding work of Sumner and Dice on *Peromyscus*, in Rensch's studies on polytypic species and on the climatic rules, and in botany by Turesson's work on ecotypes, continued by Clausen, Hiesey, and Keck. These studies were made independently of the developments in mathematical population genetics. The vigorous promotion of population thinking by systematists (see chapter 4) prepared the way for the subsequent application of these concepts to population genetics.

It was the naturalists who solved the great species problem, a problem either sidestepped altogether or answered in a typological manner by the geneticists. Darwin had concentrated on the origin of species because the problem of the origin of variation was insoluble in his day. Immediately after 1900 the genetics of variation occupied the center of the stage and the species was pushed into the background. A reversal of the roles began again in the 1930s. Almost every evolutionist realized that no one could explain the multiplication of species without first understanding what a species is. The confusion on this question is evident from Robson and Richards (1936), from various contributions to Huxley's *New Systematics* (1940), from several symposia on species and speciation in the *American Naturalist* (1940-1941), and from Goldschmidt's *The Material of Evolution* (1940, pp. 139 ff.). A strictly typological definition was still widespread and inevitably resulted in saltationist theories of speciation, not fundamentally different from that of de Vries. Dobzhansky (1937) devoted an entire chapter to the species "as natural unit" but even his account is flawed by various weaknesses. He still speaks of "forms" rather than populations; he makes no distinction between species as taxa and as categories; and he adopts a Whiteheadian definition of species, "as that stage of the evolutionary process at which the once actually or potentially interbreeding array of forms becomes segregated into two or more separate arrays which are physiologically incapable of interbreeding" (p. 312). He deliberately rejected the term "biological species" because it "conveys a wrong impression, as though there may exist a fundamental difference between morphological and biological criteria of species separation." He did not realize that the so-called morphological criteria were nearly always the manifestation of an essentialist (typological) species concept.

The majority of contemporary geneticists, such as Morgan, simply stayed away from the species problem but those who dealt with it like Haldane (1932) adopted a typological approach. Typological species definitions concentrated on species-specific intrinsic qualities, which created considerable difficulties for a gradual Darwinian speciation and encouraged a saltationist solution such as that proposed by Goldschmidt (1940).

Accustomed to closed laboratory stocks and pure lines (Mayr, 1959), the experimental geneticists, with few exceptions, were quite unaware that a populational species concept had been widely adopted by naturalists. The history of the development of the biological species concept has not yet been written but it goes back to the pre-Darwinian period, is found in some of Darwin's writings, particularly his earlier ones (Sulloway, 1979), and became steadily stronger in the post-Darwinian period (Mayr, 1957, 1963) until expressed in its total conception in the writings of K. Jordan (Mayr, 1955). This was the species concept of the ornithologists (Stresemann, Rensch) under whom I was educated, and it was widespread among those taxonomists who took an interest in evolutionary problems (see chapter 4). The biological species concept emphasizes the species as a community of populations, reproductive isolation (often based on behavioral mechanisms rather than merely on a sterility barrier), and the ecological interactions of sympatric populations that do not belong to the same species. The major intrinsic attribute characterizing a species is its set of isolating mechanisms that keeps it distinct from other species.

A large part of my 1942 book was devoted to the development of these ideas. Yet I was still confused about certain problems; in particular, I did not make a clear distinction between the species as a taxon and the species as a category. Indeed, the term taxon was not introduced into the English language literature until around 1950, and a clear discrimination between the concepts was thwarted by this terminological deficiency. Much of the writing about the species problem in the ensuing thirty years was likewise flawed by failing to make this distinction. The polytypic species, for instance, is a special kind of taxon and not, as Goldschmidt (1940) had claimed, a special category of species. After many years of argument it has now become clear, to express this in philosophical terms, that species taxa are individuals, with each member of the species being a part of the species, while the species category is a class, the members of which are the species taxa (Ghiselin, 1974; Mayr, 1976).

Speciation
The seemingly endless controversies about speciation make it quite clear that an understanding of the populational (rather than typological) nature of species was an indispensable prerequisite for understanding this process. Speciation at first sight seemed to present an insoluble conundrum for the Darwinians. It required, as Osborn (1927) and Mayr (1942) stressed, a reconciliation of Darwin's postulate of a gradual, continuous process with the fact, as was ever more clearly understood, of the sharp, bridgeless discontinuities between sympatric biological species.

The problem was indeed insoluble for those who looked for the solution at the level of genes or genotypes. One choice they had was to abandon the Darwinian postulate of gradualism, and this was done by such authors as de Vries, Bateson, and Goldschmidt. All three postulated instantaneous speciation by drastic mutations or other unknown processes. De Vries (1906) stated, "The theory of mutation assumes that new species and varieties are produced from existing forms by certain leaps." For Goldschmidt (1940), systemic mutations were variations on the same theme. He acknowledged natural selection as contributing to the adaptation of geographic races, but neo-Darwinian gradualism could not account for speciation: "The neo-Darwinian conception, which works perfectly well within the limits of the species, encounters difficulties and is not sustained by the actual facts when the step from species to species has to be explained" (p. 139). "Species formation is based upon a different type of evolutionary procedure than that of subspecific differentiation" (p. 154). "The decisive step in evolution, the first step toward macroevolution, the step from one species to another, requires another evolutionary method [that is, the origin of hopeful monsters] than that of sheer accumulation of micromutations" (p. 183). Like Charles Lyell a hundred years earlier, Goldschmidt was willing to recognize variation within the type (the species), but denied that such variation could transgress the species limits.

Bateson stated as late as 1922 that "We have no acceptable account of the origin of the species . . . The production of an indubitably sterile [with its parents] hybrid from completely fertile parents which had arisen under critical observation from a single common origin is the event for which we wait." To be sure, instantaneous speciation by polyploidy or other chromosomal processes does occur, particularly among plants, but it is not the prevailing process of speciation and is virtually unknown among higher animals. Other geneticists attempted an explanation by gradual change through splitting of a single population, so-called sympatric speciation. Such models, all of them more or less unsatisfactory, were proposed by Morgan, Bonnier (1927), and Muller. "Conditioning to new hosts is potentially an important method of [sympatric] speciation in monophagous species and in particular in plant feeders" (Mayr, 1942, p. 211), but the occurrence of such speciation is subject to so many limitations that it is surely rare, if not exceptional. Even though some geneticists mentioned the importance of geographical barriers, they did nothing with this information.

Most of these discussions made no distinction between phyletic evolution (temporal genetic changes in populations) and the multiplication of species—that is, the splitting of a phyletic line into two or more reproductively isolated lines. Those who treated speciation as an intrapopulation phenomenon overlooked that the great conundrum of speciation could be solved quite easily by approaching it through a back door. Speciation is not a phenomenon of genes or genotypes but of populations: "A new species develops if a population which has become geographically isolated from its parental species acquires during this period of isolation characters which promote or guarantee reproductive isolation when the external barriers break down" (Mayr, 1942, p. 155).

The most important conceptual advance was a clear formulation of the problem. To explain speciation it is not sufficient to explain the origin of variation or of evolutionary change within populations. It is the origin of reproductive isolation between populations that must be explained. Speciation is not so much the origin of new types but of protective devices against the inflow of alien genes into gene pools.

In retrospect, the solution of the problem of speciation was simplicity itself. The first author to have suggested geographic speciation was apparently von Buch (1825). The concept was strong in Darwin's 1837-1838 notebooks and in his 1842 and 1844 essays (Sulloway, 1979), as well as in Wallace's 1855 paper. For the further history of the concept see chapter 4 and Mayr (1963, pp. 482-488). No one has questioned the importance of geographic speciation since 1942; the only question that remains has been the relative importance of other processes of speciation, such as instantaneous speciation (for example, by polyploidy) and sympatric speciation, processes that are again quite popular (Bush, 1975; White, 1978).

The solution of the problem of speciation cleared the deck for a treatment of numerous other, previously rather neglected, problems of evolution, such as the study of the geographic variation of populations (summarized in Mayr, 1963, pp. 297-333).

Macroevolution

Another field of evolutionary biology in which major advances in thinking were made by naturalist-taxonomists is macroevolution, which deals with evolutionary phenomena above the species level. If one studies the macroevolutionary literature prior to the 1930s, such as textbooks of paleontology, or books on evolution written by taxonomists or morphologists, one is astonished to see in how few works the Darwinian viewpoint is clearly defended. Most of the numerous books on macroevolution published during this period defend either some kind of orthogenetic finalism or some form of saltationism—that is, the origin of evolutionary novelties by sudden drastic mutations in the old paleontological definition of the term mutation. Most writers considered macroevolutionary processes and causations to be of a special kind. (For more detail see chapters 4, 6, and 8.)

If paleontologists between 1925 to 1935 had been asked why they maintained the views they did, the saltationists would have responded that they were simply applying de Vriesian mutationism to their material and the orthogenesists would have said that they found evidence everywhere for progressive adaptive evolution and that Darwinian evolution by accidental mutations was contradicted by their findings. Paleontologists in that period did not form a solid front. There was an irreconcilable conflict between authors like Schindewolf, who represented a Goldschmidtian saltationism, and H. F. Osborn whose aristogenesis, even though he himself disclaimed it, was a form of orthogenesis. Some authors favored natural selection, but it would require a careful analysis to determine whether they also allowed additional evolutionary mechanisms (as did many contemporary Darwinians) or whether they considered selection as the only direction-giving mechanism. Among these authors are Dollo, Kovalevsky, Abel, Goodrich, and Matthew. At any rate, until the middle 1930s the thinking of the evolutionists in paleontology seemed to be completely separated from the thinking of those in genetics.

The bridge between the two fields was eventually built by two zoologists and one geologist, who had the all-around breadth of understanding to demonstrate that the new genetic findings had removed all the objections that had previously been raised against the application of the findings of genetics to macroevolutionary events. The three evolutionists were J. Huxley, B. Rensch, and G. G. Simpson. Unfortunately, the books of the two latter authors, although complete or nearly complete in the late 1930s, were delayed by the war. Simpson's was not published until 1944, and Rensch's not until 1947. Huxley, who wrote most of his book at the same time, was luckier because *Evolution: The Modern Synthesis* was published in 1942.

Bernhard Rensch, who still upheld certain Lamarckian concepts until 1934 or 1935, published a series of journal articles from 1939 to 1943 in which he interpreted macroevolutionary events in strictly Darwinian terms, no longer upholding any of his previous reservations (see Rensch's account in chapter 9).

More than anyone else, George Gaylord Simpson was responsible for bringing paleontology and macroevolution into the synthesis. Even though primarily trained as a geologist, he realized in the 1930s that a thorough understanding of modern population genetics was indispensable for an evolutionist. He acquired this understanding by studying the genetic literature and was able, by integrating it with his fine knowledge of the paleontological evidence, to arrive at an interpretation of macroevolutionary events in *Tempo and Mode of Evolution* (1944) that was fully consistent with the findings of the new genetics (see biographical essays). Julian Huxley, a member of the Oxford school, however, had been a thorough neo-Darwinian since his student days. His interests in the 1920s and early 1930s had been largely in the area of experimental zoology (embryology and gene physiology); apparently not until he worked on allometry and edited *The New Systematics* (1940) did he become aware of the bearing of the new genetic insights on macroevolutionary problems included in *Evolution, the Modern Synthesis* (1942). I remember discussing evolutionary problems with him for days on end during the preceding years.

I suppose it is legitimate for me to mention that I was one of the early authors to apply the concepts of the new synthesis to macroevolution. Chapter 10 of my 1942 book is devoted to the phenomena of macroevolution and their interpretation in terms of neo-Darwinism. I made a series of statements on macroevolutionary factors and principles based on the firm conviction that the mechanisms of macroevolution do not differ from those of microevolution (pp. 292-298). I summarized my viewpoint in 1942 in the statement "In conclusion we may say that all the available evidence indicates that the origin of the higher categories is a process which is nothing but an extrapolation of speciation. All the processes and phenomena of macroevolution and of the origin of the higher categories can be traced back to intraspecific variation, even though the first steps of such processes are usually very minute." Dobzhansky (1937) also states that there is no difference in the underlying mechanisms of microevolution and macroevolution but does not substantiate this observation by any detailed exposition.

The interpretation of macroevolution in terms of neo-Darwinian mechanisms consisted not merely in generalized assertions, but also in attempts to apply neo-Darwinian thinking to specific phenomena and problems. When Goldschmidt said "I cannot agree with the viewpoint of the textbooks that the problem of evolution has been solved as far as the genetic basis is concerned," particularly with respect to the origin of new structures or organs (1940, pp. 6-7), he was correct as far as the published literature up to that time. However, the publications of the ensuing years, particularly those of Rensch and Simpson, filled the gap decisively. They demonstrated that the explanation of the origin of new higher taxa, of rectilinear trends of evolution, of the acquisition of adaptations, and of the origin of new organs and structures, in terms of natural selection and the new genetics, was consistent with the observations of the paleontologists. Indeed Simpson developed an elaborate theory of macroevolution (1944, 1953).

In my own account (1942) I made two observations that may help to explain why the fossil record is not more useful. The first is that "aberrant types can be produced only in effective isolation and in rather small distributional areas . . . The probability that they will leave a fossil record is very small" (p. 297). The second is that "many of the primitive forms were inhabitants of tropical forests of which we have a particularly poor fossil record" (p. 297). The importance of ecological factors and of population structure for the explanation of evolution were very much on my mind. Haldane's *Causes of Evolution* (1932) was not well known outside England; otherwise one would also have to mention it here, for Haldane definitely attempted to explain macroevolutionary phenomena in Darwinian terms.

Are the Developments of the 1930s-1940s Rightly Considered a Synthesis?

Although Julian Huxley introduced the term synthesis for the unification of evolutionary biology in 1942, he justifies this designation only in the preface: "The time is ripe for a rapid advance in our understanding of evolution. Genetics, developmental physiology, ecology, systematics, paleontology, cytology, mathematical analysis, have all provided new facts or new tools of research: The need today is for concerted attack and synthesis. If this book contributes to such a synthetic point of view, I shall be well content."

The period of the 1930s and 1940s was indeed a time of "a rapid advance of our understanding"; the meeting of the minds achieved in this period is customarily called the "evolutionary synthesis." What do we mean by the word synthesis, and between what did this synthesis take place?

But first let me take up the question of chronology. Was it really the 1930s that saw the decisive advance? Unquestionably a few enlightened pioneers supported already in the 1920s (or even earlier) the ideas that later characterized the synthesis, particularly the Oxford group (J. Huxley, Ford, and others), Haldane, J. Grinnell (Berkeley), E. Baur (Berlin), and most likely some others. But they did not publish any books that would have exposed and promoted these ideas. They were ahead of their times, and the ideas of the synthesis actually made little headway prior to 1937, as abundantly documented by the Tübingen meeting of 1929, Morgan (1932), Robson and Richards (1936), Goldschmidt (1940), and by the additional literature I have quoted above and have referred to again in chapters 4 and 9.

Another aspect of the synthesis might also give rise to question. What was the particular missing piece of information or missing theory that had previously prevented the synthesis, the discovery of which now made it possible? No such missing piece was involved. Nearly all the individual components of the synthesis represent insights that were acquired during the 1910-1930 period (as far as the nature of genetic variation is concerned) or even before (such as the understanding of species and speciation by the naturalists).

What led to the synthesis was far more profound than the elimination of a few errors or the incorporation of a few new insights. The radically different thinking and preoccupations of the two camps of evolutionary biologists, those of the experimental geneticists and of the population-naturalists, differed in their studies of causation (proximate vs ultimate), in the level of the evolutionary hierarchy with which they were concerned, and in the dimensions they studied. They represented two very different "research traditions" (Laudan, 1977). The evolutionary synthesis was a fusion of these widely different traditions. Such an event occurs only occasionally in the history of science, as Laudan writes: "There are times when two or more research traditions, far from mutually undermining one another, can be amalgamated, producing a synthesis, which is progressive with respect to both the former research traditions" (p. 103). What happened between 1937 and 1947 was precisely such a synthesis between research traditions that had previously been unable to communicate.

As Laudan stresses, no victory of one paradigm over another was involved, as in Kuhn's theory of scientific revolutions, but rather an exchange of the most viable components of the previously competing research traditions. For this reason to state that the synthesis was merely an acceptance by the naturalists of the newer findings of genetics ignores the numerous concepts that the geneticists took over from the naturalists: population thinking, the multidimensionality of the polytypic species, the biological species concept (with the species defined as a reproductively and ecologically autonomous entity), the role of behavior and change of function in the origin of evolutionary novelties, and so on. These concepts are indispensable for understanding evolution and yet they had been originally absent from the conceptual framework of the experimental geneticists. The crucial significance of the synthesis, then, was the fusion of the widely diverging conceptual frameworks of experimentalists and naturalists into a single one. There is every justification to designate this process as a synthesis.

The Importance of Bridge Builders

When I read what was written by both sides during the 1920s, I am appalled at the misunderstandings, the hostility, and the intolerance of the opponents. Both sides display a feeling of superiority over their opponents "who simply do not understand what the facts and issues are." How could they have ever come together? Just as in the case of warring nations, intermediaries were needed, evolutionists who were able to remove misunderstandings and to build bridges between hierarchical levels. These bridge builders were the real architects of the synthesis.

What qualifications did an evolutionist require to be able to serve as a bridge builder? Finding the answer to this question was one of the objectives of our conference. A few facts, however, are evident. None of the bridge builders was a narrow specialist. They all had, so to speak, a foot in several camps. They all were willing to learn the new findings in areas of biology outside their own field of specialization. Sumner, Rensch, and Mayr accepted in their earliest papers the existence of a certain amount of soft inheritance but all three subsequently adopted the Darwinian interpretation without reservations. Additional insight can be acquired by indicating the authors or schools that were particularly active in the bridge building. The Russian school under Chetverikov included naturalists-geneticists who understood the nature of species and natural populations and applied population thinking to the new findings of genetics. Geneticists, like Ford, had the interests and the background of naturalists. Systematists, like Rensch, Mayr, and Simpson, were thoroughly familiar with the evolutionary literature and took the trouble to become acquainted with the new genetics. Zoologists, particularly Huxley, had been raised in a Darwinian milieu and were equally familiar with the work of the geneticists and of the naturalists-systematists. Botanists, such as Anderson and Stebbins, were thoroughly familiar with both the taxonomic and genetic literature.

Each of the ten or twelve people most active in the synthesis occupied a special niche. Just mentioning the names Dobzhansky, Simpson, Rensch, and Huxley makes this very obvious. Yet they all recognized the communication gap between the various evolutionary schools, and they all were aware of the rather radical changes in genetic theory that had occurred between the early mutationism of de Vries, Johannsen, and Bateson, and the new genetics of Morgan, Muller, Wright, and their contemporaries. The intriguing problem is how did a paleontologist (Simpson), a zoologist (Rensch), a systematist (Mayr), and a botanist (Stebbins) acquire their understanding of the rather fundamental conceptual change in genetics? Our discussions should shed some light on this problem.

Another question is to what extent the synthesis was completed by the end of the 1940s. Arguments continued, in fact still continue, on certain components of the synthetic theory. For instance, how homogeneous or heterogeneous are gene pools? How much of the observed genetic variability is the result of selected balancing processes and how much of it is simply "noise" produced by stochastic processes ("non-Darwinian evolution")? What is the evolutionary role of populous, contiguous populations and what is that of small peripherally isolated populations? All these problems are merely superimposed on the framework of the synthetic theory. To me at least, the basic structure of the theory arrived at in the 1940s seems in no way changed regardless how these and other questions are answered. For me the synthesis was completed in principle in the 1940s.

The Princeton Conference

Perhaps the most astonishing aspect of the evolutionary synthesis was the rapidity with which it spread through most countries except France. Several conferences on evolution had been held in the 1920s and early 1930s, like the Tübingen meeting of paleontologists and geneticists (see chapter 9), but all such meetings ended in virtually total discord. The degree of consensus among evolutionists was again tested at an international conference under the auspices of the National Research Council held at Princeton, New Jersey, January 2-4, 1947 (Jepsen, Mayr, and Simpson, 1949). The organizers of the Princeton conference made a special effort to bring together representatives of the most diverse fields, including paleontologists, morphologists, ecologists, ethologists, systematists, and geneticists of various schools. Among the participants were E. H. Colbert, D. D. Davis, Th. Dobzhansky, G. Jepsen, E. Mayr, J. A. Moore, H. J. Muller, E. Olson, B. Patterson, A. S. Romer, G. G. Simpson, W. P. Spencer, G. L. Stebbins, C. Stern, H. E. Wood, and S. Wright from the United States and E. B. Ford, J. B. S. Haldane, D. Lack, and T. S. Westoll from England.

If such a meeting had been held fifteen years earlier there would have been bitter arguments from morning to night, each participant trying to refute the claims of the others. Nothing of the sort happened at Princeton. In fact, it was almost impossible to get a controversy going, so far-reaching was the basic agreement among the participants. To test the reliability of my memory, I circulated a questionnaire among the survivors and received statements from Colbert, Ford, Moore, Patterson, Simpson, Stebbins, and Stern completely corroborating my recollections. All of them recalled an essential agreement among all the participants on the gradual mode of evolution, with natural selection as the basic mechanism and the only direction-giving force. The few arguments raised concerned aspects within the synthetic theory (such as the relative importance of population size and ecological components of speciation). One of the organizers (Jepsen) had given out the instruction that there should be "no segregation by discipline" during the social hour and at meals. As a result, groups of taxonomists, geneticists, and paleontologists, of steadily changing composition, happily engaged in mutual enlightenment. There was no evidence of any misunderstandings. It was not that the synthesis was hammered out during the Princeton conference— rather, the conference constitutes the most convincing documentation that a synthesis had occurred during the preceding decade. Furthermore, no major revision of the consensus of the 1940s became necessary during the ensuing years, only a somewhat increased sophistication in certain formulations that had been oversimplified originally. I do not claim that by 1947 every evolutionist had become a neo-Darwinian. Many members of the older generation, like Buchner, Smirnov, Weidenreich, Willis, Cannon, and others, were unable to convert. What it does demonstrate is that evolutionary biology was no longer split into two noncommunicating camps. The previously existing barrier between them had almost completely disappeared.

Was the Synthesis a Scientific Revolution?
We may ask whether the synthesis was a scientific revolution. Some scientific revolutions extended over long periods of time; other rather sudden conceptual changes fail to qualify as genuine scientific revolutions. According to Kuhn (1962), a scientific revolution is characterized by the fact that a previously prevailing paradigm is dramatically replaced by a new one. Scientific revolutions, stated in such extreme terms, occur only rarely in the biological sciences, if at all. In the case of all the more difficult problems of biology there are usually several competing theories, particularly the more complex the phenomenon. No simple replacement is possible; in most cases, individual components are modified until the competing theories have merged into a single one.

If "scientific revolution" is defined as the occurrence of something drastically new, then the synthesis certainly does not qualify as a revolution. Indeed, in its most important components, the synthesis is remarkably similar to Darwin's original theory of 1859. It was not so much the discovery of new facts that characterized the synthesis as the removal of misunderstandings. The period from 1859 to 1935 had suffered from being dominated by erroneous conceptualizations, such as essentialism (typology), a belief in an intrinsic tendency toward progress, a misunderstanding of the nature of inheritance and of mutation, a failure to understand the nature of populations and the uniqueness of individuals, and other erroneous concepts. Some, such as a belief in the existence of soft inheritance and a failure to apply population thinking consistently, had been part of Darwin's original theory. The synthesis, then, evidently was not still another revolution but simply the final implementation of the Darwinian revolution.

This conference, and others like it, are important both to the evolutionary biologist and to the historian of science. By looking at the gradual maturation of the now prevailing concepts of evolutionary biology, we can attain a far better understanding of the true nature of these concepts than merely by definitions or by attempts at empirical induction. Each concept 'is burdened with the heritage of its past and thus it often has a somewhat different meaning to representatives of different fields, let us say a developmental biologist, a paleontologist, and a population geneticist.

For the historian of science, the events during the synthesis illustrate in the most graphical manner certain phenomena and processes that in a similar manner have occurred again and again in the history of science. The resistance to new ideas, the role of terminologies, the failure of communication, the fusing together of disjunct pieces into new theories, the bridging of gaps between hierarchical levels, the role of generalizers, and many other phenomena are important not only for objective historiography but also for an understanding of the actual method of science. It is my sincere hope and optimistic expectation that this conference will make a major contribution in both respects.

References

Allen, G. E. 1968. Thomas Hunt Morgan and the problem of natural selection. *Journal of the History of Biology* 1:113-139.

Anderson, E. 1939. Recombination in species crosses. *Genetics* 24:668-698.

Babcock, E. B. 1947. The genus *Crepis* I, II. *University of California Publications, Botany*, 21, 22.

Bateson, W. 1894. *Materials for the study of variation*. New York: Macmillan.

———— 1909. *Mendel's principles of heredity*. Cambridge: Cambridge University Press.

———— 1913. *Problems of genetics*. New Haven: Yale University Press.

———— 1922. Evolutionary faith and modern doubts. *Science* 55:55-61.

Bather, F. A. 1927. Biological classification: past and future. *Proceedings of the Geological Society of London* 83:62-104.

Baur, E. 1925. Die Bedeutung der Mutationen für das Evolutionsproblem. *Zeitschrift für Induktive Abstammungs- und Vererbungslehre* 37:107-115.

Bonnier, G. 1927. Species-differences and gene differences. *Hereditas* 9:137-144.

Buch, L. von. 1825. *Physicalische Beschreibung der Canarischen Inseln*. Berlin: Königliche Akademie der Wissenschaften, pp. 132-133.

Buchner, P. 1938. *Allgemeine Zoologie*. Leipzig: Quelle und Meyer.

Bush, G. 1975. Modes of animal speciation. *Annual Review of Ecology and Systematics* 6:339-364.

Butterfield, H. 1957. *The origins of modern science*. London: G. Bell.

Cassirer, E. 1950. *The problem of knowledge*. New Haven: Yale University Press.

Castle, W. E. 1916. *Genetics and eugenics*. Cambridge, Massachusetts: Harvard University Press.

———— 1919. Piebald rats and the theory of genes. *Proceedings of the National Academy of Sciences* 5:126-130.

————— and J. C. Phillips. 1914. Piebald rats and selection. *Carnegie Institution of Washington Publication*, no. 195.

Caullery, M. 1931. *Le problème de l'évolution*. Paris: Payot.

Chapman, F. M. 1923. Mutation among birds in the genus Buarremon. *Bulletin of the American Museum of Natural History* 48:243-278.

Chetverikov, S. S. 1926. On certain aspects of the evolutionary process from the standpoint of modern genetics (in Russian). *Zhurnal Eksperimental 'noi Biologii* A2:3-54. (English translation, 1961, *Proceedings of the American Philosophical Society* 105:167-195.)

Darlington, C. D. 1940. Taxonomic species and genetic systems. In *The new systematics*, ed. J. Huxley. Oxford: Clarendon Press, pp. 137-160.

Darwin, C. 1859. *On the origin of species by means of natural selection, or the preservation of favored races in the struggle for life*. London: John Murray.

Delage, Y., and M. Goldsmith. 1909. *Les théories de l'évolution*. Paris: Flammarion.

Detlefsen, J. A. 1925. The inheritance of acquired characters. *Physiological Reviews* 5:244-278.

de Vries, H. 1906. *Species and varieties, their origin by mutation*. Chicago: Open Court.

Dobzhansky, Th. 1937a. *Genetics and the origin of species*, 1st ed. New York: Columbia University Press.

————— 1937b. The genetic nature of species differences. *American Naturalist* 71: 404-420.

————— 1970. *Genetics of the evolutionary process*. New York: Columbia University Press.

Du Rietz, G. E. 1930. The fundamental units of biological taxonomy. *Svensk Botanisk Tidskrift* 24:333-428.

East, E. M. 1910. A Mendelian interpretation of variation that is apparently continuous. *American Naturalist* 44:65-82.

————— 1916. Inheritance in crosses between *Nicotiana langsdorffii* and *Nicotiana alata*. *Genetics* 1:311-333.

Emerson, R. A., and E. M. East. 1913. The inheritance of quantitative characters in maize. *Nebraska Agricultural Experiment Station Research Bulletin* 2.

Federley, H. 1929. Weshalb lehnt die Genetik die Annahme einer Vererbung erworbener Eigenschaften ab? *Paläontologische Zeitschrift* 11:287-317.

Fisher, R. A. 1931. The evolution of dominance. *Biological Reviews* 6:345-368.

————— 1954. Retrospect of the criticisms of the theory of natural selection. In *Evolution as a process*, ed. J. Huxley et al. London: Allen and Unwin, pp. 84-98.

————— 1959. Natural selection from the genetical standpoint. *Australian Journal of Science* 22:16-17.

Ford, E. B. 1964. *Ecological genetics*. London: Methuen.

Ghiselin, M. T. 1969. *The triumph of the Darwinian method*. Berkeley: University of California Press.

————— 1974. A radical solution to the species problem. *Systematic Zoology* 23: 536-544.

Goette, A. 1875. *Entwickelungsgeschichte der Unke als Grundlage einer ver-*

gleichenden Morphologie der Wirbeltiere.

Goldschmidt, R. 1920. *Die quantitative Grundlage von Vererbung und Artbildung.* Berlin: Springer.

───── 1934. Lymantria. *Bibliographia Genetica* 11:1-180.

───── 1940. *The material basis of evolution.* New Haven: Yale University Press.

Grassé, P. P. 1973. *L'Evolution du vivant.* Paris: Albin Michel.

Grene, M. 1959. Two evolutionary theories. *British Journal of the Philosophy of Science* 9:110-127, 185-193.

Gulick, J. T. 1873. On diversity of evolution under one set of external conditions. *Journal of the Linnean Society of London, Zoology* 11:496-505.

Haldane, J. B. S. 1932. *The causes of evolution.* London: Longmans, Green.

───── 1938. The nature of interspecific differences. In *Evolution*, ed. G. R. de Beer. Oxford: Clarendon Press, pp. 79-94.

Hartmann, M. 1947. *Allgemeine Biologie*, 3d ed. Jena: G. Fischer.

Hertwig, R. 1927. *Abstammungslehre und neuere Biologie.* Jena: G. Fischer.

Heslop-Harrison, J. W. H. 1920. Genetical studies in the moths of the genus Oporobia. *Journal of Genetics* 9:195-280.

His, W. 1874. *Unsere Körperform und das physiologische Problem ihrer Entstehung.* Leipzig: F. C. W. Vogel.

Hubbs, C. 1940. Speciation of fishes. *American Naturalist* 74:198-211.

Hull, D. 1973. *Darwin and his critics.* Cambridge, Massachusetts: Harvard University Press.

Huxley, J., ed. 1940. *The new systematics.* Oxford: Clarendon Press.

───── 1942. *Evolution, the modern synthesis.* London: Allen and Unwin.

Jenkin, F. 1867. The origin of species. *North British Review* 46:149-171.

Jepsen, G. L., E. Mayr, and G. G. Simpson, eds. 1949. *Genetics, paleontology, and evolution.* Princeton: Princeton University Press.

Johannsen, W. 1915. Experimentelle Grundlagen der Deszendenslehre. Variabilität, Vererbung, Kreuzung, Mutation. *Kultur der Gegenwart* III, 4, vol. 4:497-660.

───── 1923. Some remarks about units in heredity. *Hereditas* 4:133-141.

Jollos, V. 1934. Inherited changes produced by heat-treatment in *Drosophila melanogaster. Genetica* 16:476-494.

Jordan, D. S. 1905. The origin of species through isolation. *Science* 22:545-562.

Kammerer, P. 1924. *The inheritance of acquired characteristics.* New York: Boni and Liveright.

Kellogg, V. L. 1907. *Darwinism to-day.* New York: Henry Holt.

Laudan, L. 1977. *Progress and its problems.* Berkeley: University of California Press.

Loeb, J. 1916. *The organism as a whole from a physico-chemical viewpoint.* New York: G. P. Putnam's Sons.

Lotsy, J. P. 1916. *Evolution by means of hybridization.* The Hague: M. Nijhoff.

MacDowell, E. C. 1917. Bristle inheritance in Drosophila. II. Selection. *Journal of Experimental Zoology* 23:109-146.

Mayr, E. 1942. *Systematics and the origin of species.* New York: Columbia University Press.

———— 1955. Karl Jordan's contribution to current concepts in systematics and evolution. *Transactions of the Royal Entomological Society of London* 45-66.

———— 1957. *The species problem*. Washington: American Association for the Advancement of Science.

———— 1959. Where are we? *Cold Spring Harbor Symposia* 24:1-14.

———— 1961. Cause and effect in biology. *Science* 134:1501-1506.

———— 1963. *Animal species and evolution*. Cambridge, Massachusetts: Belknap Press of the Harvard University Press.

———— 1976. Is the species a class or an individual? *Systematic Zoology* 25:192.

———— 1978. Origin and history of some terms in systematic and evolutionary biology. *Systematic Zoology* 27:83-88.

Moorhead, P. S., and M. M. Kaplan, eds. 1967. *Mathematical challenges to the neo-Darwinian interpretation of evolution*. Philadelphia: Wistar Institute Monograph, no. 5.

Morgan, T. H. 1923. The bearing of Mendelism on the origin of species. *Scientific Monthly* 16:237-247.

———— 1932. *The scientific basis of evolution*. New York: W. W. Norton.

Nilsson-Ehle, H. 1909. Kreuzungsuntersuchungen an Hafer und Weizen. *Lunds Universitets Årsskrift*, n.s., ser. 2, vol. 5, no. 2.

Noble, G. K., and M. E. Jaeckle. 1928. [On the origin of arboreal frogs.] *Journal of Morphology and Physiology* 45:259-292.

Nordenskioeld, E. 1928. *The history of biology*. New York: Knopf (originally published 1920-1924).

Osborn, H. F. 1927. Speciation and mutation. *American Naturalist* 61:5-42.

Pattee, H. H., ed. 1973. *Hierarchy theory*. New York: George Braziller.

Patterson, B. 1949. Rates of evolution in taeniodonts. In *Genetics, paleontology and evolution*, ed. G. L. Jepsen, E. Mayr, and G. G. Simpson. Princeton: Princeton University Press.

Payne, R. 1920. Selection for increased and decreased bristle number. *Anatomical Record* 17:335-336.

Plate, L. 1913. *Selektionsprinzip und Probleme der Artbildung*, 4th ed. Leipzig and Berlin: W. Engelmann.

———— 1925. *Die Abstammungslehre*. Jena: G. Fischer.

Plough, H. H., and P. T. Ives. 1935. Induction of mutations by high temperature in Drosophila. *Genetics* 20:42-69.

Popper, K. R. 1972. *Objective knowledge*. Oxford: Clarendon Press.

Poulton, E. B. 1903. What is a species? *Proceedings of the Entomological Society of London* lxxvi-cxvi.

Provine, W. B. 1971. *The origins of theoretical population genetics*. Chicago: University of Chicago Press.

———— 1979. Francis B. Sumner and the evolutionary synthesis. *Studies in the History of Biology* 3:211-240.

Rensch, B. 1929. *Das Prinzip geographischer Rassenkreise und das Problem der Artbildung*. Berlin: Borntraeger.

———— 1939. Typen der Artbildung. *Biological Reviews* 14:180-222.

———— 1947. *Neuere Probleme der Abstammungslehre*. Stuttgart: Enke.

Robson, G. C. 1928. *The species problem*. Edinburgh: Oliver and Boyd.

———— and O. W. Richards. 1936. *The variation of animals in nature*. London: Longmans, Green.

Rothschild, W., and K. Jordan. 1903. Lepidoptera collected by Oscar Neumann. *Novitates Zoologicae* 10:492 [491-500].

Schindewolf, O. H. 1936. *Paläontologie, Entwicklungslehre, und Genetik*. Berlin: Borntraeger.

Semon, R. 1912. *Das Problem der Vererbung "erworbener Eigenschaften."* Leipzig: W. Engelmann.

Simpson, G. G. 1944. *Tempo and mode in evolution*. New York: Columbia University Press.

Strasburger, E., et al. 1919. *Lehrbuch der Botanik*, 14th ed. Jena: G. Fischer.

Stresemann, E. 1926. Übersicht über die Mutationsstudien I-XXIV und ihre wichtigsten Ergebnisse. *Journal für Ornithologie* 74:377-385.

Sturtevant, A. 1918. An analysis of the effects of selection. *Carnegie Institution of Washington Publication* 264:1-68.

Sulloway, F. J. 1979. Geographic isolation in Darwin's thinking: the vicissitudes of a crucial idea. *Studies in the History of Biology* 3:23-65.

Sumner, F. B. 1932. Genetic, distributional, and evolutionary studies of the subspecies of deer mice (*Peromyscus*). *Bibliographia Genetica* 9:1-106.

Tschulok, S. 1922. *Deszendenzlehre*. Jena: G. Fischer.

Wallace, A. R. 1855. On the law which has regulated the introduction of new species. *Annals and Magazine of Natural History* 16:184-196.

———— 1889. *Darwinism*. London: Macmillan.

Weidenreich, F. 1929. Vererbungsexperiment und vergleichende Morphologie. *Paläontologische Zeitschrift* 11:275-286.

Weismann, A. 1883. *Über die Vererbung*. Jena: G. Fischer.

White, M. J. D. 1978. *Modes of speciation*. San Francisco: W. H. Freeman.

Wilson, E. B. 1909. The cell in relation to heredity and evolution. In *Fifty years of Darwinism*, ed. anonymous [T. C. Chamberlin]. New York: Henry Holt, pp. 92-113.

Wright, S. 1967. The foundations of population genetics. In *Heritage from Mendel*, ed. R. A. Brink. Madison: University of Wisconsin Press, p. 246.

———— 1968. *Evolution and the genetics of populations*. Chicago: University of Chicago Press, vol. 1.

Zirkle, C. 1941. Natural selection before the "Origin of Species." *Proceedings of the American Philosophical Society* 84:71-124.

PART ONE

Different Biological Disciplines
and the Synthesis

1 Genetics

Advances within the field of genetics were clearly of basic importance to the evolutionary synthesis. Indeed, the most widespread view among geneticists since 1940 is that the evolutionary synthesis occurred because ideas developed by geneticists were incorporated into the thinking of systematists, field naturalists, paleontologists, cytologists, and embryologists. This view suggests that the great weakness of Darwin's theory of evolution was his faulty conception of the mechanism of heredity ("provisional hypothesis of pangenesis"), based in part on an assumption of blending inheritance. Mendel's theory of heredity perfectly filled the gap in Darwin's thinking, thus leading directly to population genetics and the evolutionary synthesis (see chapter 14 for a more detailed exposition).

Ernst Mayr's views as a systematist clearly differ in emphasis from this view favored by geneticists. Although attributing much importance to advances made by geneticists, Mayr also emphasized the contributions made by systematists, paleontologists, and field naturalists in the prologue to this volume. Wishing to represent contrasting views, Mayr invited L. C. Dunn to assess the general relation of genetics to evolution during the synthesis period, drawing from his vast knowledge of the history of genetics, and R. C. Lewontin to address the specific problem of the role of the theoretical population genetics of Fisher, Wright, and Haldane in the evolutionary synthesis.

Unfortunately Dunn died shortly before the conference just as he was beginning to write out his views on the evolutionary synthesis. The brief notes that Dunn completed before his death (available at the American Philosophical Society), his transcript from the Columbia University Oral History Project, and his *Short History of Genetics* (1965) all indicate that he thought that the evolutionary synthesis resulted from the exportation of developments within genetics to other fields of evolutionary biology, especially to systematics and paleontology.

The interaction of genetics with evolutionary theory in the period of the evolutionary synthesis is more complex, however. Many different schools of genetics research existed. Significant differences in outlook and conclusions existed not only between research groups, but sometimes also within them. Even when most geneticists could agree about the interpretation of a genetic phenomenon of importance in evolution, researchers in the other fields that were unified in the synthesis often were unaware of or uninterested in the results of genetics research. An adequate view of the contribution of genetics to the evolutionary synthesis must rest on careful, detailed historical research into specific developments.

The bulk of genetics research was conducted in the United States, but important centers were also located in Great Britain, Germany, the Netherlands, the Soviet Union, and elsewhere, as the following list of genetics research and teaching centers in the mid-1920s indicates.

Columbia University — Morgan, Wilson
Students: Muller, Sturtevant, Bridges, Altenburg, Metz, Payne, Weinstein, D. E. Lancefield

Bussey Institution of Harvard University — Castle, East
Students of Castle: Wright, Dunn, Little, MacDowell, Reed, Snell, Snyder, Gates, Whiting
Students of East: Jones, Sax, Brink, Stadler

Cornell University — R. A. Emerson, Sharp
Students: Anderson, McClintock, Beadle, Rhoades, Demerec, S. Emerson, Randolph

Johns Hopkins University — Jennings, Pearl
University of California, Berkeley — Babcock, Clausen
University of Wisconsin — Cole, Brink
University of Texas — Muller, Patterson, Stone
University of Missouri — Stadler
University of Michigan — A. F. Shull
Department of Genetics, Carnegie Institution of Washington, Cold Spring Harbor — Davenport, Blakeslee, MacDowell, Belling
Connecticut Agricultural Experiment Station — Jones, Dunn, Landauer
Oxford University — Julian Huxley, Ford
John Innes Horticultural Institution — J. B. S. Haldane, Darlington
Rothamsted Experimental Station — Fisher
University of Edinburgh — Crew
Kaiser-Wilhelm Institut für Biologie (Berlin-Dahlem) — Goldschmidt, Stern (Baur, Correns, and Nachtsheim were also in Berlin-Dahlem in the mid-1920s)

Koltsov Institute, Moscow	Koltsov, Chetverikov,
	Timoféeff-Ressovsky
Institute of Applied Botany, Leningrad	Vavilov
University of Leningrad	Philipchenko

Most of these centers of genetics research, not just the first three, produced a larger second generation of geneticists who were active and influential during the thirties and forties. Many in this generation started new centers of genetics research. Some of them were already in influential positions by the mid-1920s, such as Muller, Stadler, Jones, and Dunn.

How the first generation of geneticists focused on genetics as a research field is a fascinating problem that has so far received little attention. The intellectual pedigrees in Sturtevant's *History of Genetics* (1965) are often useful though sketchy for the second and succeeding generations of geneticists, but do not reveal the origins of the first generation. Davenport took his doctorate with Mark at the Zoological Laboratory of the Museum of Comparative Zoology at Harvard; later Jennings and Castle worked with both Davenport and Mark. None of them studied heredity until after receiving their degrees for different reasons. East was trained as a chemist, and later worked with Hopkins on selection experiments with corn at the Illinois Agricultural Experiment Station. This experience apparently turned his attention to genetics.

The membership lists at the fifth and sixth international congresses of genetics yield a very rough idea of the size of the genetics research community. The fifth congress in 1927 in Berlin had 903 paid members, with 533 from Germany, 64 from the USSR, 61 from the United States, and 45 from Great Britain (*Verhandlungen des V. Internationalen Kongresses für Vererbungswissenschaft*, 1928). The sixth congress held in 1932 in Ithaca had 823 paid members, with 693 from the United States, 49 from Canada, 15 from Great Britain, and only 11 from Germany and 5 from the USSR (*Proceedings of the Sixth International Congress of Genetics*, 1932). Geographic location of the congress obviously was a determining factor in the enrollment figures, as was the depression.

Communication between geneticists working on similar problems in the same genetics specialty was reasonably good. The international congresses every five years enabled geneticists to have personal contact; many older geneticists hold vivid memories of the interchanges at these congresses. Only a few journals were published so that geneticists could keep up with almost all related studies. In the United States, the primary journals were *Genetics*, *American Naturalist*, and the *Journal of Heredity*. England had the *Journal of Genetics*, Germany the *Zeitschrift für Induktive Abstammungs- und Vererbungslehre*, Holland *Bibliographia Genetica* and *Resumptio Genetica*, and Sweden *Hereditas*. Genetics research was also published in the proceedings of many academies (such as the National Academy of Sciences in the United States), or in such journals as the *Journal of Experimental Zoology, Journal of Mammalogy,* or *Cytologia.* Exchange of journal reprints was a highly developed means of communication. Most of the genetics research groups had access to not only the journals, but also one or more extensive reprint collections, which were generally arranged by author rather than field.

Most geneticists in universities taught general courses in genetics; as a result, they were forced to keep up with new research in many areas of genetics. Communication, however, clearly fell off rapidly outside specific fields of research. Often geneticists kept abreast of other fields in genetics only through their colleagues.

Geneticists in the late 1920s and early 1930s generally paid little attention to the problem of genetics in relation to evolution. The most prominent exceptions were, of course, Fisher, Haldane, and Wright. Others very interested in evolution included Goldschmidt and Baur in Germany; Huxley and Ford in England; Castle, East, and Jennings in the United States; and Chetverikov in Russia. The professional journals before the evolutionary synthesis focused mainly on laboratory or agricultural genetics rather than on evolution in nature.

Many of the results of laboratory or agricultural genetics were applied to the problem of evolution in nature during the synthesis period. In most cases this transfer could be done theoretically. Darwin extended the idea of artificial selection to the idea of natural selection in his first four chapters of *On the Origin of Species,* at a time when he had no experimental evidence of evolution by natural selection in nature. Similarly, the new genetic theory of evolution in nature was worked out well in advance of knowledge of the actual genetics of natural populations. For example, Sewall Wright wrote nothing on evolution before 1925, when he wrote the first draft of his famous paper, "Evolution in Mendelian Populations" (1931). Wright's "shifting balance" theory of evolution was based on his analysis of the inheritance of coat color and inbreeding in laboratory populations of guinea pigs and on the methods of selective breeding he had developed earlier. He simply assumed that natural selection, like a breeder familiar with modern genetics, would operate by the most effective mechanisms (Wright, 1978). As in Darwin's case, Wright's rather bold assumption proved very fertile.

Genetics made many other contributions to the evolutionary synthesis as the following brief list indicates.

Variability and its inheritance. The rediscovery of Mendel's theory of heredity in 1900 ushered in a period of vigorous research into the patterns of inheritance of distinct variations. Mendelian ratios were found in many organisms, from peas to flies to humans. Johannsen's research on "pure lines" of beans led to his important distinction of genotype and phenotype (Churchill, 1974). Jennings discovered that minute differences between pure lines of *Paramecia* were strictly hereditary (1908); by 1912, Morgan and his students had found that tiny differences in *Drosophila melanogaster* were inherited according to Mendel's laws of segregation (Morgan, Sturtevant, Muller, and Bridges, 1915). Nilsson-Ehle (1909) and East (1910) demonstrated that Mendelian multiple-factor inheritance could explain almost continuous arrays of variability. Muller (1927) and others showed how X rays and some chemicals could cause mutations that were inherited in the same way as naturally occurring mutations. Mendel's simple laws of segregation were modified by more complex patterns of heredity revealed by breeding experiments. Hereditary factors such as melanism in the peppered moth and blood group genes in man were discovered and later used to analyze evolutionary change.

Artificial selection in the laboratory. Opponents of Darwin's gradual view of evolution by natural selection, such as Galton, de Vries, Johannsen, and Bateson, had claimed that selection of small fluctuating differences could not move a population beyond original limits of variability. Selection experiments by Castle on rats (Castle and Wright, 1916) and Sturtevant (1918) and Payne (1918) on *Drosophila* proved that selection of small differences was far more powerful than the mutationists believed.

Physiological genetics. Careful experiments by Goldschmidt (1921) on sex determination in the moth *Lymantria* and by Castle and Wright on guinea pigs (1916) revealed clearly that genes have pleiotropic effects. Gene interaction and position effect were important research topics for geneticists by the late 1920s.

Theoretical population genetics. Pearl, Jennings, and Wright invented quantitative theories of inbreeding. Fisher partitioned phenotypic variance into environmental and genetic components, and the genetic component further into components caused by additive, epistatic, and dominance effects (1918). Fisher, Haldane, and Wright showed theoretically how selection in combination with factors such as mutation rate, migration rate, assortative mating, and effective population size could explain evolution in nature without the need of other hypotheses. The inheritance of acquired characters was no longer a theoretical necessity for evolution; no convincing evidence for it appeared, and the concept was dropped by most geneticists.

Cytogenetics. De Vries's theory of evolution by mutation in *Oenothera* was refuted resoundingly. Genes were located with some assurance into linkage groups, then into linear arrangements, and finally on the chromosomes. The evidence was sufficient to convince most geneticists by 1920. Chromosomal rearrangements such as inversions, translocations, and deletions were discovered. Inversions in particular were used by Sturtevant and Dobzhansky by the mid-1930s to study changes in the gene pools of *Drosophila pseudoobscura*. Botanists used cytogenetics (for example, polyploidy) to trace phylogenies.

Agricultural genetics. The great advances in animal and plant breeding as a result of the rise of genetics can be seen by comparing Davenport's *Principles of Breeding* (1907) with Jones's *Genetics in Plant and Animal Improvement* (1925). Two developments in agricultural genetics were important in evolutionary theory during the synthesis period. First, the concepts developed in the analyses of inbreeding and outbreeding were later applied to problems of effective population size and genetic diversity in natural populations. The development of Sewall Wright's theory of evolution is understandable only in the light of his practical and theoretical work on inbreeding in guinea pigs. Second, breeders became quite sophisticated in their use of variance analysis to develop effective patterns of artificial selection. For example, the selection experiment at the Maine Agricultural Experiment Station to increase fecundity in fowl produced no positive results and convinced Raymond Pearl in 1909 that Darwinian selection could not work in nature (Pearl and Surface, 1909). Variance analysis indicated that fecundity in fowl had a low heritability. Using techniques of sib-selection instead of mass selection, a selection experiment on fecundity in fowl was started in the mid-1930s at the University of California, Berkeley, and produced remarkable positive results. Heritability estimates were essential for understanding selective pressures upon phenotypic characters in natural populations.

Genetics of natural populations. Goldschmidt was the first geneticist to study the genetics of geographical races seriously in his work on *Lymantria* (1933). He believed that geographic races of *Lymantria* generally differed in a relatively few Mendelian characters. The meticulous work of Sumner on geographic races of the deer mouse *Peromyscus* revealed that the differences between geographic races were not simple Mendelian characters, but continuously varying quantitative characters. Only after ten years of rejecting the theory did he adopt the view that the differences were best explained by multiple-factor Mendelian inheritance (Provine, 1979). In the 1920s, work by Chetverikov and his students showed many Mendelian recessive mutants in natural populations of *Drosophila*. Sturtevant and Dobzhansky began their serious work on genetic variation in *D. pseudoobscura* in the mid-1930s. Geneticists who studied natural populations of plants included Baur, Blakeslee, Müntzing, and Babcock.

These discoveries in the field of genetics were certainly not immediately accepted within the genetics research community. Nor did most geneticists see their application to evolution in natural populations. Workers in other fields, notably systematics and paleontology, were quite slow to appreciate the importance of these advances in genetics. Books published as late as Mayr's *Systematics and the Origin of Species* (1942) and Simpson's *Tempo and Mode in Evolution* (1944) still had a major influence in bringing about the evolutionary synthesis.

Many other discussions and writings about the role of genetics were produced at the conference. The correspondence files and transcripts of the tape recordings of sessions contain valuable information. All of these documents are available at the library of the American Philosophical Society. W.B.P.

References

Castle, W. E., and S. Wright. 1916. *Studies of inheritance in guinea pigs and rats.* Carnegie Institution of Washington, 195.

Churchill, F. B. 1974. William Johannsen and the genotype concept. *Journal of the History of Biology* 7:5-30.

Davenport, E. 1907. *Principles of breeding.* New York: Ginn.

Dunn, L. C. 1965. *A short history of genetics.* New York: McGraw-Hill.

East, E. M. 1910. A Mendelian interpretation of variation that is apparently continuous. *American Naturalist* 44:65-82.

Fisher, R. A. 1918. The correlation between relatives on the supposition of Mendelian inheritance. *Transactions of the Royal Society of Edinburgh* 52:399-433.

Goldschmidt, R. 1921. The determination of sex. *Nature* 107:780-784.

———— 1933. Lymantria. *Bibliographia Genetica* 11:1-186.

Jennings, H. S. 1908. Heredity, variation, and evolution in Protozoa: 2. *Proceedings of the American Philosophical Society* 47:393-546.

Jones, D. F. 1925. *Genetics in plant and animal improvement.* New York: Wiley.

Morgan, T. H., A. H. Sturtevant, H. J. Muller, and C. B. Bridges. 1915. *The mechanism of Mendelian heredity.* New York: Holt.

Muller, H. J. 1927. Artificial transmutation of the gene. *Science* 66:84-87.

Nilsson-Ehle, H. 1909. Kreuzungsuntersuchungen an Hafer und Weizen. *Lunds Universitets Årsskrift*, n.s., ser. 2, vol. 5, no. 2.

Payne, F. 1918. An experiment to test the nature of the variations on which selection acts. *University of Indiana Studies* 5:1-45.

Pearl, R., and F. Surface. 1909. Is there a cumulative effect of selection? *Zeitschrift für Induktive Abstammungs- und Vererbungslehre* 2:257-275.

Proceedings of the Sixth International Congress of Genetics. 1932. Menasha, Wisconsin: Brooklyn Botanic Garden.

Provine, W. B. 1979. Francis B. Sumner and the evolutionary synthesis. *Studies in the History of Biology* 3:211-240.

Sturtevant, A. H. 1918. *An analysis of the effects of selection*. Carnegie Institution of Washington publ. no. 264.

Verhandlungen des V. internationalen Kongresses für Vererbungswissenschaft. 1928. Two-volume supplement to *Zeitschrift für Induktive Abstammungs- und Vererbungslehre.* Leipzig: Borntraeger.

Wright, S. 1931. Evolution in Mendelian populations. *Genetics* 16:97-159.

———— 1978. The relation of livestock breeding to theories of evolution. *Journal of Animal Science* 46:1192-1200.

Theoretical Population Genetics in the Evolutionary Synthesis

Richard C. Lewontin

Addressing the role of the theoretical work of Fisher, Wright, and Haldane in the evolutionary synthesis of the 1930s and 1940s is difficult, indeed impossible, because of the confusion between the history of science and the sociology of science. Or put another way, the problem is perhaps the difference between what Karl Popper calls objective knowledge and subjective knowledge (1965). The perceptions of those engaged in the synthesis of evolutionary ideas differed significantly from the actual state of knowledge that existed on the theoretical plane. Much of the theoretical work was not in fact incorporated into the thinking of biologists.

Almost all biologists are either scornful, hostile, or fearful of mathematical theory, sometimes with good reason. As a result, the theoretical work, especially as it becomes mechanically complex, becomes truly difficult for biologists to incorporate into their thinking. I think the evolutionary synthesis actually did occur without a great deal of direction from theory. Nevertheless, a tremendous amount of understanding and synthesis of evolutionary ideas could have been derived chiefly from the theoretical works of Fisher and Wright. These insights were explicit in the writings of Fisher and Wright, but simply unavailable to most biologists for reasons of literacy.

Contributions of Sir Ronald A. Fisher

In 1918 Sir Ronald Fisher published "The Correlation between Relatives on the Supposition of Mendelian Inheritance," which laid the foundations for all of current biometrical genetics—the branch of genetics devoted to analysis of phenotypes by calculating means, variances, and covariances and other statistical properties of continuously varying phenotypes without specific reference to the frequencies of genes. Fisher's paper was an immense advance, because he considered not one gene but two. That may seem trivial to the biologist, but by considering two genes simultaneously Fisher could estimate the importance of interactions between the genes and generalize to many genes, especially if the interaction turned out to be small. Fisher first showed what is now a regular routine of biometrical geneticists—the partition of the variation of phenotype into not only genetic and nongenetic, but the genetic variation into additive, dominance, and epistatic components. The heritability studies and indeed the whole biometrical tradition utilized by the agricultural geneticist not dealing with gene frequencies come from Fisher.

Yet the outcome of this important paper still remains to be incorporated properly into the thinking of evolutionary biologists who come from a different tradition with the gene frequency as the focus of interest. There are two theoretical structures: the theoretical structure in which everything is dealt with in terms of phenotype, and the theoretical structure in which everything is dealt with in terms of the frequencies of genes. They appear to be independent theoretical structures addressing two different kinds of variation that are linked to each other reciprocally, because developing the theory of phenotype and phenotypic transformations as Fisher did in 1918 must begin with genes. Then the genes get lost in the shuffle. They get absorbed into mysterious parameters like the heritability or the average effect. On the other hand, Mendelian population genetics —that is, dealing with gene frequencies—is impossible unless something about the phenotype of the genotypes—their fitnesses —can be asserted. But those phenotypic transformations get lost in the shuffle because they are assumed constant; a given genotype has a given fitness.

Ledyard Stebbins mentions how little his professors at Harvard College in the 1920s understood about genetics in relation to evolution. In 1951 I was taught by Professor Henry Bryant Bigelow of Harvard that genetics was a passing fad. At the same time I was told by Professor Jeffries Wyman that there was no sense in going to population genetics because Dobzhansky had milked it dry. And when I first learned biometrical genetics as a young assistant professor in North Carolina, I discovered to my horror that my colleagues there, who were all plant and animal breeders, did not know the corpus of the standard Mendelian population genetics. They were solving all the same problems as I was, using different variables, different parameters, different notions—yet they all came to the same conclusions. There were two traditions. The synthesis has not brought those traditions really well together, not even to this day. Systematics and paleontology deal in phenotypes, yet they continue to use the Mendelian population genetics of single genes of fixed effect as a theoretical base for explanation, which only shows consistency and not entailment. As an evolutionary geneticist, I do not see how the origin of higher taxa are the necessary consequence of neo-Darwinism. They are sufficiently explained, but they are not the necessary consequences. The body of knowledge being built in a field must not be confused with the sociology of that knowledge in a scientific community. One of our difficulties in understanding the evolutionary synthesis is the constant confusion between these two issues.

Differences in the Work of Fisher and Wright

The fundamental difference between the work of Fisher and Wright is best exemplified by the titles of their famous great works. Fisher's book, *The Genetical Theory of Natural Selection* (1930), was concerned with phyletic change, the change in the characteristics of a population in time. He was anxious, of course, to show that the suppositions of Mendelism and the supposition of slight differential fitness of genotypes would lead to evolution and could also lead in some cases to the stabilization of variation. Indeed, Fisher is best known among theoreticians not for his theory of balanced polymorphism but for his fundamental theorem of natural selection: "the rate of increase in fitness of any organism at any time is equal to its genetic variance in fitness at that time" (1930, p. 35). This theorem says nothing about balance, only evolutionary progress. It predicts the rate of change in fitness over time. It caused a split between the theoreticians and the naturalists who followed Fisher. Fisher's fundamental theorem has no role to play in the school of ecological genetics of Ford (1964), which is chiefly concerned with balanced polymorphism. For the biometrical geneticist and even for modern theoreticians, including James F. Crow and Motoo Kimura, the beauty of Fisher's work is his fundamental theorem of natural selection and its extensions.

Fisher did not cope with the conflict between the utilization of variation by natural selection, which he brilliantly showed would occur even for very small selection coefficients, and the destruction of variation that accompanied that utilization. A consequence of Fisher's fundamental theorem is that genetic variation will disappear from a population as natural selection operates, the primary problem in Darwinism. If we believe that species arise by the transformation of variation within populations into variation between populations, we must explain the renewal of variation. Wright explicitly faced this problem in his paper, "Evolution in Mendelian Populations" (1931), a more ambitious title than Fisher's book. He tried to encompass the whole of evolution, not simply the question of natural selection. In the first few pages Wright set forth a *Bauplan* for an evolutionary theory: some evolutionary forces preserve variation and others destroy variation.

The essential and new feature that Wright brought into the theory of evolution really was the role of random processes. Most important, he realized that random processes played a significant part in the process of speciation. I do not find in Fisher's writings any theoretical work leading to an understanding of how species are formed, nor do I find in Fisher anything that explains the third of that triad of evolutionary processes: extinction. Phyletic evolution (adaptation), speciation, and extinction are the three processes of evolution. Fisher demonstrated that Mendelism allows us to understand the first. Only in the synthetic work of Wright do we get hints (and only hints) of the way in which both speciation and extinction can flow mechanically from the processes of modulation of variation. This is the role of chance in evolution. For Wright, chance modifies and softens the directive effects of selection so that populations may find themselves with a genetic constitution different from the end-point to which selection is driving them at any particular moment. If populations find themselves at a different genotypic composition than selection is driving them toward, then extinction may occur. This is almost extinction by bad luck.

Wright, unlike Fisher, based his evolutionary synthesis on the importance of gene interactions. Wright always identified himself primarily as a developmental geneticist. He said to me once that the mathematical work was really a diversion from his first love, which was guinea pigs. Wright's papers provide immensely complicated diagrams of the interactions of all the coat color genes in guinea pigs. He introduced his new four-volume work on evolution (1968-1978) with this kind of developmental genetic manipulation. In 1931, Wright only briefly mentioned that the creative importance of genetic drift is that it causes other genes besides the one being studied to change their frequencies. In this way the selective forces on the gene change because interaction between genes is important in determining fitness. Wright's formulations are not restricted to single genes. The equations do indeed involve single genes, but at the same time the parameters, such as selection coefficients, are said to be changing genetically so that the population explores the field of gene frequencies. The exploration of the field of gene frequencies in Wright's synthetic theory depends on the fact that populations are not restricted to a single value of gene frequency by natural selection because the interaction between the genes forced them to go from one adaptive peak to another.

This insight of Wright, that speciation depends on changes in the genetic background of genes and upon changes in the interactions between genes, has not really been incorporated into subsequent theoretical investigations because of its great mathematical difficulty and complexity. Thus, theory has had a bad name among natural, historical, and biological workers as being simplistic bean-bag genetics, solely concerned with single genes. This reaction has hidden the essential contributions of Wright in 1931 and of Fisher in 1918. They were interested in coping with gene interactions from the very beginning.

I cannot define the role of Wright's perceptions in actually molding the the synthesis of the late 1930s because that is a subjective question not really available from the historical record. Occasional works like Dobzhansky's *Genetics and the Origin of Species* (1937) do make explicit reference to Wright's theory. Of course, oral history is just as unreliable as written history because people justify themselves in retrospect and often change their perspectives.

Theoretical Population Genetics and the Evolutionary Synthesis

At this moment, theoretical population genetics plays a remarkably small role in the synthetic attempts of modern evolutionary geneticists to understand what they see. I am constantly amazed at the rubbish that people write about their interpretations of natural historical observations and experimental observations. To this day they have not incorporated the insights of Fisher, Wright, and Haldane in the early 1930s into their thinking. There are no natural historical observations on the statics of gene frequencies—that is, on the patterns of spatial distribution of the frequencies of genes in species—which enable us unambiguously to distinguish among competing hypotheses of their control. There are prejudices, desires, and axes to be ground. But Wright's 1931 paper shows that the stochastic synthetic theory is such that nearly all observations can be explained by a great variety of hypotheses. Pseudoparameters appear often in stochastic theory, for example the product of population size and selection coefficients, or the product of population size and migration rates. Competing hypotheses must be judged on the basis of a guess about the product of a relatively large but unknown number, the effective size of a population, and a relatively small but unknown number, the migration rate, or the selection coefficient, or the mutation rate for genes, which makes it extremely difficult to distinguish between various hypotheses. What marks the schism between the Wrightian and Fisherian schools of theory and the schools of natural historical and experimental work that follow them is notions about the sizes of these products.

Fisher, for example, showed in his work that dominance would evolve, even though by a second-order selection process, provided there were any advantage to hiding the recessive effects of genes. Wright's response was that chance elements were too strong to allow such a weak selection process to be effective. That is, Wright never contradicted Fisher on the *eventual* efficacy of small selection coefficients, but he claimed that we could not wait for those selection coefficients because the fluctuations of environment and the effects of finiteness of populations were such that random forces would intervene before the deterministic force of weak selection coefficients carried the process to completion.

LERNER: I certainly don't deny the tremendous importance of Fisher's 1918 paper. But people don't remember that in 1917 Wright published a paper in the Washington Academy of Sciences in which he partitioned variance in guinea pigs between environmental and genetic factors. In effect, without knowing the name, he used an analysis covariance.

LEWONTIN: Fisher's 1918 paper is important not simply because of its distinction between environmental and genetic variance, but his partition of the genetic variance into its components.

Curiously, the science of inbreeding in animals and plants is clearly dominated by the Wrightian scheme, because inbreeding always deals with the probability of homozygosity—that is, with gene frequency. But in dealing with progress under selection for continuous variation we depend on Fisher's scheme that talks about rates of change of means and of variance.

Wright's notion that random processes allow the exploration of the field of gene frequencies, which a deterministic force does not allow, and the consequent idea of multiple possible equilibria are extremely important. For example, this view explains the apparent diversity of adaptive answers to the same problem. It frees the evolutionist from having to ask such questions as why did some Ceratopsian dinosaurs have three horns and some have two, and some have a frill and some not? Wright's theory does not require an explanation of adaptive difference between those ornamentations. They are just the alternative adaptive peaks, all providing a defense mechanism for an herbivore. This extremely important element in Wright's theory does not appear either in Haldane or in Fisher.

DARLINGTON: Why did Wright and Fisher both think entirely in terms of a few gene mutations? Why in fact did neither of them ever consider the interaction of gene differences with structural changes in chromosomes? Every combination of genes is liable to be affected by not only such changes as deletions, which of course can occur in an astronomically greater variety than gene mutations, but also inversions, which affect the relation of any two genes on the same chromosome. The occurrence of inversions along with gene mutations enormously increases the role of chance in variation. Wright was compelled to think of small populations, but every inversion creates a small population in respect of the segment that is inverted. It at once becomes a unique piece of the population. The result, of course, is the same with polyploidy and many other chromosome changes. They create a role for chance in variation that is far greater than Wright was ever able to adduce from talking about actual small and relatively unimportant populations. The really important small populations are the little bits of chromosomes that are populations within which recombination cannot occur and that are therefore isolated from the rest of the species with chromosomes that are not inverted.

LEWONTIN: What importance did Wright assign to the organization of genes on the chromosome and recombination? Here he simply had the wrong intuition. I discussed this with him at the Cold Spring Harbor Symposium in 1955. His intuition was based on a famous paper by Hilda Geiringer (1944) in which she showed that genes would eventually come to be randomly associated with each other at a rate that depended on the amount of recombination. Provided there were any recombination at all, no matter how small, this random association would eventually arise. Wright then thought, incorrectly as it turned out, that even in the presence of natural selection the random association would occur—that is, that because of the randomizing effect of recombination in the end the genes would be at random with respect to each other, even though natural selection was operating on them. However, when the theory is worked out, natural selection will keep genes in nonrandom association if there is little enough recombination, as in inversion heterozygotes.

DOBZHANSKY: What genetic investigation during the thirties and forties, and perhaps in the fifties, would have been impossible or unlikely to be undertaken without the influence of Fisher, Wright, and Haldane? In other words, to what extent did this theoretical, mathematical work in your opinion guide or inspire pedestrian experimental work? Do you think that the fancy mathematics of Kimura, Lewontin, Crow, and others will also guide such experimental work in the future? That is, do you believe that this theoretical work has a value independent of experimental work or do you regard this theoretical work as an instrument designed to guide experimental work?

LEWONTIN: First, theoretical work can directly inform experimental work. For example, in the series of papers on the genetics of natural populations Dobzhansky attempted to understand the breeding structure of

Drosophila pseudoobscura, using the theoretical relationship between mutation rates and allelism of lethals. In those papers, he attempted to get an insight into the breeding structure of the species. Dobzhansky could not have written those papers without Sewall Wright; indeed he could not have written them without Wright's 1931 paper. I hesitate to use this example, because you characterized the experimental work as pedestrian and I don't regard this one as pedestrian. If you ask me in a broader context how much field population genetics work has depended directly for its direction on theoretical work, the answer is very little.

What is the use of all of this mathematical theory? Is it simply "mental masturbation"? Theory generally should not be an attempt to say how the world is. Rather, it is an attempt to construct the logical relations that arise from various assumptions about the world. It is an "as if" set of conditional statements. If the mutation rate is this, if the selection coefficients are that, if the migration is this, if the population size is that—then you may expect to see the following. That kind of theory acts as a guide for perplexed experimentalists. When they observe something about the statics of gene frequency, differences among populations, the rates of change of gene frequencies, or the heterozygosities, then they can make correct inferences about the causal mechanisms involved. The modern synthesis is incomplete precisely because experimentalists and natural historical geneticists even today are explaining their observations as if those explanations were clearly proved by the observations. The theory shows, however, that those observations cannot be used to distinguish between causal hypotheses. For example, the statics of gene frequency are the observations most available to us. If an evolutionist is not as fortunate as E. B. Ford is to have fossil or subfossil populations of snails, he must only look at present patterns of geographic variation to find heterozygosity and generally a small amount of differentiation of frequencies of genes from one population to another. The observations of heterozygosity and the rather great similarities in gene frequencies between populations of *D. pseudoobscura*, for example, can be interpreted as evidence for strong heterotic selection maintaining the same genes in stable equilibrium everywhere the species exists. Without knowing the details of the ecology of the population, I don't know how one could say anything more specific than that.

Equally well, the observations could be explained by saying that the genes are not selected, that they are segregating because of recurrent mutation, and that the reason the populations have such similar gene frequencies is because of enough migration between neighboring populations that the product of effective population size and migration rate is about one. If the effective population size is ten thousand, the migration rate need only be one in ten thousand. We cannot use the observations to distinguish between those hypotheses. The most egregious example I can think of is a paper (F. M. Johnson, 1971) describing populations of *D. ananassae* on islands in the South Pacific. Johnson observed that the gene frequencies were very similar on islands of the same archipelago but very different in places very far apart. The most obvious conclusion is that these genes were neutral. The frequencies are similar in the close islands of the archipelago because there is migration between them, but of course if the populations are separated by a thousand miles of ocean, very little migration could occur and so the frequencies would drift apart. But Johnson's conclusion was the opposite. He said his data indicated selection as the cause of differentiation. The reason for the similarity within an archipelago is that the environments of islands close together are similar while islands five hundred miles apart have very different environments. The role of theory is to say which hypothesis generates which set of observations. The observations should be used in the light of the theory to distinguish among the hypotheses. Theory tells us at the present moment that we have not yet found the observations sufficient to distinguish among the hypotheses.

DOBZHANSKY: Does theoretical, mathematical investigation make sense unless it leads to experiments or interpretations?

LEWONTIN: Obviously not. After all, we are concerned with a contingent world and not with the world of poetry.

BOESIGER: I agree with Lewontin that comparison of gene frequencies in populations living close to each other and in populations far from each other does not by itself permit discrimination between selection and neutral genes. But many other experiments exist. For example, placing two genotypes in a population cage and leaving it until equilibrium is reached, as Dobzhansky has done, indicates that the same kind of equilibrium is reached in each trial. Thus we can conclude that selection is operating.

LEWONTIN: I agree in principle that if you observe repeated selections always going in the same direction, selection is the prime suspect for the change. Unfortunately, if the experiments are not well designed, the result can be caused by linkages with deleterious genes. The point I was trying to make in response to Dobzhansky was not that in principle our science is incapable of progress, but that the function of theory is to map out the consequences of various hypotheses to discriminate between them. What I claim as a historical fact about the synthesis period is that most of the theory was not incorporated correctly into the thinking of most experimentalists. Even now experimentalists are not using available theory to help them make distinctions between hypotheses.

MAYR: Does this situation arise because the experimentalists simply are insufficiently familiar with the theory? Or is it because they feel that even if they understood the whole theory, they would not know how to design experiments that would settle these points?

LEWONTIN: I don't understand how that is relevant to my point. If existing theory tells you that the experiments that you have designed are incapable of distinguishing between two hypotheses, then you should know that. You should not continue to perform experiments whose interpretation is necessarily ambiguous.

MAYR: What percentage of published work in population genetics would be wiped out by your statement?

LEWONTIN: A very healthy percentage.

MAYR: Wasn't the evolutionary synthesis possible on the basis of a very minimal agreement that there are small genetic changes and that the phenotypes produced by even very small genetic changes may differ and probably usually differ in their selective values? These two minimal assumptions allow interpretation of almost anything in the evolutionary phenomena that an ordinary, naturalistic evolutionary biologist would observe. Were the far more sophisticated mathematical elaborations of Fisher and Wright necessary for these very simpleminded minimal assumptions that permitted most of the subsequent synthesis? I am not arguing that Fisher and Wright should have stopped doing what they did. I merely suggest that perhaps we should not complain about the failure of application of the sophisticated and advanced theoretical analyses, because for the purposes of the evolutionary biologists these very advanced theoretical analyses were superfluous. For them, even minimal rather elementary calculations were sufficient.

LEWONTIN: I really disagree. Do you accept my characterization of the chief role of theory, at the present historical period of our science, as dictating what you can validly infer from your observations and what you cannot? If so, I can think of very little in the theories of Fisher and especially of Wright and even of that much-maligned gentleman Kimura that is irrelevant to the question of what can be inferred from the observations. Although we regard their work as an arcane and perhaps baroque manifestation of mathematics, the theoretical population geneticists are trying to ask really practical questions. If the pattern of migration of individuals from one population to another is of a particular sort, how much differentiation can we expect, given different amounts of selection? Very difficult mathematics are required to answer it. I cannot read half of the mathematics. However, almost always after the mathematics is done the answer comes out in a very simple form that the experimentalist usually can read if he's willing to ignore the machinery and look only at the conclusions. We can use the products of the mathematical machinery to help us decide whether our experiments tell us what we want to know. A great deal more theory still needs to be done on problems of estimation and hypothesis testing. All of us have published results in which we have not faced the question about the limits of error on the numbers we get. We talk about estimating rates of migration, selection coefficients, and population sizes, but too few papers put a standard error on those numbers. The calculated standard errors are frighteningly large, which indicates another function of theory: to tell you how to construct an experiment that gives you estimates of the things you want to know within a reasonable limit.

References

Dobzhansky, Th. 1937. *Genetics and the origin of species*. New York: Columbia University Press.

Fisher, R. A. 1918. The correlation between relatives on the supposition of Mendelian inheritance. *Transactions of the Royal Society of Edinburgh* 52:399-433.

———— 1930. *The genetical theory of natural selection*. Oxford: Oxford University Press.

Ford, E. B. 1964. *Ecological genetics*. London: Methuen.

Geiringer, H. 1944. On the probability theory of linkage in Mendelian heredity. *Annals of Mathematical Statistics* 15:25-57.

Johnson, F. M. 1971. Isozyme polymorphisms in *Drosophila ananassae*. Genetic diversity among island populations of the South Pacific. *Genetics* 68:77-95.

Popper, K. R. 1965. *The logic of scientific discovery*. New York: Harper Torchbook. See especially section 8, Scientific objectivity and subjective conviction.

Wright, S. 1917. The average correlation within subgroups of a population. *Journal of the Washington Academy of Sciences* 1:532-535.

———— 1931. Evolution in Mendelian populations. *Genetics* 16:97-159.

———— 1968-1978. *Evolution and the genetics of populations*. Chicago: University of Chicago Press, 4 vols.

2 Cytology

The role of the field of cytology in the evolutionary synthesis is difficult to assess. Morgan and Wilson had from the beginning of the *Drosophila* work emphasized cytological aspects of research in genetics. Thus *The Mechanism of Mendelian Inheritance* (1915, 1922) by Morgan and his students Sturtevant, Muller, and Bridges; *The Physical Basis of Heredity* (Morgan, 1919); and *The Theory of the Gene* (Morgan, 1926) emphasized the chromosome theory of heredity. Even Morgan's *Critique of the Theory of Evolution* (1916) contained much cytological analysis. Yet as Weinstein points out in his comment in this section, Sturtevant, who began to work with Morgan and Wilson in 1909 and who did much pioneering work on the chromosome theory of heredity, argued in 1939 that the development of cytology had actually been more or less apart from that of genetics until Darlington's work in the early 1930s. Darlington and Carson shed much light on cytology's small contribution to the study of evolution before the 1930s.

Two contributions of cytology to the evolutionary synthesis before 1930 are, however, important. The work of Renner, Cleland, and many others showed conclusively that the so-called mutations in *Oenothera*, which de Vries had used as the experimental evidence for his influential mutation theory, were in fact not mutations at all. De Vries's theory of discontinuous evolution lost much of its former influence as a direct result of this cytological refutation of his evidence. As Mayr has pointed out in the prologue, the wide acceptance of de Vries's theory was a major obstacle to the evolutionary synthesis. As Darlington points out, none of the *Oenothera* workers before 1930 developed an evolutionary perspective that they interpreted with the cytological work on *Oenothera*.

Cytology also made a very general but significant contribution to the understanding of the physical basis of inheritance. In 1900 a systematist interested in evolution in natural populations had no compelling reason to believe in the material particles hypothesized in the prevalent theories of heredity, particularly those of Weismann and de Vries. By 1930, however, cytological analysis had established clearly that genes, the determinants of Mendelian heredity, lay linearly on chromosomes. Moreover, chromosome complements of some closely related species were shown to be consistently and observably different. In short, the field of cytology was becoming ripe for synthesis with a systematic and evolutionary perspective.

During the 1930s, cytology literally exploded into the synthetic view of evolution. Two developments were crucial. Darlington presented a comprehensive and unified view of chromosome behavior (1932). Although many particulars were later revised in Darlington's theory, it furnished him the possibility of applying an evolutionary perspective to cytology. The result was his theory of the evolution of genetic systems, which was enormously influential in leading evolutionary biologists to an understanding of genetic recombination in natural populations. Dobzhansky used cytological techniques in his systematic and evolutionary analysis of natural populations in his 1937 book, *Genetics and the Origin of Species,* and in his "Genetics of Natural Populations" series. Both of these developments are examined by Darlington and Carson in this chapter. Weinstein, a Morgan student who published very important work on the mechanics and geometry of crossing-over, has also contributed to this section.   W.B.P.

The Evolution of Genetic Systems:
Contributions of Cytology to Evolutionary Theory

C. D. Darlington

The central problem of cytology and genetics in 1926, as perceived by my teacher, Frank Newton, and later by me was how to discover the universal rules underlying meiosis in plants and animals. At that time, there were two rival schools of interpretation of meiosis. The one favored by E. B. Wilson assumed side-by-side pairing of chromosomes (parasynapsis). The other was the end-to-end pairing (telosynapsis) that was favored by many plant cytologists, especially by the *Oenothera* school of R. R. Gates. The John Innes Horticultural Institution (whose first director, Bateson, died in 1926) had a governing body dominated by two men: E. W. MacBride, an embryologist who was a staunch defender of Paul Kammerer and an obsessed Lamarckian; and J. B. Farmer, a cytologist who invented the story of *Primula kewensis* having doubled its chromosome number by crosswise splitting of the mythical continuous spireme, a feature of telosynapsis.

Newton's prime object was to try to establish the theory of side-by-side pairing on which Morgan's coordination of Janssens' crossing-over theory and *Drosophila* breeding had been based. Newton was convinced about side-by-side pairing from his own work on diploid tulips. To clinch this matter he had the idea that we should look at meiosis in triploids. He was looking at triploid tulips and I was looking at triploid and tetraploid hyacinths when Newton died in 1927. I was left alone to carry the whole program—alone apart from the moral support of J. B. S. Haldane who had arrived at John Innes in 1927. The observations on the triploids and tetraploids carried me much further than ever I or Newton imagined I should go when we started trying to prove synapsis and test Janssens' theory.

To save the telosynapsis theory, it had been suggested that when chromosomes appeared to be coming together at the prophase of meiosis the threads were just the chromatids of one chromosome and not the two homologous chromosomes. The telosynaptic theory held that there was no side-by-side pairing. All the pairing was end-to-end. Newton wanted to examine a triploid to see if the three chromosomes came together. If so, it would prove side-by-side pairing. This was very simple. The three chromosomes could be seen coming together, and we proved parasynapsis to our own satisfaction (Newton and Darlington, 1927).

Out of this work many unexpected things came to light. First the three chromosomes always came together in pairs, never in threes. In fact I hesitated for a month or two because the threads change partners rather often and when they do so the unpaired third chromosome is dragged around the other two in a spiral on account of the spiral torsion shown in all chromosome pairing preparatory to crossing-over. This drawing round of the third chromosome made it look as though perhaps all three were sometimes associated together. Finally my tetraploids were decisive in resolving that question. I found that the fourth chromosome was always paired. There was no question of the pairs drawing a singleton around; they were always symmetrically and separately in pairs of pairs.

Going back to the triploid, it would be perfectly simple if, as John Belling had claimed, all three chromosomes remain together in later stages. But they do not. These chromosomes at metaphase, the latest stage, often fall apart, leaving the third chromosome unpaired. What I observed was that this third chromosome remains paired with the other two only if it has formed a chiasma. I therefore looked at all the chromosome configurations that occurred in these triploids and I found that all chromosome association at the metaphase of meiosis was conditioned by chiasmata. I then tried very precisely to define these chiasmata. Janssens' definition, I found, was not a definition, but rather a lot of descriptions of chiasmata. I proposed to define the chiasma as always an exchange of partner among four chromatids. There was no question of a "total chiasmatypy" or "partial chiasmatypy"; there was no question of chromosomes ever sticking together; the chromosomes often look as though they are sticking together but when they touch at the ends I said this was a "terminal chiasma" and was always the result of the movement of an original chiasma to an end.

In this way I had enormously simplified the description of meiosis, but I also found myself making the assumption that all chromosome association in my examples must be by chiasmata. This assumption, I concluded from looking at the literature, might be properly applied to meiosis in all sexually reproducing organisms. It was a very wide generalization, which caused great indignation and exposed me to several years of bitter controversy.

That, however, was only one part of the story. By saying that all the attraction between chromosomes was in pairs and by saying that, after crossing-over in the diplotene stage, the attraction was still only between pairs of chromatids (the meaning of the chiasma theory of pairing), I concluded that attraction was between pairs of threads at all stages of meiosis. That is exactly what one sees with the reproduction of chromosomes in mitosis. So on the principle of Occam's razor, I was confronted with the hypothesis that meiosis and mitosis were on the same mechanical plane of an attraction in pairs at all stages. No attraction of chromosomes occurred at meiosis that did not occur at mitosis. I dismissed the idea that perhaps the phenomenon was a sexual attraction or something mysterious or vitalistic or designed or Lamarckian. This new position also aroused indignation.

On top of all this, I was faced with the problem of crossing-over. Were the chiasmata the result of crossing-over? Before I had established the consequences of chiasmata, Newton died. He had been opposed to reaching any conclusions about crossing-over. But when I described our work, I decided I must make up my mind about crossing-over, now a doubly difficult problem. If I said that chiasmata are universal in meiosis and crossing-over was universally the cause of chiasma formation, what was I going to do about male *Drosophila* with no crossing-over? This situation was embarrassing for a young man without any established appointment. I was already embattled with all of the cytologists and botanists. If I took on all the Drosophilists as well, life would be difficult (Darlington, 1979c).

My courage failed me and I put off facing this problem in 1929. But in 1930 I took my hyacinth preparations with their indisputable configurations at pachytene, and with them I showed that a chiasma can take place between two exchanges of partners. Then I knew that crossing-over must have occurred. It was inconceivable that having formed the pachytene configuration the chromatids could reshuffle to form chiasmata: they must break. This was the critical configuration. Of course, my colleagues and I later found many other types of critical configuration including the great system of inversion crossing-over relationships.

I was compelled now to face the problem of universal chiasmata and universal crossing-over. Putting aside male *Drosophila* for the moment, I saw that these universal generalizations involved three issues. The first issue was that of reductionism. I was supposing that the whole mechanism of mitosis and meiosis in chromosome behavior could be explained in physicochemical terms without any of the vitalistic impulses and essences that people had so long assumed. This position enabled me to look forward, and I immediately proceeded to get physicists and chemists interested in chromosome structure (as described by Olby, 1974). I had thus arrived at the point that Muller had reached speculatively some years earlier. We agreed very nicely on the reductionist issue. The idea that pairing and reproduction were both associations of chromosomes in twos was what Muller and I thought was so significant from the reductionist point of view.

The second great issue was the genetic one. Crossing-over was now on my view coextensive with sexual reproduction. But in 1930 we knew about crossing-over and linkage in perhaps only a dozen species of plants and animals. So when I said that chiasmata were universal and that they resulted from crossing-over, I had explicitly made a generalization that was enormously wider. We now could say, if this generalization was true, that recombination by crossing-over was coextensive with sexual reproduction. That wasn't a generalization that could have been made on the basis of *Drosophila*. It was implied by the Drosophilists, by the claim that the laws of heredity and crossing-over had been discovered from *Drosophila*. But it was a big, speculative step. I had made that step by shifting the argument to the level of chiasmata, which were now visible or on my view were to be inferred in hundreds of species all the way from amoeba to man. The problem was put on a universal footing.

The third great issue was evolution. I argued that meiosis had arisen from mitosis at a single step without any vague or mysterious intervention. It had to be an evolutionary process, a mutation if you like. On my view it had to have survived and spread and governed the whole of evolution itself, by an evolutionary process, by its selective advantage. I didn't want to appear too original; I therefore made it clear that long before me, Weismann had proposed in 1887 that sexual reproduction with meiosis had arisen by an evolutionary process.

In a letter to *Nature* of September 10, 1887, Weismann said, "Fertilization is no longer an unknown impulse given to the egg cell by the entrance of a spermatozoon, but it is simply the union of the germ plasmas of two individuals." Replacing the idea of an "impulse" with a mechanical event was a revolutionary idea whose significance was clouded by the dust and storm of controversy about Weismann's views.

Weismann also wrote that "there are no *essential* merely individual differences between the nuclear substance of the spermatozoon and that of the ovum." I would call this a reductionist point of view. There is no essence and no mysterious sexual impulse. The chromosomes that come together are the same kinds of things, maternal and paternal indistinguishably.

The third quotation from Weismann's letter is, "Sexual propagation must confer immense benefits upon organic life. I believe that such beneficial results will be found in the fact that sexual propagation may be regarded as a source of individual variability, furnishing material for the operation of natural selection." I don't remember how carefully I studied Weismann in 1930 but this last phrase captures my idea of the evolution of genetic systems. The hint was obvious to me; it gradually became obvious to other people (first of all to Muller). This idea meant that heredity evolved—an evolution of the hereditary mechanism—an evolution of mitosis and meiosis. I had opened up an immense field of inquiry. I had to find out how the genes could control the chromosomes, how the chromosomes could control themselves at the same time that they were controlling the organism. I first discussed this problem in 1932. Formally anybody can repeat the idea and learn and write it down, but few have grasped its implications, as I think, for biology.

I assumed, of course, that the entire process by which heredity evolved was itself a result of natural selection. This assumption went against the "evergreen" superstition of Lamarckism. The idea of use and disuse, the idea of the effects of the environment producing meiosis from mitosis was rejected. The advantage of meiosis is something that only appears in the next generation. Indeed it appears only in the course of evolution itself. These things destroy Lamarckism straightaway; meiosis and recombination cannot be adaptations to the environment. They are adaptations to the environments of future generations. No Lamarckian has ever honored me by rebutting my suggestion that this in fact is how sexual reproduction works (Darlington, 1979a).

In June 1932 my chapter leading toward my book *The Evolution of Genetic Systems* appeared. I attended the Sixth International Congress of Genetics in Ithaca in August 1932, where I heard Fisher discuss the same question (1932). Fisher had an aphoristic turn of style and he improved on what appears in print. He said, "Others have considered the bearing of the theory of heredity on evolution" (and when he said "others" he meant Wright and Haldane); "I am going to consider the bearing of evolution on heredity—I am going to reverse the accepted order of discussion." He was unaware that I had just published an account arriving at exactly this conclusion.

The odd thing is, and we never discussed this, I have no idea of whether Fisher ever thought of the chromosomes from this point of view at all. His views on the evolution of heredity, particularly the evolution of dominance, formed a beautiful theory that had immediate practical application to many genetic problems. But his use was on a different level from my use of evolution of heredity because I was considering the whole behavior of genes and chromosomes. This wide field of study has occupied me and many others ever since. It will continue to be of interest because it is impossible to talk about chromosomes now without thinking of their reciprocal relations with heredity. Heredity controls them and they control heredity; they evolve and heredity evolves with them. This point of view I think is strange to the naturalist who by tradition concentrates his attention on the organism. A chromosome man might say that a naturalist is obsessed with organisms, but after all we ourselves are organisms. It is difficult for us to look at the organism from inside, from the point of view of chromosomes. But if you have spent as much time looking at chromosomes as at organisms, you may see that the reversal of the point of view is possible, and that meiosis was the means for exposing variation to natural selection as Weismann thought.

The crux of this means of exposing variation lay in crossing-over, its frequency, and its distribution. I will discuss one example of how this problem appeared. Fisher had just published his book on the genetical theory of natural selection (1930). He had pointed out the existence of so much crossing-over, more than evolution seemed to require. Now from my point of view, there was a simple—not a complete but apparently complete—answer to this question. I said the chromosomes required chiasmata to pair, and the chiasmata must arise by crossing-over; therefore more crossing-over exists than would suit the needs of evolution. That is the way it seemed to me at first. Later I came to a different conclusion.

Three months later when I arrived at the California Institute of Technology where I wanted to look at meiosis in *Drosophila*, Dr. Dobzhansky kindly lent me his admirable preparations of *D. pseudoobscura*, which fortunately has better chromosomes than *D. melanogaster*. I found a very extraordinary thing. I had already supposed that there was something different about meiosis in the male *Drosophila*, but what I found on my interpretation of Dobzhansky's slides was a new kind of meiosis in which, so far as the autosomes were concerned, it was possible to do without chiasmata. Sex chromosomes still had them in a particular way—reciprocal chiasmata, as I believed—whose effect was that there was no visible crossing-over.

For the other chromosomes, the autosomes, it was different from an evolutionary point of view. The chromosomes of the males, that is, of half the flies, had done away with chiasmata; the fly had thus done away with crossing-over in the male sex. This form of meiosis was clearly derivative; it was not the original form because the original form characteristic of the whole breadth of organisms was found in a female. The essential form (which was that crossing-over should continue in the species) was kept going in the female.

At first I thought that *Drosophila* was just neatly doing what Fisher wanted it to do, reducing the frequency of crossing-over in the species by half. It was only thirty years later that it occurred to me that *Drosophila* was in effect doing something much cleverer than that. It was providing two systems of heredity, two breeding systems, in the same species— what I have called two-track heredity. The female broke up the chromosomes into genes, produced the recombinations, and sometimes produced a good recombination. The male (normally about half her offspring were bound to be male) could stereotype any combination that was good so that the lineage could shift from recombination to stereotyping and back again in successive generations, always subject to natural selection. This system seems to me to be the most beautiful example of the use of the chromosomes for exposing variation to natural selection by doing it in the same species in two entirely different ways. As we now know, this kind of two-track heredity occurs in plants as well as very widely in animals (Darlington, 1973).

Oenothera, my next object of study, had given rise to a false theory of evolution, as well as a false theory of mutation and a false theory of heredity and chromosome behavior. Indeed, if we talk about scientific research blocking the progress of a science, *Oenothera* was the imperishable example. And what a collection of people it picked up on the way! At the head was Hugo de Vries, confronting the world with his dogmatic notion of evolution. Following him came the trail of lesser men led by Reginald Ruggles Gates, with the words "mutation" and "telosynapsis" echoing down the years. He was one of the obstacles I had to overcome, no less because his friend Farmer was a governor of the John Innes.

My problem in 1928 was to show that *Oenothera* was entirely consonant with my two chiasma theories as well as with *Drosophila* genetics. The *Oenothera* people had argued that plants might be quite different from animals; *Oenothera* might have one set of rules and *Drosophila* another. I had to show that they had the same rules; that a system of interchange, parasynapsis, and chiasmata (terminalized) could produce the rings in *Oenothera* which would be inherited in exactly the way that Cleland had demonstrated to me at the Fifth International Congress of Genetics in Berlin (1928). Although I had been told by Newton and Karpechenko that a good deal that was claimed for *Oenothera* was probably bogus, when I saw Cleland's slides I realized that they were perfectly genuine. There were the extraordinary rings. It was easy enough to explain one ring of four chromosomes arising from the examples of interchange in Belling's explanation of *Datura*; but how could a ring of fourteen chromosomes arise? It meant that five successive interchanges must have occurred. And as Haldane pointed out to me, every additional one must have had a selective advantage over its predecessor. It was fairly obvious to me but I was glad to see that Haldane accepted the idea. This was a notable illustration of Fisher's aphorism that natural selection is a means of generating events of extreme improbability. These extremely improbable rings had been generated, accumulated, and directed by natural selection.

Otto Renner, whose work on *Oenothera* was so important, always took a static viewpoint and never developed an evolutionary point of view. Blakeslee's approach was like that of Renner. But for my purposes I had to adopt an evolutionary viewpoint. I have explained the contrast in these points of view in my obituary of Renner (Darlington, 1962).

I have said that *Oenothera* generated frauds; but it also generated Renner who extinguished the frauds. Renner conducted his beautiful breeding analysis of *Oenothera* over a period of ten years, examining in addition the viability of embryo sacs, pollen grains, and embryos. He was doing microbial genetics twenty years before the microbes themselves emerged. He demonstrated the selection and competition among the embryo sacs produced by segregation. And by crossing *Oenothera* he was the first to demonstrate segregation at meiosis (Renner, 1919, 1921). His formula can be expressed in one phrase: the idea of the complex. Each species of *Oenothera* was a hybrid of two complexes, the *velans* and *gaudens*, or alpha and beta as they were sometimes called. But Renner was not concerned with the evolution of these complexes; he was concerned with their elucidation in contemporary physiological and genetic terms.

My problem was to explain at the same time the origin of the chromosome pairing system, of the complexes, and of the mutations in both. Where were Renner's complexes located? The idea that Belling had for *Datura* was an interchange caused by breakage of the chromosomes at what we now call the centromere; he thought the chromosomes broke there, as indeed they did in *Datura*, and that the two arms were chromosome segments. But I thought the interchanges took place at an intercalary position anywhere along the arm, like chiasmata or crossing-over or Muller's chromosome breakage in *Drosophila*.

If the chromosomes broke at different places, then the interchanges with a ring of six would yield a certain standard constitution. Some segments in the middle would never have a chance of crossing-over with one another. They would always be committed to one or other of the complexes. These "differential segments" then must be, in my view, the seat of Renner's complexes. If crossing-over does occur between them, as Renner found in breeding and as I found cytologically, the large ring will break down into a smaller ring and the complex will then disintegrate.

By this view, strictly on parasynaptic, chiasmatic, and Drosophilitic assumptions, I had explained the origin of the complexes and their breakdown in what Renner and de Vries called half-mutants, which arose in *Oenothera lamarckiana*.

What then about the first problem with *Oenothera*: how were the interchanges favored by natural selection? Some heterozygous advantage was necessary—and here we come back to Muller's balanced lethals and Ford's balanced polymorphism—in the hybrid condition of the high ring-forming species. To test this hypothesis, I tried to produce rings in species of *Campanula persicifolia*, and I got rings as high as twelve chromosomes. Then I found that if rings with six chromosomes were self-fertilized, those rings bred true; but if they were crossed with their first cousins or second cousins also with rings of six that were different in general genetic constitution, then the rings broke down. In other words, inbreeding favored and stabilized hybridity.

About then American colleagues, Cleland particularly, made it clear that after the ice retreated *Oenothera* had moved across the United States and in doing so had lost its cross-fertilizing mechanism. One outbreeding species had become many inbreeding species, a general property of species that are advancing on their frontiers. They are liable to be forced into inbreeding, which is an agent of genetic isolation. Inbreeding creates selection favoring the preservation of hybridity (Darlington, 1956).

So in the end we come back with *Oenothera* to something closely related to a balanced polymorphism, but at a gametic level. Thus the hybrid and mutating *Oenothera* fits in several fundamental respects with the ordinary workaday theory of heredity and variation. However, it has been subjected to quite extraordinary selective stresses as a result of its chromosomes being suited to interchange and having an ability to generate the strange physiological compensating mechanisms that Renner discovered.

In both *Drosophila* and *Oenothera* the mechanism of heredity, the shape of the chromosomes, even the number of the chromosomes, their life cycles and their reproductive organizations, their breeding behavior (outbreeding or inbreeding, sexually differentiated or hermaphrodite) are all related in a system whose parts are mutually adapted and adaptively connected. This is the idea of the genetic system as I have developed it.

The mutual adaptiveness of all the different aspects in the genetic system introduces too many dimensions for the mathematician to cope with. When I suggested this to Fisher, he never appeared to hear. I think he was quite right not to hear, because no one can cope mathematically with all the effects of selection when the mechanism controlling selection and the unit that is being selected are both themselves being changed by selection. This kind of situation introduces difficulties for those who think of natural selection in terms of organisms. I am the most bigoted selectionist in the world; but I noticed early these contradictions in the traditional organismal theory. Their resolution, however, lies outside the present discussion (Darlington, 1979a,b,c).

References

Cleland, R. E. 1928. The genetics of Oenothera in relation to chromosome behavior, with special reference to certain hybrids. *Verhandlungen des V. internationalen Kongresses für Vererbungswissenschaft*. Leipzig: Borntraeger, pp. 554-567.

Darlington, C. D. 1932. *Recent advances in cytology*. London: Churchill, chap. 16, The evolution of genetic systems.

———— 1939. *The evolution of genetic systems*. Cambridge: Cambridge University Press.

———— 1956. Natural populations and the breakdown of classical genetics. *Proceedings of the Royal Society of London* B 145:350-364.

———— 1962. Otto Renner: 1885-1960. *Biographical Memoirs of the Royal Society of London* 7:207-220.

———— 1973. The place of the chromosomes in the genetic system. *Chromosomes Today* 4:1-13.

———— 1979a. A diagram of evolution. *Nature* 276:447-452.

———— 1979b. The chromosomes as feedback systems in evolution. *Kybernetes* 8:275-284.

———— 1979c. Morgan's crisis. *Nature* 278:786-787.

Fisher, R. A. 1930. *The genetical theory of natural selection*. Oxford: Oxford University Press.

———— 1932. The evolutionary modification of genetic phenomena. *Proceedings of the Sixth International Congress of Genetics*. Menasha, Wisconsin: Brooklyn Botanic Garden, pp. 165-172.

Newton, W. C. F., and C. D. Darlington. 1927. Meiosis in a triploid tulip. *Nature* 120:13.

Olby, R. C. 1974. *The path to the double helix*. London: Macmillan.

Renner, O. 1919. Zur Biologie und Morphologie der männlichen Haplonten einiger Oenotheren. *Zeitschrift für Botanik* 11:305-380.

———— 1921. Heterogametie im weiblichen Geschlecht und Embryosackentwicklung bei den Oenotheren. *Zeitschrift für Botanik* 13:609-621.

Weismann, A. 1887. Letter, *Nature* 36:607-609. (This letter is an abstract of Weismann's essay, On the number of polar bodies and their significance in heredity, *Essays upon heredity*, trans. and ed. E. B. Poulton, S. Schönland, and A. E. Shipley. Oxford: Oxford University Press, 1889, pp. 331-384.)

Cytology in the T. H. Morgan School

Alexander Weinstein

Before Morgan began his work on *Drosophila*, the original purpose of which was to produce mutations, a detailed study of the chromosomes of the fly was made by Nettie M. Stevens (1907, 1908). She found the material difficult. In 1907 she wrote, "The chromosomes of *Drosophila ampelophila* have proved to be much more puzzling than those of any other forms yet studied by the author." This did not prevent Morgan from choosing *Drosophila* for his experiments, in part presumably because he was still skeptical of the importance of chromosomes in heredity, and in part because of the advantages of *Drosophila* for breeding work—small size, short life cycle, and large number of offspring.

Stevens considered her results "not very satisfactory," especially in the male. In 1914 the chromosomes of both sexes were successfully worked out by Bridges, and he proved that the distribution of X chromosomes in heredity precisely parallels the distribution of sex-linked genes both in normal cases and in cases of nondisjunction. In 1921 he extended the proof to cases of nondisjunction of the fourth chromosome. In 1921 he also proved the parallelism of genes and chromosomes in triploids. And in 1922 Lillian V. Morgan proved the parallelism in strains with attached X's.

The chiasmatype theory of Janssens in 1909 and subsequent years raised questions that were regarded as still not definitively settled in the 1920s; these and the cytological implications of the work on *Drosophila* were discussed in Morgan's fly room, but they could not be studied cytologically in *Drosophila* because the material was not sufficiently favorable.

In 1913-1914, when I took Wilson's course in cellular biology, he said that Janssens had not proved that a chiasma represents an actual exchange between chromosomes. This agnostic position was still held by Wilson in the third edition of his book, *The Cell*, published in 1925. Morgan's theory of crossing-over, of course, had originally been suggested by Janssens' interpretation of chiasmata. But *The Mechanism of Mendelian Heredity*, in both the first edition (1915) and the second (1923), said that the strepsinema stage, when the chromosomes are condensed and chiasmata were reported by Janssens, was not the only stage at which exchanges might be supposed to occur; and cited as another possibility the earlier leptotene stage, when the chromosomes are twisted about each other but are not yet condensed. Muller (1916) argued that the leptotene was the more likely stage because the accuracy of crossing-over is more compatible with the thinness of the chromosomes at leptotene than with their thickness at strepsinema; and the same opinion was expressed by Morgan in *The Physical Basis of Heredity* (1919). Both editions of *The Mechanism of Mendelian Heredity* also said that it is unnecessary to suppose that an exchange between chromosomes takes place at every point at which they lie across each other.

Janssens' figures in 1909 showed each strepsinema chromosome split lengthwise, so that there were four strands. He thought that generally each chiasma involves only two strands, so that they become crossovers while the two others remain noncrossovers. Less frequently, he thought, the two other strands also recombine at the same level, so that four crossovers result.

The question whether crossing-over takes place at a four-strand stage (when each chromosome has already split lengthwise) or at a two-strand stage (before the split) was discussed in *The Mechanism of Mendelian Heredity* (1915) and also by Bridges (1916) and by Muller (1916) in their theses. A decision could not be reached because ordinarily only one chromatid remains in the egg, the three others going out into the polar bodies. But when there is nondisjunction, two strands remain in the egg; in a few of these cases it was found that one was a crossover and the other a noncrossover at the same level. Because only one of these two strands had been involved in the exchange, at least one additional strand must have been present when the exchange occurred. This inference was in agreement with Janssens' scheme, according to which four strands are present at the time of exchange, because each chromosome has already split lengthwise. But there was an alternative possibility: that the chromosomes had not yet split but that nevertheless more than two strands were present because the egg was not an ordinary diploid but a triploid or tetraploid. Such eggs would be rare, but so were the exceptional individuals containing one crossover and one noncrossover chromosome; they numbered only fifteen. It was therefore impossible to decide between the two alternatives; although the split-chromosome interpretation was recognized as a possibility, the theory of crossing-over was developed at that time entirely on a two-strand basis.

The situation changed in 1925 when, in experiments with attached X's, high nondisjunction, and triploids, individuals with one crossover and one noncrossover chromosome were found to occur not as rare exceptions, but frequently and regularly (Anderson, 1925; Bridges and Anderson, 1925). This discovery meant that four strands are present when crossing-over occurs; but it was found that only two of the four strands cross over at any one level. It then became necessary to reconstruct the theory of crossing-over on a four-strand basis. This had not been done when the *Drosophila* group moved permanently from Columbia in June 1928. I reconstructed the theory that summer.

The reconstruction required the overcoming of two obstacles not present in the two-strand theory. If there are only two strands when crossing-over occurs, the result is two chromosomes that are complements of each other; and when these chromosomes divide lengthwise, each forms two identical sister strands complementary to the two sister strands of the other chromosome. Hence, of the three strands that go out into the polar bodies, one is identical with the one in the egg, the two others complementary to it. However, if crossing-over takes place when four strands are already present, crossing-over may have occurred at more than one level, and the chromosome remaining in the egg may have been involved in only some of the exchanges, or in none of them. This difference made it necessary to devise a method of calculating the constitutions of the three chromatids that do not remain in the egg.

There was still the second obstacle to be overcome. If crossing-over occurs at a two-strand stage, there are as yet no sister strands, and the question whether there is crossing-over between them does not arise. But if each chromosome is already split before crossing-over takes place, the question of possible crossing-over between sister chromatids must be considered. It cannot be answered directly, because the parts interchanged would be identical, and hence the constitution of the strands involved would remain unchanged. But the occurrence of a sister-strand exchange could be detected if it interferes with crossing-over between nonsister strands. Calculations showed that exchanges between sister chromatids either do not occur or do not interfere with exchanges between nonsister chromatids. But a chromatid may cross over with either chromatid of the other chromosome; and which two chromatids cross over at one level does not influence which two cross over at other levels.

Thus the four-strand theory made it possible to reconstitute the tetrad; to discover how the four chromatids are arranged during crossing-over; to analyze interference, coincidence, and internode frequency on a four-strand basis; and to deduce the kinds of offspring, and their frequencies, in attached-X females and in triploids (Weinstein, 1928, 1932, 1936).

The presence of more than two strands during crossing-over, whether because of previous lengthwise splitting of chromosomes or because of polyploidy or polysomy, raised the question whether the strands conjugate in twos or in groups of more than two. Each view was supported by some evidence. In *Drosophila* triploids, Bridges and Anderson (1925) found that a chromosome may cross over with a second at one level and with the third at another level; they concluded that all three conjugate together along their entire lengths. But calculations that I made on the basis of this assumption on Redfield's 1930 data of linkage in triploids gave frequencies of hexads that were negative and hence impossible. On the other hand, no negative frequencies resulted if only two chromosomes conjugate and the third remains unmated. This finding indicates that a chromosome conjugates with only one other, at least for long stretches, though it may change partners when there is a long distance between chiasmata. This interpretation found support in Bridges and Anderson's own data, where, as Muller pointed out, change of partners is relatively rare, and when it does occur, is absent from long stretches of chromosomes.

In 1928 genetic analysis had far outstripped cytological observation. Anderson's study of crossing-over in attached X's showed that the end at which they were attached was the one that had been arbitrarily called the right end. But because the different parts of a chromosome ordinarily look alike, it was not possible to correlate specific parts with specific genes and thus test the validity of genetic maps cytologically. And because the two chromosomes of a pair ordinarily look alike, it was not possible to test cytologically the exchange of parts between them.

In 1929 Muller and Painter, in a study of deletions and translocations produced by exposure of *Drosophila* to radiation, showed that when genes are deleted from a linkage group, the chromosome of that group is shortened; and when genes are translocated to another linkage group, the chromosome of that group is lengthened. The shortening and the lengthening are not necessarily proportional to the lengths of the deleted or translocated sections on the linkage maps. But, as had been found in deletions and translocations in nonirradiated flies, the deleted and the translocated genes are always genes that are consecutive on the genetic maps; and this shows that the order of the genes on the maps is correct.

In 1931 cytological proof of crossing-over was obtained in maize by

Harriet B. Creighton and Barbara McClintock and in *Drosophila* by Curt Stern. In each case a pair of chromosomes was used that differed by visible morphological traits at or near each end, and also differed genetically at two loci. It was found that whenever there was crossing-over between the two genes, the visible traits were also exchanged. This proved that genetic crossing-over results from exchanges of pieces of chromosomes. It also proved that in each chromosome the two genes were located between the two visible markers; but as the visible markers were at or near the ends of the chromosomes, the determination of the positions of the genes was not very precise.

Precise cytological determinations of the positions of genes in chromosomes were made possible by studies of the giant chromosomes that are found in the salivary glands and certain other tissues in some *Diptera*. These chromosomes, although observed and studied since 1881, were not known to be chromosomes until 1933, when their chromosomal nature was proved by Heitz and Bauer in *Bibio* and by Painter in *Drosophila*. In these giant chromosomes there are numerous transverse lines and bands, which are arranged at different intervals in different parts and which therefore make it possible to distinguish the parts from one another. Painter, by using deletions, translocations, and inversions, located specific genes accurately in specific parts of chromosomes. Very detailed maps of this kind for all four entire chromosomes were made by Bridges (1935, 1938, 1939); and maps of parts of chromosomes have been made by others.

Of the cytological questions that were considered by the *Drosophila* workers in the 1910s and 1920s, some were subsequently treated by Darlington (1932). Some of his interpretations were similar to theirs, some were different. Darlington's book was regarded by some biologists as too speculative and not in agreement with the evidence. But Darlington's ideas were accepted by Sturtevant and Beadle in their *Introduction to Genetics* (1939). Later, however, they changed their minds, as is indicated in the preface to the second edition of their book in 1962, though Darlington's name is not mentioned in that passage. And in 1965 Sturtevant, in his *History of Genetics*, went into some detail about why he had changed his mind, citing facts and arguments that had been brought forward by other biologists as inconsistent with Darlington's theory (pp. 78-79).

References

Anderson, E. G. 1925. Crossing over in a case of attached-X chromosomes in Drosophila melanogaster. *Genetics* 10:403-417.

Bridges, C. B. 1914. Direct proof through non-disjunction that the sex-linked genes of Drosophila are borne by the X-chromosome. *Science* 40:107-109.

———— 1916. Non-disjunction as proof of the chromosome theory of heredity. *Genetics* 1:1-52, 107-163.

———— 1921a. Genetical and cytological proof of non-disjunction of the fourth chromosome of Drosophila melanogaster. *Proceedings of the National Academy of Sciences* 7:186-192.

———— 1921b. Triploid intersexes in Drosophila melanogaster. *Science* 54:252-254.

———— 1935. Salivary chromosome maps. *Journal of Heredity* 26:60-64.

———— 1938. A revised map of the salivary gland X-chromosome. *Journal of Heredity* 29:11-13.

———— and E. G. Anderson. 1925. Crossing-over in the X chromosomes of triploid females of *Drosophila melanogaster*. *Genetics* 10:418-441.

———— and P. N. Bridges. 1939. A new map of the second chromosome. *Journal of Heredity* 30:475-477.

Creighton, H. B., and B. McClintock. 1931. A correlation of cytological and genetical crossing over in *Zea mays*. *Proceedings of the National Academy of Sciences* 17:492-497.

Darlington, C. D. 1932. *Recent advances in cytology*. London: Churchill.

Heitz, E., and H. Bauer. 1933. Beweise für die Chromosomennatur der Kernschleifen in den Knäuelkernen von Bibio hortulanus. *Zeitschrift für Zellforschung und mikroscopische Anatomie* 17:67-82.

Janssens, F. A. 1909. La théorie de la chiasmatypie. *La Cellule* 25:387-411.

Morgan, L. V. 1922. Non-criss-cross inheritance in Drosophila melanogaster. *Biological Bulletin* 42:267-274.

Morgan, T. H. 1919. *The physical basis of heredity*. Philadelphia: Lippincott.

————, A. H. Sturtevant, H. J. Muller, and C. B. Bridges. 1915, 1923. *The mechanism of Mendelian heredity*. New York: Holt.

Muller, H. J. 1916. The mechanism of crossing-over. *American Naturalist* 50:193-221, 284-305, 350-366, 421-434.

———— and T. S. Painter. 1929. The cytological expression of changes in gene alignment produced by X rays in Drosophila. *American Naturalist* 63:193-200.

Painter, T. S. 1933. A new method for the study of chromosome rearrangements and the plotting of chromosome maps. *Science* 78:585-586.

———— and H. J. Muller. 1929. Parallel cytology and genetics of induced translocations and deletions in Drosophila. *Journal of Heredity* 20:287-298.

Redfield, H. 1930. Crossing-over in the third chromosomes of triploids of *Drosophila melanogaster*. *Genetics* 15:205-252.

Stern, C. 1931. Zytologisch-genetische Untersuchungen als Beweise für die Morgansche Theorie des Faktorenaustausches. *Biologisches Zentralblatt* 51:547-587.

Stevens, N. M. 1907. The chromosomes of Drosophila ampelophila. *Proceedings of the Seventh International Zoological Congress*.

———— 1908. A study of the germ cells of certain Diptera with reference to the heterochromosomes and the phenomena of synapsis. *Journal of Experimental Zoology* 5:359-374.

Sturtevant, A. H. 1965. *A history of genetics.* New York: Harper & Row.

———— and G. W. Beadle. 1939. *An introduction to genetics.* Philadelphia: Saunders. (2d ed., 1962; New York: Dover.)

Weinstein, A. 1928. Four-strand crossing-over. *Anatomical Record* 41:109-110.

———— 1932. A theoretical and experimental analysis of crossing over. *Proceedings of the Sixth International Congress of Genetics* 2:206-208.

———— 1936. The theory of multiple-strand crossing-over. *Genetics* 21:155-199.

Wilson, E. B. 1925. *The cell in development and heredity.* New York: Macmillan.

Cytogenetics and the Neo-Darwinian Synthesis

Hampton L. Carson

An attempt to study the evolution of living organisms
without reference to cytology would be as futile as
an account of stellar evolution which ignored spectroscopy.   —*J. B. S. Haldane, 1937*

At the beginning of this century, cytology and genetics were wholly disparate sciences. Mendelism arose in studies that grew from the tradition of plant breeders such as Kölreuter and Gärtner. There was no experimental cytological approach to inheritance. The three "rediscoverers" of Mendel—Correns, de Vries, and Tschermak—worked in the same noncytological frame.

Traditionally, the cytology of inheritance (karyology) was defined as a branch of developmental and reproductive biology; it emerged into the twentieth century with the traditions of a sedate and properly inductive, descriptive science, as the result of the elegant and sophisticated chromosome studies of Pfitzner, Flemming, Van Beneden, Strasburger, and Boveri. These workers, including Hertwig, had indeed established a firm basis for cellular inheritance by 1900. There was a general awareness of this discovery, but the science still remained inseparable from developmental biology.

The chromosome theory of heredity, in its modern form, however, burst upon the world in a single sentence, possibly added as an afterthought, at the end of Sutton's 1902 paper on the chromosomes of the grasshopper *Brachystola magna:* "I may finally call attention to the probability that the association of paternal and maternal chromosomes in pairs and their subsequent separation during the reducing division as indicated above may constitute the physical basis of the Mendelian law of heredity."

Boveri came to the same conclusion independently in the same year and a year later Sutton expanded on what was to become one of the greatest unifying concepts in biology—that is, the genes are indeed carried on the chromosomes. Sutton proposed the basic formula 2^n for the number of combinations resulting from random segregation, thus building on Weismann's original idea.

A clear statement, however, was not enough to immediately integrate these two diverse approaches. The stranglehold of development on karyology needed to be loosened so that genetical cytology could emerge as a science in its own right. During the first decade of the century, classical cytology continued to be pursued, in traditional nineteenth-century style, by Wilson, Robertson, McClung, Montgomery, and Carothers. The influence of eminent developmentalists like Boveri and Baltzer was still very strong. The flurry of activity of the neo-Mendelians was largely overlooked by cytologists and no strong attempts at integration were apparent.

The merger of cytology and genetics into a single science was the major contribution of the *Drosophila* work of the Morgan school. As has been documented in a number of historical accounts (for example, Sturtevant, 1965), the fly room at Columbia University gave forth a rapid succession of fundamental discoveries: sex-linked inheritance, linkage, crossing-over, and, finally, nondisjunction as proof of the chromosome theory of heredity.

The union of genetics with classical cytology was painful. Janssens' chiasmatype theory could have served as an early bridge between the two cultures but much criticism was leveled against it and, in turn, in a number of laboratories the *Drosophila* work was viewed with considerable misgivings and suspicion. Eventually, however, nuclear cytology and genetics were integrated by 1920, aided and confirmed by the early work of Allen, Winge, and Winkler on plants.

After 1920 it is difficult to separate cytogenetics from any other kind of genetics. Accordingly, a separate cytological contribution to the evolutionary synthesis that emerged in the 1930s cannot be discerned. Nuclear cytology had simply been converted into one of a growing number of powerful genetic techniques.

Nevertheless it may be asked to what extent the proliferating cytogenetic research of the 1920s had an evolutionary flavor and was moving toward the synthesis. Plant cytogenetics in particular showed an explosive development in the mid-1920s. The tentative partial explanations of polyploidy by Winkler and Winge in 1916 and 1917 were the forerunners of the coming into flower of plant cytology. Renner and Cleland started the final clarification of the bizarre complexities of *Oenothera* that had puzzled de Vries and Gates before 1910. Blakeslee and Belling plunged into the *Datura* work, which in retrospect seems curiously nonpopulational. Heitz began work on *Melandrium* and Babcock and Navashin took up the study of *Crepis*, which also lacked populational emphasis. A few years later came the work on *Nicotiana* (Goodspeed and R. E. Clausen) and *Zea* (Creighton, McClintock, and Beadle).

The explosion of research in this field at first strongly stressed polyploidy. Kihara and Ono coined the terms autopolyploidy and allopolyploidy in 1926; Karpechenko synthesized *Raphanobrassica* in 1927; and in 1929 came Newton and Pellew's classic study, the allopolyploid origin of *Primula kewensis*.

During the decade of the 1920s chromosomal studies led toward interpretations of the origin of cultivated plants. A survey of these papers however, shows almost no attempt at synthesis. The words "evolution," "phylogeny," or "origin" are almost never used. In a 1928 paper, however, Darlington remarked concerning his work on *Prunus:* "The work was undertaken partly for its pure cytological interest and partly in an attempt to throw light on the origin of the cultivated forms, their relationships and genetic behavior." Having said this, the author's interest shows through and "pure cytological interest" rules the later discussion and conclusion.

Interest in biogeographic cytology was rudimentary; the populational approach continued to lag and there was a curious lack of interest in organisms in a state of nature. Helwig's study of *Circotettix* in 1929 was one of the first to deal with intraspecific chromosome variability. It was perhaps influenced by Seiler's earlier paper on psychid moths.

Only in the 1930s did Huskins and Anderson publish their respective studies on *Spartina* and *Iris*. I ascribe this lack of interest in evolution to several things, mostly to the excited preoccupation of many workers with exploiting the breakthrough stemming from polyploidy and recombinational genetics. In the rush to apply the new methods and do experiments, few stopped to reflect on the larger picture. When beautiful new kinds of data are unfolding before one's eyes, this display tends to dampen enthusiasm for a broader outlook. The reductionist approach was very important in the case of cytology and it carried its own reward.

A second inhibiting force was the rejection of the Darwinian approach by Morgan and Bateson (see Allen, 1968). Powerful personalities have a way of setting fashion in science, as elsewhere. Indeed, a sterile thought can, under eloquent advocacy, rule a generation. Most of Morgan's young colleagues, nevertheless, disagreed with him, as became clear in the 1930s, but during the 1920s their views were not articulated.

For cytogenetics, the preoccupation with detail did not break until 1932 when it was done almost simultaneously by Haldane and Darlington. I mention Haldane first because *The Causes of Evolution* has a much greater intellectual sweep than *Recent Advances in Cytology*, although the treatment is almost semipopular. *Causes* is most famous, of course, for its remarkable mathematical appendix, which confirms Fisher's population genetics and then is extended with a special Haldanian flair. But where else in that era can we find integrated and facile discussions of the evolutionary implication of the *Drosophila* chromosome studies, allopolyploidy in *Primula*, and Darwinian fitness? Here Haldane neatly conjoins Darwin and Mendel, Fisher and Wright, Newton and Kihara. In the evolutionary context, Haldane deals for the first time with inversions and translocations, polyploidy and hybridization. The paleontological record is woven into the argument. The synthesis had begun in earnest.

Darlington's book is no less remarkable in its abrupt departure from traditional thinking. The book pulls together the disparate karyology of the preceding decade. The data are organized in a provocative manner under deductive headings that jolt the reader and shatter the classical conservative approach to cytology. The book is at once imaginative and brash; speculations are freely made without apology.

Perhaps it was the shock that Darlington's book produced which prevented it from serving the evolutionary synthesis in an immediate and direct way. It was the first articulation of a new science, a strange and difficult kind of visual chemistry with rules only dimly perceived. Even Haldane, who wrote the introduction, cautioned: "It is improbable that all the conclusions of Dr. Darlington's last chapter are correct . . . Nevertheless I am convinced that [it] is a prolegomenon to every future theory of evolution."

The last chapter that Haldane refers to is titled "The Evolution of Genetic Systems." This remarkable essay is found only in the first edition (1932) of *Recent Advances*. It was omitted from the 1937a edition and was expanded and published as a separate book under the same title in 1939. The 1932 chapter is an important milestone. Most relevant to the present inquiry into cytogenetics is the concept that the differences in the chromosomal system are themselves subject to evolutionary processes. Darlington's view of this is neo-Darwinian: "The essential Darwinian assumptions of variation followed by natural selection hold good . . . The principles of change, shown by observations of chromosome behavior, . . . show that the changes are never adaptive to the conditions that produced them but that adaptation arises by selection, most of the original changes having been disadvantageous."

There are indications, in these early discussions by Darlington, that his view of selection was essentially as a normalizing process and it was only later that these ideas were refined. However, a new phase of evolutionary cytogenetics had been born. That chromosomes could be used to construct phylogenetic schemes through comparative studies of chromosome numbers and polyploids had been perceived during the mid-1920s and articulated in the broad context by Haldane. Darlington's view of the genetic system as a generator of variability on which evolutionary processes could operate brought an entirely new dimension to the subject. Crystallized in the essay "The Biology of Crossing-Over" (1937b), it emancipated cytogenetics from dependence on the use of karyotypes and chromosome numbers as an adjunct of systematics. The subject had now become dynamic and it opened a new avenue to the causal analysis of the evolutionary process.

The statements in Darlington's book might be supposed to have had an immediate stimulating effect on the research of the time, but the book was such a mixture of fact and fancy that the inductively operating cytologists of the day were highly suspicious of it, if indeed, they could understand it.

One of the theories that immediately became a major issue was Darlington's treatment of chiasmatypy and its relation to crossing-over and disjunction. At the Sixth International Congress of Genetics at Ithaca, held the same year (1932) that Darlington's book appeared, Karl Sax led the chorus of criticism that arose in response to the Darlingtonian ideas. Those controversies preoccupied cytologists for at least another decade. It is my opinion that their arguments served as a largely inadvertent smokescreen that hid the novelty and potential usefulness of the concept of the evolution of the genetic system.

The culminating event in the integration of cytogenetics into synthetic evolutionary thought came in October 1936, when Dobzhansky delivered a series of lectures at Columbia University; they were published in 1937 as the book *Genetics and the Origin of Species*. The role of this work in the evolutionary synthesis does not have to be elaborated by me; the excitement that it immediately engendered throughout biology defies simple description and is worthy of historical study in its own right. Nowhere, however, was its effect greater than in cytogenetics. Dobzhansky wrote from a background knowledge of genetic and chromosomal variability, both within and between species. His approach was strongly populational and he was the first to consider the supergene inversions as related to geographic distribution. This strongly contrasted with the Belling-Blakeslee-Babcock approach. Because he had an equally broad background in biogeography and selection theory, Dobzhansky was peculiarly well equipped to frame the basic synthesis in a simple eloquent statement.

I was a student at the University of Pennsylvania when both Darlington's and Dobzhansky's books arrived in the laboratory. Graduate students are, in the long run, the most sensitive targets for new ideas. The older members of this strongly cytological department received the Darlington book with stiff attitudes of outrage, anger, and ridicule. The book was considered to be dangerous, in fact poisonous, for the minds of graduate students. It was made clear to us that only after we had become seasoned veterans could we hope to succeed in separating the good (if there was any) from the bad in Darlington. Those of us who had copies kept them in a drawer rather than on the tops of our desks.

Reception of the Dobzhansky book was quite the opposite. As I have suggested, it caused excitement among graduate students and senior staff alike, although by the time I left Pennsylvania in 1943, no genetics and evolution course had yet appeared in the curriculum. Indeed, it was only after sitting in Anderson's remarkable "Genetics and Natural History" course at Washington University (St. Louis) that Harrison Stalker and I put together our first joint course, "Modern Theories of Evolution."

The long-term effects of the Dobzhansky book can be discerned in research trends to the present day. One immediate effect was the birth of a major branch of *Drosophila* work dealing with cytogenetics, evolution, and natural populations pioneered at the University of Texas and presaged by the paper of Patterson, Stone, and Griffen (1940) on the evolution of the *virilis* group. A few years later another book of enormous influence, traceable ultimately to Haldane's influence, appeared. *Animal Cytology and Evolution* by White (1945) brings together three cytological approaches: those of Belar, Dobzhansky, and Darlington. It drew large quantities of cytological information into the evolutionary literature. For the first time, White provided the evolutionary framework for the integration of such subjects as Metz's work on *Sciara*, that of the Schraders on coccids, and of Matthey and Makino on chromosome number. White weaves together theories on the origin of parthenogenesis and polyploidy including many of the things touched on by Muller's writings but never really synthesized.

White's book was followed in 1950 by Stebbins' monumental *Variation and Evolution in Plants*, which finally synthesized all of the loose ends of the enormously diverse works on plants that had appeared in the twenty years after the beginning of Babcock, Turesson, and Gustafsson.

Stebbins' imaginative and far-reaching mind was already evident in his first papers on *Antennaria* (1932) coming only a year or so after Babcock was still inquiring whether evolution was a continuous or discontinuous process.

This brief period in the history of science is especially interesting because it shows how advancing understanding tends to unify knowledge by the removal of the artificial walls that man erects as he fumbles along. In this case, the first wall to come down was the distinction between cytology and genetics. But even this new unified science, about which we were congratulating ourselves, was soon perceived to be walled off from evolutionary theory. That wall came down rather slowly and painfully. At this juncture we would do well to sharpen our perception to identify barriers as yet only dimly seen.

References

Allen, C. E. 1919. The basis of sex inheritance in *Sphaerocarpos*. *Proceedings of the American Philosophical Society* 58:298-316.

Allen, G. E. 1968. Thomas Hunt Morgan and the problem of natural selection. *Journal of the History of Biology* 1:113-139.

Anderson, E. 1936. The species problem in *Iris*. *Annals of Missouri Botanical Garden* 23:457-509.

Babcock, E. B. 1924. Genetics and plant taxonomy. *Science* 59:327-328.

——— and M. Navashin. 1930. The genus *Crepis*. *Bibliographia Genetica* 6: 1-90.

Baltzer, F. 1910. Über die Beziehung zwischen dem Chromatin und der Entwicklung und Vererbungsrichtung bei Echinodermenbastarden. *Archiv für Zellforschung und mikroskopische Anatomie* 5:497-621.

Bateson, W. 1909. *Mendel's principles of heredity*. Cambridge: Cambridge University Press.

Beadle, G. W. 1930. Genetical and cytological studies of Mendelian asynapsis in *Zea mays*. *Cornell University Experiment Station Memoirs* 129.

Belar, K. 1928. *Die cytologischen Grundlagen der Vererbung*. Berlin: Borntraeger.

Belling, J. 1927. The attachment of chromosomes at the reduction division in flowering plants. *Journal of Genetics* 18:177-205.

Blakeslee, A. F. 1922. Variations in *Datura* due to changes in chromosome number. *American Naturalist* 56:16-31.

Boveri, T. 1909. Die Blastomerenkerne von *Ascaris megalocephala* und die Theorie der Chromosomenindividualität. *Archiv für Zellforschung* 3:181-268.

Carothers, E. E. 1913. The Mendelian ratio in relation to certain orthopteran chromosomes. *Journal of Morphology* 24:487-511.

Cleland, R. E. 1922. The reduction divisions in the pollen mother cells of *Oenothera franciscana*. *American Journal of Botany* 9:391-413.

Correns, C. 1900. G. Mendels Regel über das Verhalten der Nachkommenschaft der Rassenbastarde. *Berichte der deutschen botanischen Gesellschaft* 18:158-168.

Creighton, H. B., and B. McClintock. 1931. A correlation of cytological and genetical crossing over in *Zea mays*. *Proceedings of the National Academy of Sciences* 17:492-497.

Darlington, C. D. 1928. Studies in *Prunus*. *Journal of Genetics* 19:213-256.

——— 1932. *Recent advances in cytology*, 1st ed. Philadelphia: Blakiston.

——— 1937a. *Recent advances in cytology*, 2d ed. Philadelphia: Blakiston.

——— 1937b. The biology of crossing-over. *Nature* 140:759.

——— 1939. *The evolution of genetic systems*. Cambridge: University Press.

Darwin, C. 1859. *On the origin of species by means of natural selection, or the preservation of favored races in the struggle for life*. London: John Murray.

de Vries, H. 1900. Sur la loi de disjonction des hybrides. *Comptes rendus de l'Académie des Sciences* (Paris) 130:845-847.

——— 1903. *Die Mutationstheorie*. Leipzig: Veit.

Dobzhansky, Th. 1937. *Genetics and the origin of species*. New York: Columbia University Press.

Fisher, R. A. 1930. *The genetical theory of natural selection*. Oxford: Clarendon Press.

Flemming, W. 1882. *Zellsubstanz, Kern und Zelltheilung*. Leipzig: F.C.W. Vogel.

Gärtner, C. F. von. 1849. *Versuche und Beobachtungen über die Bastarderzeugung im Pflanzenreich*. Stuttgart.

Gates, R. R. 1909. The stature and chromosomes of *Oenothera gigas* De Vries. *Archiv für Zellforschung* 3:525-552.

Goodspeed, T. H., and R. E. Clausen. 1927. Interspecific hybridization in *Nicotiana*. *University of California Publications in Botany* 11:117-125.

Gustafsson, A. 1935. Studies on the mechanism of parthenogenesis. *Hereditas* 21:1-112.

Haldane, J. B. S. 1932. *The causes of evolution*. New York: Harper.

——— 1937. Foreword. In *Recent advances in cytology* by C. D. Darlington. Philadelphia: Blakiston, pp. v-vi.

Heitz, E. 1925. Beitrag zur Cytologie der Melandrium. *Zeitschrift für wissenschaftliche Biologie* 1:241-259.

Helwig, E. R. 1929. Chromosomal variations correlated with geographical distribution in *Circotettix verruculatus* (Orthoptera). *Journal of Morphology* 47:1-36.

Hertwig, O. 1874. Das Problem der Befruchtung und der Isotropie des Eies, eine Theorie der Vererbung. *Jenaische Zeitschrift* 18:276-318.

Huskins, C. L. 1931. The origin of *Spartina townsendii*. *Genetica* 12:531-538.

Janssens, F. A. 1909. Spermatogenèse dans les Batraciens. *La Cellule* 25:387-411.

Karpechenko, G. D. 1927. Polyploid hybrids of *Raphanus sativus* L. x *Brassica oleracea*. *Bulletin of Applied Botany* (St. Petersburg) 17:305-410.

Kihara, H. 1930. Genomanalysis bei Triticum und Aegilops. *Cytologia* 1:263-284.

——— and T. Ono. 1926. Chromosomenzahlen und systematische Gruppierung der Rumex-Arten. *Zeitschrift für Zellforschung und mikroskopische Anatomie* 4:475-481.

Kölreuter, J. G. 1766. *Vorläufige Nachricht von einigen das Geschlecht der Pflanzen betreffenden Versuchen und Beobachtungen, Nebst Fortsetzungen* in Ostwald's *Klassiker,* 1893.

McClung, C. E. 1905. The chromosome complex of orthopteran spermatocytes. *Biological Bulletin* 9:304-340.

Makino, S. 1932. An unequal pair of idiochromosomes in the tree cricket *Oecanthus longicauda* Mats. *Journal of the Faculty of Science, Hokkaido University* VI 2:1-36.

Matthey, R. 1932. Les chromosomes et la systematique zoologique. *Revue suisse de zoologie* 39:229-237.

Mendel, G. 1866. Versuche über Pflanzen-Hybriden. *Verhandlungen des Naturforschenden Vereins in Brünn* 4:3-47.

Metz, C. W. 1938. Chromosome behavior, inheritance and sex determination in *Sciara. American Naturalist* 72:485-520.

Montgomery, T. H. 1901. A study of the chromosomes of germ cells of *Metazoa. Transactions of the American Philosophical Society* 20:154-236.

Morgan, T. H. 1903. *Evolution and adaptation.* New York: Macmillan.

Muller, H. J. 1940. Bearings of the *Drosophila* work on systematics. In *The new systematics,* ed. J. Huxley. Oxford: Clarendon Press, pp. 185-268.

Newton, W. C. F., and C. Pellew. 1929. *Primula kewensis* and its derivatives. *Journal of Genetics* 20:405-467.

Patterson, J. T., W. S. Stone, and A. B. Griffen. 1940. Evolution of the *virilis* group in *Drosophila. University of Texas Publications* 4032:218-250.

Pfitzner, W. 1881. Ueber den feineren Bau bei der Zelltheilung auftretenden fadenförmigen Differenzierungen des Zellkerns. *Morphologisches Jahrbuch* 7:289-311.

Renner, O. 1925. Untersuchungen über die faktorielle Konstitution einiger komplexheterozygotischer Oenotheren. *Bibliographia Genetica* 9:1-168.

Robertson, W. R. B. 1915. Chromosome studies. *Journal of Morphology* 26:109-141.

Sax, K. 1932. The cytological mechanism for crossing-over. *Proceedings of the Sixth International Congress of Genetics* 1:256-273.

Schrader, F., and S. Hughes-Schrader. 1926. Haploidy in *Icerya purchasi. Zeitschrift für wissenschaftliche Zoologie* 128:182-200.

Seiler, J. 1923. Geschlechtschromosomenuntersuchungen an Psychiden. *Zeitschrift für induktive Abstammungs- und Vererbungslehre* 31:1-99.

Stalker, H. D. 1940. Chromosome homologies in two subspecies of *Drosophila virilis. Proceedings of the National Academy of Sciences* 26:575-578.

Stebbins, G. L., Jr. 1932. Cytology of *Antennaria* I. normal species. *Botanical Gazette* 94:134-151.

——— 1950. *Variation and evolution in plants.* New York: Columbia University Press.

Strasburger, E. 1875. *Ueber Zellbildung and Zelltheilung.* Jena: G. Fischer.

Sturtevant, A. H. 1965. *A history of genetics.* New York: Harper & Row.

Sutton, W. S. 1902. On the morphology of the chromosome group in *Brachystola magna. Biological Bulletin* 4:24-39.

Tschermak, E. 1900. Ueber künstliche Kreuzung bei *Pisum sativum*. *Berichte der deutschen botanischen Gesellschaft* 18:232-239.

Turesson, G. 1922. The species and the variety as ecological units. *Hereditas* 3:100-113.

Van Beneden, E. 1876. La maturation de l'oeuf, la fécondation et les premières phases du développement embryonnaire des mammifères. *Bulletin de l'Académie Royale de Belgique* 41:1160-1205.

Weismann, A. 1892. *Essays on heredity*, trans. A. E. Shipley and S. Schönland. Oxford: Oxford University Press.

White, M. J. D. 1945. *Animal cytology and evolution*. Cambridge: Cambridge University Press.

Wilson, E. B. 1896. *The cell in development and inheritance*, 1st ed. New York: Macmillan.

Winge, O. 1917. The chromosomes: their number and general importance. *Compte rendu des travaux du Laboratoire de Carlsberg* 13:131-275.

Winkler, H. 1916. Über die experimentelle Erzeugung von Pflanzen mit abweichenden Chromosomenzahlen. *Zeitschrift für Botanik* 8:417-531.

Wright, S. 1932. The roles of mutation, inbreeding, crossbreeding and selection in evolution. *Proceedings of the Sixth International Congress of Genetics* 1:356-366.

3 Embryology

Darwin once expressed the opinion that embryology had provided him with the best evidence for evolution. Haeckel and his followers took up this hint and used the "law of recapitulation" to establish phyletic lineages. The result was a flowering of comparative embryology from the 1860s through the 1880s.

The basic facts of embryology, Darwin claimed in *On the Origin of Species*, all "can be explained on the view of descent with modification" (1859, p. 443). Darwin's language here is precise and crucial. Wherever possible, he claimed that phenomena could be explained by the mechanism of natural selection, not merely the looser "descent with modification." In fact, neither Darwin nor any of the great comparative embryologists could meaningfully synthesize embryology with evolution by natural selection, because the connections between hypothetical germinal material and the process of development were too vague to command the interest of serious biologists. Comparative embryology thus remained a purely descriptive science.

When *Entwicklungsmechanik* began to flourish in the late 1880s, comparative embryology went into decline. The new causal embryology, which attempted to explain development in strictly physicochemical terms, soon provided important new insights into the mechanics of development. But the problem of the germinal control of development was still beyond analysis. The experimentalists who pursued the *Entwicklungsmechanik* approach had little appreciation or concern for evolution in natural populations, and by the time Mendelism arose as a science in the 1900-1910 decade, embryologists did not see their science as closely connected to the study of heredity or the mechanism of evolution. Although all the embryologists were evolutionists, very few embryologists before the 1930s, with the exception of the Oxford school (Goodrich, Huxley, de Beer), endorsed natural selection as the primary mechanism of evolution. Many prominent embryologists actively minimized the importance of natural selection, even as late as the 1940s and 1950s.

The role of embryology in the evolutionary synthesis raises many questions. Why did embryologists resist accepting the idea of natural selection for so long? How important were geneticists who worked on gene action and developmental genetics in bringing together the fields of genetics and embryology? Were the books of Schmalhausen (1949) and Waddington (1957) the first major works envisioning the synthesis of embryology with genetics and evolution, or did such a synthesis really require knowledge of the molecular biology of the 1950s and 1960s? What contribution, if any, did embryology make to the evolutionary synthesis of the 1930s and 1940s? The two essays in this section address all of these and related questions, particularly on the evolutionary thinking of embryologists in Germany and England.   w.b.p.

References

Darwin, C. 1859. *On the origin of species.* London: Murray.
Schmalhausen, I. I. 1949. *Factors of evolution.* Philadelphia: Blakiston.
Waddington, C. H. 1957. *Strategy of the genes.* London: Allen and Unwin.

Embryology and the Modern Synthesis in Evolutionary Theory

Viktor Hamburger

Did embryology and, more specifically, experimental embryology assist in the creation of the modern synthesis during the thirties and early forties, or, on the contrary, was it a retarding element? The major works that embody the modern synthesis by Huxley (1942), Dobzhansky, Fisher, Haldane, Mayr, Simpson, and Wright hardly mention embryonic development. This omission is somewhat strange because Huxley was sufficiently interested in growth and development to write two books on this topic: one on relative growth (1932), and a text of experimental embryology with de Beer (1934). His book on evolution clearly shows the dominance of Huxley the naturalist over Huxley the embryologist. Obviously, the modern synthesis had a strong foundation in Mendelian and population genetics and its mathematical treatment, in ecology, and in field studies of speciation. It could well afford to dispense with embryology, although its implications for genetics and evolution were recognized by the founders of the modern synthesis. Conversely, the contemporary leading books on experimental embryology, by Schleip (1929), Spemann

(1936), and Weiss (1939), did not include considerations of evolution. The modern synthesis did not receive assistance from contemporary embryologists.

Could one imply that the leading experimental embryologists of the 1920s and 1930s, such as Harrison, Spemann, Lillie, Conklin, Dalcq, and Child, whose names were held in considerable esteem among most biologists, delayed the synthesis by opposition or indifference? They were all evolutionists and they all conceded the effectiveness of natural selection, at least to some extent. But many had misgivings about a key dogma of the modern synthesis; namely, the claim that natural selection is the sole explanation of all adaptations; and Lamarckist ideas were by no means dead. However, with few exceptions, such as Dalcq's somewhat later publications (1949, 1951) and some remarks in Spemann's autobiography (1943, pp. 156 ff., 272), there was little public discussion and no open opposition by this group. In fact, I believe that few embryologists in the 1920s and 1930s were aware of the emergence of a new synthesis in evolutionary theory. In Spemann's "reprint room," where his colleagues and *Doktoranden* often met during tea hour in the late twenties and early thirties, there was a lively, continuous dialogue. Spemann and his colleague, Fritz Süffert, an expert on adaptive coloration in butterflies and moths, often discussed selectionist versus Lamarckist explanations of such complex adaptations and their embryological implications, with Süffert on the selectionist side and Spemann inclined to Lamarckism. I do not remember any extension of the discussions to speciation or to the particular issues involved in the evolutionary synthesis. I can document only Harrison's awareness of the latter: "The development of modern genetics, the experimental study of the origin of mutations and the new mathematical theory of natural selection are hopeful signs of the applicability of exact methods to the study of evolutionary processes" (1937, p. 7). T. H. Morgan wrote *Embryology and Genetics* (1934), but I omit his case because a special session is devoted to him. Indeed, I believe that some leading embryologists had a retarding influence on the modern synthesis.

The lack of interest of experimental embryologists in evolutionary problems can be traced back to the founder of experimental embryology, Wilhelm Roux. Before him, for once, and only for a short time, embryology, genetics, and evolution had been united in a complete synthesis in Weismann's *The Germ-Plasm: A Theory* (1893). But this superb intellectual feat soon foundered; its foundations broke down, partly under the impact of work of the early experimental embryologists. Roux, a student of the major German prophet of evolution, Haeckel, and with impeccable credentials as a selectionist (he had extended selectionism to the explanation of adaptive structures within the organism in a book published in 1881) broke away from Haeckel in the matter of recapitulation. He founded experimental embryology or *Entwicklungsmechanik* in the 1880s as a deliberate countermove against Haeckel's categorical verdict that phylogeny is the sufficient cause of ontogeny, and that there is nothing else to explore in this matter. Roux's decisive move from ultimate or remote to proximate causes (following that of His), and the concomitant introduction of the experimental method as the indispensable tool for the analysis of proximate factors in embryonic development, started the alienation of embryological from evolutionary thinking.

By the 1920s and 1930s, experimental embryology and genetics both had accomplished a major breakthrough, experimental embryology through the achievements of Harrison and Spemann and their schools, and genetics through the Morgan school. Both fields were deeply absorbed in their own problems and took little notice of each other. The embryologists were involved in the study of epigenetic mechanisms, such as induction, gradient fields, and morphogenetic movements. Evolutionary considerations turned up rarely, as for instance in the mistaken dichotomy of "mosaic" versus "regulation" eggs.

To some extent, the evolutionists were aware of the role of embryology in evolutionary theory—to be more specific, of the fact that phenotypes, which are the target of selection, are the end result of developmental processes which in turn are the manifestation of gene activity. As Huxley says, "Any originality which this book may possess lies partly in furthering Fisher's ideas and partly by stressing the fact that a study of genes during development is as essential for an understanding of evolution as are the study of mutation and selection" (1942, p. 8). Unfortunately, his intent was not fulfilled in the writing of his book; only a short chapter deals with heterogonic growth. In a recent publication Mayr states: "The fact that fitness is determined by the phenotype is the reason for the extraordinary evolutionary importance of the developmental processes that shape the phenotype" (1970, p. 108). I assume that he was aware of this notion at a much earlier date. Of the founding fathers, only Wright was actively engaged in studies of developmental genetics. His brief résumé of 1934 goes beyond generalities. It illustrates the role of specific genes in growth processes and pattern formation and the implication of such manifestations for evolution.

I do not imply a criticism of the originators of the modern synthesis for their neglect of developmental genetics. On the contrary, I would assert that it has always been a legitimate and sound research strategy to relegate to a "black box," at least temporarily, wide areas that although pertinent would distract from the main thrust. No great discoveries or conceptual advances are possible without this expediency. Von Frisch would probably never have achieved what he did if he had allowed himself to be sidetracked by worrying over a dance center in the cerebral or thoracic ganglia of the honeybee!

Nevertheless, the modern synthesis as formulated at the time was incomplete without a chapter dealing with the effects of selection on the gene-controlled variability of developmental processes. This type of theoretical consideration was then actually in the making in the work of the Russian academician, Schmalhausen. Unfortunately, the Second World War interrupted communications between the East and the West. Schmalhausen's book, *Factors of Evolution*, which I believe offered one of the most succinct expositions of the problem and important contributions to its solution, did not become known to English readers until 1949.

Alienation between Experimental Embryology and Genetics

Before a chapter on evolutionary changes in embryonic development could be written, it was necessary to develop a new borderline field, a developmental or physiological genetics. In other words, the concepts and methods had to be created that would lead to an understanding of the role of genes and their products in embryonic development. The beginnings of such a synthesis of embryology and genetics can be traced to Boveri and Driesch at the end of the last century. But the actual analysis of gene action beginning with the study of the development of mutants and the application of the transplantation method did not come to fruition until the 1930s. It was accomplished by a younger generation, against the background of strong skepticism from at least some of the leading experimental embryologists of the older generation who were at the zenith of their accomplishments, power, and influence in the 1920s and 1930s. Some went as far as to assert a fundamental incompatibility between concepts, goals, and methods of the two fields; they saw, in principle, no chance of a meaningful amalgamation. Obviously, if this viewpoint had prevailed, there would have been no place for developmental genetics nor for a consideration of development in evolutionary thinking.

The expression of these ideas, which are presented succinctly in Lillie's essay (1927), coincides in time with the ripening of the modern synthesis. Morgan's book *Embryology and Genetics* illustrates the ambiguity of this situation. He wrote, "The story of genetics has been so interwoven with that of experimental embryology that the two can now, to some extent, be told as a single story . . . It is possible to attempt to weave them together in a single narrative" (1934, p. 9). The story goes that after the publication of the book, Morgan asked a prominent visitor what he thought of it. The visitor frankly responded that he could not find a synthesis of the two fields; whereupon Morgan, tongue in cheek, asked "What does the title say?"

The roots of the uncompromising attitude of leading experimental embryologists can be found in two major divisive issues. The first was based on the preoccupation of the geneticists with the nucleus and of the embryologists primarily with the cytoplasm. The bias of the embryologists dates back to the famous experiment of Driesch (1891) in which he isolated the first two blastomeres of the sea urchin egg, resulting in two complete larvae. The experiment demonstrated the equivalence of the blastomere nuclei, the regulative capacity of the egg, and an interaction of the two blastomeres in normal development. All three points were found to have general validity and far-reaching implications. It was argued that if every cell is in possession of the same complete genome, progressive differentiation must result from cytoplasmic differentials. The argument was reinforced by numerous impressive studies of egg structure and cytoplasmic prelocalization of organs, both in invertebrate and vertebrate eggs, and experimental proof of the importance of cytoplasmic differences for progressive differentiation. Lillie speaks of the "almost universally accepted genetic doctrine today that each cell receives the entire complex of genes. It would therefore appear to be self-contradictory to attempt to explain embryonic segregation by behavior of the genes, which are *ex hypothesi* the same in every cell" (1927, p. 13). Lillie further states:

> I do not know of any sustained attempt to apply the modern theory of the gene to the problem of embryonic segregation. As the matter stands, this is one of the most serious limitations of the theory of the gene considered as a theory of the organism. We should of course be careful to avoid the implication that in its future development the theory of the gene may not be able to advance into this unconquered territory. But I do not see any expectation that this will be possible, *even in principle*, as long as the theory of the integrity of the entire gene system in all cells is maintained. If this is a necessary part of the gene theory, the phenomena of embryonic segregation must, I think, lie beyond the range of genetics (pp. 14-15, italics added).

And Harrison wrote ten years later: "The prestige of success enjoyed by the gene theory might easily become a hindrance to the understanding of development by directing our attention solely to the genome, whereas cell movements, differentiation, and, in fact, all developmental processes are actually effected by the cytoplasm" (1937, p. 9).

Another issue formed an even more formidable obstacle to mutual understanding. A basic tenet of embryology has always been the structural and physiological unity, the individuality, of the embryo that continues through life. Lillie says: "The germ is physiologically integrated as an individual at all stages" (1927, p. 4). And the same theme pervades the entire lifework of Child, who was very influential at that period. Progressive differentiation, while creating an increase in complexity, proceeds within the confines of the individuum. This statement is not an abstraction but an expression of the epigenetic nature of development, of the internal inductive, regulative, and feedback mechanisms that integrate and synchronize developmental processes. (The embryologist has always been concerned with the individual embryo, moving from there to lower levels of organization, whereas one of the key elements of the modern synthesis was the conceptual shift from the individual to populations—another ground for alienation.)

To complicate matters, there are unitary subsystems with the same properties as the whole, which we define operationally as "morphogenetic fields." For instance, if in the tail bud stage of a urodele amphibian embryo half of the limb-forming mesodermal disc, or one half of the optic vesicle is removed, the residual group of seemingly undifferentiated cells restores the whole, and an organ of normal size and structure ensues. How would one handle such phenomena of regulation with the genetic concepts and methods available at that time? Lillie states: "Individuation is clearly an [internal] environmental relationship mediated through the cytoplasm, not through the nucleus" (1927, p. 13). Dalcq, in his review essay of Simpson and Cuénot (1951), puts the dilemma more succinctly:

> The cytoplasm of the egg is by itself an organized system along general lines, endowed with a pattern which the genetic system lacks. Moreover, it integrates the activity of numerous constituents in such a way that it is capable of regulation. Since Driesch's discovery [the experiment of regulation after blastomere isolation] the embryologists are forced to uphold more or less explicitly the notion of "configuration globale" [Ganzheit, whole]. This notion, so intimately tied to a pattern, is lacking in the system of concepts used by geneticists and notably in the synthetic theory. The latter is based on a particularistic, atomistic viewpoint which neglects, despite everything, this other factor which resides in the totality of the organization (p. 135; my translation).

Even though Dalcq may have been somewhat behind the times with his atomistic notion of the genome, the embryologists of that period can hardly be blamed for their failure to recognize the discrete units of the genome, which, moreover, were identical in each cell, as critical factors in the continuous flow of epigenetic development. Many contemporaries of Lillie shared his pessimistic outlook: "The progress of genetics and physiology of development can only result in a sharper definition of the two fields, and any expectation of their reunion (in a Weismannian sense) is in my opinion doomed to disappointment" (1927, p. 18).

Spemann had a more positive attitude. Coming from Boveri, he realized the close relationship of problems of localization and activation of hereditary factors. In his 1924 address to the German Society of Genetics, he tried to find common ground between the two fields. But he was no more successful than others in attaining a real synthesis. His general statement that "the activation of the genome does not occur by autonomous segregation of the hereditary factors but under far-reaching interaction between the parts, hence epigenetically" (1924, p. 78) was no more constructive than that of Wilson. But eight years later Spemann and Schotté (1932) reported the classical experiment of xenoplastic transplantation between anurans and urodeles that, though designed originally to deepen the understanding of embryonic induction, actually opened up problems of profound evolutionary significance. This aspect was taken up by Baltzer and his student E. Hadorn and their coworkers who made substantial contributions to what may be called evolutionary embryology (Baltzer, 1952).

To ensure that I have not overstated the amount of polarization, I returned to the book that we then considered as the ultimate arbiter in all such matters: Wilson's *Cell in Development and Heredity* (1925). The superb chapter, "Development and Heredity," is a succinct discussion of cytoplasmic organization with a survey of all experimental evidence for its role in development and an equally lucid presentation of nuclear organization. But, again, the synthesis is limited to the insight that all developmental processes including egg organization are controlled by genes and to the general statement that "heredity is effected by the transmission of a nuclear preformation which in the course of development finds expression in a process of cytoplasmic epigenesis" (p. 1112). If one of the most profound minds of his time with a full command of both genetics and embryology bypassed the problem of gene action in development, then obviously the embryologists of this generation were not ready to come to the aid of the architects of the new synthesis.

Beginnings of Developmental or Physiological Genetics

To overcome the impasse, it was necessary to make a fundamental conceptual shift from the antithesis nucleus versus cytoplasm to the idea of nucleocytoplasmic interactions in development. Driesch anticipated this notion as early as 1894. His hypothesis was based on the then-established premises that the nucleus is the bearer of heredity, that all nuclei in an embryo are equivalent and "totipotent," and that the egg cytoplasm has an organized structure created under the control of the oocyte nucleus. He postulated that the nucleus would effect chemical changes in its cytoplasmic environment by a "fermentative effect," in a way that would preserve its completeness or totality. "Determinative substances would not be released directly by the nucleus but originate in the cytoplasm under the control of the nucleus" (p. 88). The chemical changes in the cytoplasm, in turn, would cause the release of other specific fermentative effects in the nucleus leading to the next step in cytoplasmic differentiation, and so forth. This model, with its emphasis on the chemical nature of the interaction, on the indirect way in which genes control cytoplasmic differentiation, and on the notion of feedback between nucleus and cytoplasm, has a modern ring. But Driesch hardly made an impact on his contemporaries. His ideas were forgotten even by his friend Morgan with whom he spent several winters at the Stazione Zoologica in Naples; they certainly discussed nucleus and cytoplasm. Four decades later, Morgan (1934) presented the reciprocal feedback notion as his own novel solution to the problem of nucleocytoplasmic interaction.

A younger generation, with open minds, had to make a new beginning. Among them, Goldschmidt was the dominant figure. He became one of the founders of physiological genetics and remained its most forceful promoter. Beginning around 1915 with the study of sex determination and determination of pigment patterns in larvae of the moth *Lymantria*, he synthesized extensive experimental data and imaginative theoretical ideas in a quantitative theory of gene action, which for the first time placed the physiological role of genes as determinants of developmental processes in the center of the stage. Like Driesch who spoke of "ferments," Goldschmidt postulated that because genes have enzyme properties, their influence is of a chemical nature. A key element in his theory is the assumption that genes operate by controlling rates or velocities of developmental processes and different alleles represent quantitative differences of gene activity. It follows that they produce variations in speed of development. As in Driesch's theory, a necessary corollary of the enzymatic nature of gene action is the assumption of specific substrates located in the cytoplasm, hence the postulate of cytoplasmic organization and a chain of specific gene-cytoplasm interactions. Because he considered the entire genome as operating in an integrated fashion, some of the difficulties that Lillie and Dalcq found insuperable were overcome. From the beginning, he emphasized the importance of physiological genetics for evolutionary theory; his book, *The Material Basis of Evolution* (1940), is a synthesis of his ideas.

I am not concerned with the speculative and controversial nature of some of his novel ideas, nor with the fact that his "rate-gene" theory was soon superseded by the biochemical genetics initiated by Beadle, Ephrussi, and Tatum. The historical fact remains that he broke new ground by supplementing the contemporary genetics, which was essentially transmission genetics, with a dynamic conception of the gene as a physiological agent controlling developmental processes. By creating this link, he opened a meaningful dialogue between some embryologists and some geneticists. I and other embryologists became interested in genetics in the 1920s largely by reading Morgan's book, *The Physical Basis of Heredity* (1919), which became available to us in German in Nachtsheim's translation in 1921. And we became avid readers of Goldschmidt's books of 1920 and 1927; they opened our eyes to the challenging idea that genes as factors in development could be incorporated into our experimental-embryological thinking. In historical perspective, Boveri in his merogony experiment of 1896 had already combined experimental embryological and genetic methods and thoughts. The design of cross-fertilization of enucleated eggs with foreign sperm to elucidate nuclear-cytoplasmic compatibility had been taken up in the school of Boveri's student, Baltzer (1940) and by some others. At any rate, through Goldschmidt, Baltzer, Herbst, and others, we were immunized early against the notion of an antagonism between experimental embryology and genetics. In the late 1930s, the burgeoning field of physiological genetics had already gained foothold. At the First Growth Symposium (1939, ed. Berrill), which, like its successors, was supposed to bring us up to date on research frontiers in growth and development and brought together animal and plant embryologists, geneticists, microbiologists, biochemists, and physiologists, two of the ten presentations (by Stern and Waddington) dealt with the role of genes in development.

Before embryology could make a meaningful contribution to evolutionary theory, the basic mechanisms of gene action and the concepts derived from them had to be worked out by the endeavors of developmental geneticists. Some of those concepts proved to be useful for evolutionary theory, particularly those developed in the period before 1940, which corresponded to the formative years of the modern synthesis.

The concept of heterogonic growth (Huxley, 1932) is closely related to Goldschmidt's concept of rate genes. The significance of heterogonic growth for evolution has been discussed by Huxley (1942) and Goldschmidt (1940). If a single gene controls differential growth rates in different parts of the organism, then a mutation can bring about a multiplicity of changes while preserving relative proportions, thus "lightening the burden of natural selection." Both authors point out implications for an understanding of neoteny, metamorphosis, vestigial organs, and extinction resulting from excessive growth of specialized structures.

The concept of "norm of reaction," which is significant in this context implies that the phenotypic expression of a gene is dependent on variables in the internal and external environment during development. Each gene has a potential range of expression, the extent of the range being characteristic of each gene. Modifications of the phenotypes resulting from this developmental plasticity are important raw materials for natural selection. Closely related is the concept of sensitive periods in development: restricted phases in a developmental process or a metabolic sequence are particularly susceptible to disturbances. Each developmental process has its specific sensitive period or periods. Chemical teratogens, extreme temperature shocks, X rays, or other agents produce malformations of specific structures such as eyes or legs, or their components, when applied at the pertinent sensitive period, but at no other time. The link to genetics was Goldschmidt's discovery (1935) that malformations produced in this way often have a striking resemblance to abnormal mutants. He produced "phenocopies" of several mutants of *Drosophila* by heat shock applied at the appropriate sensitive period. The implication is that phenocopying agents and the phenocopied gene interfere with the same developmental process or metabolic sequence; as a result, sensitive periods would represent phases of determinative gene actions. Although the hope that in this way one might locate the site of primary gene action was thwarted by the consideration that a sensitive period may be anywhere along the path from primary gene effect to the phenotypic end product and not necessarily at its beginning, this point does not detract from the importance of sensitive periods for evolutionary considerations.

The concept of pleiotropic or polyphenic expressions of single genes in different organs or metabolic processes helps to simplify the complexity of the problem of genetic control of integrated developmental processes. Penetrance is the frequency (in percentages) of the phenotypic expression of a gene. Penetrance below 100 percent indicates that intrinsic or extrinsic factors set a threshold for gene manifestation. Many factors are involved in the lowering of penetrance, such as the genetic background, including modifier genes, quantitative variations in gene products, or variations in environmental factors. But again any mechanism that introduces variability in gene-controlled developmental processes gives selection a foothold.

The term "pattern genes" does not refer to a special category but to a variety of genes that control the differentiation of integrated structural patterns. The older experimental embryologists such as Dalcq found insuperable difficulties in reconciling the atomistic configuration of the genome with the holistic and regulative aspects of epigenetic development and its morphogenetic fields and gradients. The demonstration of single genes that control determinative processes resulting in complex structural patterns, such as the pigment banding pattern in the wings of moths, or the bristle pattern on the thorax of *Drosophila*, or the toe patterns in forelegs and hindlegs of guinea pigs, goes a long way toward mitigating this misconception.

The operation of pattern genes is possible only in epigenetic development with its built-in plasticity. In only a few cases was a detailed analysis of pattern-gene action possible. One of the earliest and most detailed studies was the investigation of the pigment bands on the wings of the meal moth *Ephestia kühniella* by Kühn and Henke between 1929 and 1936. The pattern consists of alternating light and dark bands of different widths, with a bilateral symmetrical arrangement on each wing. The building blocks are scales formed as flattened outgrowths of single epidermal cells. Scales are classified according to differences in shape and pigmentation. Defect and heat shock experiments were used to establish the details of the determination process of the whole pattern, in which the spreading of streams of determining agents (probably diffusible substances) plays a major role. These processes of organizing the wing surface in bands that are, of course, invisible occur in the pupal epidermis long before the actual differentiation of the scales takes place. Mutants were found that modify the "wild-type" pattern by modifying or slowing down the spreading of the determination streams, thus creating abnormal total patterns; other genes modify pigmentation. The essential point is the genetic control of supercellular invisible patterns which in turn determine the fate of the individual scale-cells (Kühn, 1936).

Later, Stern (1954) was able to carry the analysis of several pattern genes to considerably greater depth by the ingenious use of genetic mosaics on a particularly favorable and much simpler system, the bristle pattern on the thorax of *Drosophila*.

Meanwhile, a discovery of far-reaching consequences had been made in Kühn's laboratory. His student, Caspari, working with a pleiotropic gene in *Ephestia* that affects the pigmentation of eyes, skin, testis, and brain, transplanted larval mutant testis into larval coelom of wild-type individuals, and vice versa, and found pigment changes in the host eyes, or host testis, respectively. This study provided the first evidence of diffusible gene-produced substances (Caspari, 1933; Kühn, Caspari, and Plagge,

1935). Shortly thereafter, and independently, Beadle and Ephrussi (1936), using a similar transplantation technique with larval eye discs, discovered diffusible eye-pigment-determining substances in *Drosophila*. These discoveries ushered in the era of biochemical genetics. At the same time, they demonstrated a new mechanism by which primary gene effects could be amplified.

The Missing Chapter

It was clear to the founders of the modern synthesis that embryological considerations had to be incorporated in evolutionary theory, for the simple reason that the only way by which genes can produce the material for natural selection—that is, phenotypic variability—is in their capacity as controlling agents and modifiers of developmental processes. But this chapter had not been written. It would have taken a biologist with very broad interests, who would be familiar with genetics, speciation, evolution and at the same time knowledgeable in experimental embryology and the intricacies of epigenetic development to write it. The only person of this rank at the time was Schmalhausen. In the foreword to his book, *Factors of Evolution* (1949), Dobzhansky says: "The book of I. I. Schmalhausen advances the synthetic treatment of evolution starting from a broad base of comparative embryology, comparative anatomy, and the mechanics of development. It supplies, as it were, an important missing link in the modern view of evolution" (p. ix). In fact, Schmalhausen had been active in research in all these fields.

Among the younger generation, Waddington in England had independently developed ideas similar to those of Schmalhausen during the war. He was also eminently qualified to provide the missing link; he had been engaged in research in experimental embryology and genetics and written basic books in both fields. His book, *The Strategy of the Genes* (1957), in which his ideas on the relationships of genetics, experimental embryology, and evolution are synthesized, could be considered a draft of the "missing chapter."

Three issues, or trends of thought, illustrate the kind of material that I think might have become part of the missing chapter of "evolutionary biology." First, the epigenetic mechanisms that have been elucidated by experimental embryology take a great burden off the genome in its role as controlling agent of developmental processes. The activity of a single gene operating at a particular focal structure can have widespread secondary effects through the mediation of diffusible gene products, hormones, inductions, and other epigenetic or physiological mechanisms. I have mentioned pattern genes. If the segregation process in the distal part of the limb bud subdivides the mesodermal field into six instead of five units, a polydactylous phenotype emerges. This result can be accomplished by a single mutation; how the gene operates is unknown. But each digit is complete in its complex structure. Neurogenetic mechanisms provide innervation for the extra digit, without requiring direct involvement of the "polydactyly gene."

The induction of the embryonic eye by the underlying mesoderm triggers a sequence of subsequent inductions: the optic vesicle once formed induces the lens when it contacts the overlying ectoderm, and the lens, in turn, induces cornea formation. A single gene, weakening the inductive capacity of the mesoderm, or a gene producing anophthalmia, thus initiates specific, remote structural changes. For a time it was fashionable to design pedigrees of indirect gene effects illustrating such ramifications. Epigenetic mechanisms reduce the number of genes required for the production of the structural and physiological complexity of the phenotype by amplifying the primary gene effects.

Second, the remarkable regulative properties of most eggs and embryonic subsystems have intrigued and puzzled embryologists since Driesch discovered this phenomenon in 1891. The regulation of isolated blastomeres can be described as an adaptive feature, but it is doubtful whether even a harmoniously proportioned dwarf larva has a good chance of survival. However, regulation can make sense in the context of natural selection as a stabilizing agent. Schmalhausen strongly emphasized this point and elaborated on this theme. He considered the regulative capacity of undifferentiated systems (such as morphogenetic fields) as a major agency in the stabilization of the phenotype (1949, p. 221). Generally speaking, both the genome and the differentiation processes must be balanced and integrated to guarantee the survival of the embryo as a whole; any mutational or other change of the genome that disturbs the equilibrium requires a buffering device. In epigenetic development, the capacity for regulation, in the broadest sense, is one of the most effective means to accomplish this feat. The embryologist Dalcq had a somewhat bizarre idea: he envisaged "ontomutations," that is, mutational radical alterations in very early stages, as, for instance, mutants affecting egg structure or gastrulation, and considered them as the basis of the origin of higher taxa. And he argues that such drastically modified systems could survive only by the stabilizing capacity for regulation (1949). Schmalhausen and Waddington clearly realized that the stabilizing effect of regulation or other epigenetic mechanisms can serve in two different situations: in so-called stabilizing or normalizing selection that aims at retaining the status quo in a population when conditions remain constant; or as buffers against perturbations of developmental processes in so-called dynamic or directional selection that aims at shifting the phenotype in adaptation to changing conditions. Waddington speaks of canalization rather than stabilization of developmental processes.

Third, both Schmalhausen and Waddington are particularly interested in a process that Schmalhausen calls "autonomization" and Waddington, more aptly, "genetic assimilation." The evolutionist is familiar with the following situation. Suppose an environmental change occurs in a well-adapted population. It will result in the selection of extreme variants from the range of nonheritable modifications of the phenotypes, which are best adapted to the new conditions but still within the norm of reaction of the unchanged genome. If the altered condition prevails over a long period (long in evolutionary terms), mutations or recombinations that guide development in the direction of the better-adapted phenotype will occur and become the target of positive selection. In this way, in the course of time, the well-adapted phenotype will be "assimilated" in the genome, giving the impression of inheritance of acquired characters. Cornification of the foot pads in terrestrial vertebrates that are formed already in the embryo is a common example.

Schmalhausen makes an additional important point. The process of genetic assimilation makes the developmental system independent of the environmental conditions or of external stimuli that originally were required to operate continually for the purpose of selection, as long as the better-adapted variant was a nonheritable modification. The shift to the better-adapted phenotype by the creation of a new, balanced genetic system—that is, internalization of developmental determining factors—contributes significantly to the stabilization of the phenotype. Generally speaking, genetic assimilation is an effective stabilizing device in the sense that it makes the development of the better-adapted phenotype independent of environmental factors.

The relations of experimental embryology, genetics, and evolutionary theory during the first half of this century were much too complex to be treated adequately in this brief essay. For instance, I have not dealt with the problem of the embryonic and evolutionary origin of highly specialized adaptations, such as concealing coloration that many embryologists found difficult to explain solely in terms of natural selection from small-step mutations. But here we enter the borderline field of science and philosophy, or Weltanschauung, which is beyond the scope of our discussion.

I am grateful to Dr. Garland E. Allen and Dr. Jane Oppenheimer for helpful suggestions.

References

Baltzer, F. 1940. Über erbliche letale Entwicklung und Austauschbarkeit artverschiedener Kerne bei Bastarden. *Naturwissenschaften* 28:177-206.

———— 1952. Experimentelle Beiträge zur Frage der Homologie. Xenoplastische Transplantationen bei Amphibien. *Experientia* 8:285-297.

Beadle, G., and B. Ephrussi. 1936. The differentiation of eye pigments in *Drosophila* as studied by transplantation. *Genetics* 21:225-247.

Berrill, N. J., ed. 1939. *First symposium on development and growth.*

Boveri, Th. 1895. Über die Befruchtungs- und Entwicklungsfähigkeit kernloser Seeigel-Eier und über die Möglichkeit ihrer Bastardierung. *Roux's Archiv für Entwicklungsmechanik der Organismen* 2:394-443.

Caspari, E. 1933. Über die Entwicklung eines pleiotropen Gens bei der Mehlmotte, *Ephestia kühniella*. *Roux's Archiv für Entwicklungsmechanik der Organismen* 130:354-381.

Dalcq, A. 1949. *L'apport de l'Embryologie causale au Problème de l'Evolution.* Portugaliae Acta Biologica, Serie A. Goldschmidt Volume, pp. 367-400.

———— 1951. Le Problème de l'Evolution, est-il près d'être résolu? *Annales de la Société Royale zoologique de Belgique* 82:117-138.

Driesch, H. 1891. Entwicklungsmechanische Studien I: Der Wert der beiden ersten Furchungszellen in der Echinodermen-Entwicklung. *Zeitschrift für wissenschaftliche Zoologie* 53:160-178.

———— 1894. *Analytische Theorie der organischen Entwicklung.* Leipzig: Engelmann.

Goldschmidt, R. 1920. Die quantitativen Grundlagen von Vererbung and Artbildung. *Roux' Vorträge und Aufsätze* 24.

———— 1927. *Physiologische Theorie der Vererbung.* Berlin: Springer.

———— 1935. Gen und Ausseneigenschaft. *Zeitschrift für Induktive Abstammungs- und Vererbungslehre* 69:38-131.

———— 1940. *The material basis of evolution.* New Haven: Yale University Press.

Harrison, R. G. 1937. Embryology and its relations. *Science* 85:369-374.

Huxley, J. S. 1932. *Problems of relative growth.* London: Methuen.

———— 1942. *Evolution: the modern synthesis.* London: Allen and Unwin.

———— and de Beer, G. R. 1934. *The elements of experimental embryology.* Cambridge: Cambridge University Press.

Kühn, A. 1936. Versuche über die Wirkungsweise der Erbanlagen. *Naturwissenschaften* 24:1-10.

————, E. Caspari, and E. Plagge. 1935. Über hormonale Genwirkungen bei *Ephestia kühniella*. *Nachrichten von der Gesellschaft der Wissenschaften zu Göttingen* 2:1-30.

Lillie, F. R. 1927. The gene and the ontogenetic process. *Science* 66:361-368.

Mayr, E. 1970. *Populations, species, and evolution.* Cambridge, Massachusetts: Harvard University Press.

Morgan, T. H. 1919. *The physical basis of heredity.* Philadelphia: Lippincott.

———— 1934. *Embryology and genetics.* New York: Columbia University Press.

Roux, W. 1881. *Der Kampf der Theile im Organismus*. Leipzig: Engelmann.

Schleip, W. 1929. *Die Determination der Primitiventwicklung*. Leipzig: Akademische Verlagsanstalt.

Schmalhausen, I. I. 1949. *Factors of evolution*. Philadelphia: Blakiston.

Spemann, H. 1924. Vererbung und Entwicklungsmechanik. *Naturwissenschaften* 12:65-79.

———— 1936. *Experimentelle Beiträge zu einer Theorie der Entwicklung*. Berlin: Springer. (1938 English translation: *Embryonic development and induction*. New Haven: Yale University Press.)

———— 1943. *Forschung und Leben*, ed. F. W. Spemann. Stuttgart: Engelhorn.

———— and O. Schotté. 1932. Über xenoplastische Transplantation als Mittel zur Analyse der embryonalen Induktion. *Naturwissenschaften* 20:463-467.

Stern, C. 1954. Two or three bristles. *American Scientist* 42:213-247.

Waddington, C. H. 1957. *The strategy of the genes*. London: Allen and Unwin.

Weismann, A. 1893. *The germ-plasm: a theory of heredity*, trans. W. N. Parker and H. R. Ronnfeldt. London: Walter Scott.

Weiss, P. 1939. *Principles of development*. New York: Holt.

Wilson, E. B. 1925. *The cell in development and heredity*, 3d ed. New York: Macmillan.

Wright, S. 1934. Genetics of abnormal growth in the guinea pig. *Cold Spring Harbor Symposium on Quantitative Biology* 2:137-147.

The Modern Evolutionary Synthesis and the Biogenetic Law

Frederick B. Churchill

When historians survey the diversity of research promoted by biologists at the onset of the 1930s, they find it hard to disentangle those progressive elements which contributed to the emergence of the evolutionary synthesis from the recalcitrant ones which resisted the new developments. The heterogeneous area of embryology is a particularly good example of a jumble of interlocking aspirations and convictions. It covered an extraordinary range of clearly definable and seemingly quite independent subfields. Each in isolation possessed its own specific goals and working assumptions; collectively the spectrum encompassed by them all cut across a grid of scientific norms and methods broad enough to include descriptive natural history at one side and experimental biochemistry at the other.

Given the extraordinary diversity in studies that come under the general rubric of embryology, it would be impossible to survey the science as it contributed to or confronted the modern synthesis in the 1930s. It might be possible but scandalous to generalize from a few case studies to the field as a whole. Under these circumstances the tools of an intellectual historian of science can be useful. When confronted with a complex forest of research, our instincts tell us to dig deeply and narrowly with the hopes of striking the tap roots that nourish the trees that are too tangled for our comprehension. A double case study can illustrate the relevance of these assertions for this conference.

First, however, we should dispose of two loosely connected conceptual issues of long standing in embryology that were "hot" topics at the end of the nineteenth century but by 1910 were all but dead issues. The first of these was the age-old dispute between preformation and epigenesis; the second was the eternal debate between vitalism and mechanism. Their ghosts certainly lingered into the 1930s, but I cannot see at the moment that they played any role in the subject before us. I won't mention them again.

Two other nineteenth-century preoccupations of embryologists were very much in the center of concern during the 1930s: the germ-layer theory and the biogenetic law. Both were connected because both assumed an embryological criterion of homology. (See Wilson, 1905, for an account of the first two preoccupations and Baxter, 1974, for a discussion of all four.)

The outlines of the history of the germ-layer theory are easily accessible in Russell's classic book, *Form and Function* (1916) and a paper by Jane Oppenheimer (1940). Pander recognized three layers in the chick; von Baer generalized the layers (with him, four) to all vertebrates; Kowalevsky earned the credit for extending the generalization to a large variety of invertebrate groups; Haeckel and Lankester earned the notoriety of taking a generalization in descriptive embryology and molding it into a phylogenetic fiction. Oppenheimer's historical sketch carries the germ-layer theory into the period of its decline and rejection in the twentieth century. She describes some of the early experiments of Spemann, Mangold, Holtfreter, Hörstadius, and others that repeatedly indicated that the specificity of the germ-layers was not absolute; that when a tissue with a given presumptive fate was transplanted to a different location its actual fate could be very different—in fact, that of a different germ-layer derivative. Mangold had reviewed the implications of these transplantation experiments for the germ-layer theory in 1923 so it is worth noting that when Oppenheimer wrote her sketch in 1940 the issue was still alive and that she was really writing in the role of a biologist for biologists. "The only conclusion that can be maintained, as a result of these experiments that have been enumerated," she exhorted the members of her profession, "is that the doctrine of the absolute specificity of the germ-layers as enunciated in the last century *must* be abandoned" (1940, p. 286). Gavin de Beer also brought up the subject in 1938: "But the important point to notice," he wrote after reviewing recent transplant experiments, "is that *structures can owe their origin to different organizers without forfeiting their homology*" (1938, p. 69). At the time of the modern synthesis, the germ-layer doctrine positing an embryological criterion for homology was very much a focus of debate for embryologists. The morphological formalism implied by the germ-layer theory was giving way to causal analysis and experimental realism.

The other nineteenth-century preoccupation is the biogenetic law formulated by Haeckel to explain the apparent parallelism between the increasing complexity in the phylogenetic scale and embryological development. Brief histories of this doctrine are also easy to find: again Russell (1916) remains the most satisfactory older text; recent studies by Ospovat (1976) and Gould (1977) indicate the ease with which biologists and historians confounded von Baer's laws and Haeckel's doctrine; Maienschein's examination of cell-lineage studies (1978) demonstrates a lesser-known assault on the biogenetic law. The two aspects of Haeckel's biogenetic law relevant here are his claim that ontogeny recapitulated the adult form of preceding ancestors in the phylogenetic sequence, and his unambiguous and repeated insistence that phylogeny caused ontogeny. It is well known that His (1874), Goette (1875), and Rauber (1880), among others, challenged the causal implication of the law soon after its enunciation, and Russell leaves the impression that the rise of *Entwicklungsmechanik* in the 1880s rechanneled the quest for the explanation of ontogenetic events into a mechanical and physiological mode and away from Haeckel's historical obsession. This direction appears to be logically accurate; it comes as a surprise to find out that the law had not been sufficiently repudiated in embryological circles. Oscar Hertwig (1906) and Adam Sedgwick (1909) singled it out for special criticism in the first decade of the century. In 1928 in his role as president of the zoological section of the British Association for the Advancement of Science, Walter Garstang delivered a well-known attack against the law; his argument centered on his own demonstration of the functional adaptations of larval forms. Garstang's lead was followed by many other disclaimers: Joseph Needham (1930) denied recapitulation of the biochemical processes of development, Waldo Shumway (1932) explained how modern genetics and experimental embryology militated against Haeckel's law, and Gavin de Beer in 1930 presented a comprehensive review and rejection of the recapitulation doctrine. The historian finds sections in standard embryology textbooks that single out the biogenetic law for a special attack. MacBride seems to be the only embryologist by the 1920s who seriously maintained the causal as well as descriptive aspects of the law. The paleontologists, in fact, were much more attracted to the recapitulation concept than embryologists. I find MacBride's justifications, which, as readers of Koestler also know, included positive assertions about the inheritance of acquired characteristics, intellectually vacuous (1914, 1917). I cannot comprehend how his views could have been taken seriously by contemporary embryologists when they were submitted—let alone their open-armed reception by a humanist in the 1970s. From the number of attacks on it, the biogenetic law was nevertheless viewed by its vilifiers as worthy of their efforts, which perhaps suggests that scientific generalizations that are both easy to comprehend and comprehensive in scope are indeed difficult and time-consuming to filter out from the great sump of common wisdom. It may also be that the undue preoccupation with this dying, if not already slain, foe reflected a curious misunderstanding about the life sciences on the part of the victors.

A number of very different areas of embryology militated against Haeckel, including the refined details of larval forms of Garstang, the "inducers" and "organizers" of Spemann's school, the biochemical approach of Needham, and even the study on relative growth rates pursued by Ford and Huxley and exploited by de Beer. Embryologists as a community were eager to dispatch the biogenetic law in both its descriptive and causal meanings not once, not twice, but over and over again at the time when the modern synthesis was brought together by geneticists and students of natural populations.

Two domains of theory making and discussion, the germ-layer theory and the biogenetic law, both preoccupations of nineteenth-century embryologists, thus spilled over into the 1930s as vital concerns. This double case study—I can barely resist calling it the case of the Oxford twins, but this would conceal the important differences I wish to bring out—embraces Julian Huxley (Needham, 1975) and Gavin de Beer (*Nature*, 1972) who was fifteen years Huxley's junior. Both were trained at Oxford; Huxley was a lecturer for a short time at Balliol before teaching at Rice Institute in Texas. He returned to England at the outbreak of World War One when de Beer was an undergraduate at Magdalen. After the war Huxley was a fellow at New College until 1925 when he became a professor of zoology at King's College, London. De Beer became a fellow at Merton College upon his graduation and in 1926 was appointed Jenkinson Lecturer in Embryology. Their first collaborative effort appeared in 1923 in a paper on differential inhibition (published appropriately in Goodrich's *Quarterly Journal of Microscopical Science*). In 1930 de Beer published the first edition of his classic review of the recapitulation doctrine, *Embryology and Evolution*; two years later Huxley published his *Problems of Relative Growth*, (1932), which drew together a wide range of information including ten years of his own examinations of differential growth. Both he and de Beer used this material in their discussions concerning the connections between ontogeny and phylogeny. In 1934 the two men brought out their monumental text on *The Elements of Experimental Embryology*, a masterful synthesis that drew together the enormous amount of material which had been produced by that time. In recognition of the two great centers of experimental embryology of the day, they dedicated their book to Ross G. Harrison and Hans Spemann.

Huxley's study on relative growth rates was to a great extent stimulated by D'Arcy Thompson's *Growth and Form*. I also suspect that the biometricians served as a model during the ten years Huxley collected data on how the growth rates of its parts affected the form of the organism's whole. Experimental embryology it certainly was not, but then Huxley made no pretensions at uncovering anything more than what he called "empirical laws," which were to find their physical explanations in terms of the mechanisms unveiled by the developmental physiologists. He felt he could demonstrate that the differential growth rates were fairly constant throughout the life of an organism and that these ratios were "associated with a growth-gradient culminating in a growth centre" (1932, p. 243).

Huxley felt these empirical laws had something to say about evolutionary problems. First, he argued that relative growth must be explained within the framework of classical genetics. Because he had recently collaborated with Ford on a study of the amphipod *Gammarus* in which they examined different pigmentation rates of the eye through crossbreeding experiments, it is a fair guess that Huxley was sensitive to the implications genetics held for the physiology of form. Huxley was also impressed by Goldschmidt's endeavor to explain sex determination in moths in terms of "rate genes," a term invented by Huxley (1932, pp. 229-234). Second, Huxley argued that what had previously passed for recapitulation in Haeckel's terms could easily be explained in terms of growth-rate genes; that is, if a gene caused accelerated growth, a given trait would express itself earlier in the life cycle and the biologists would find progressive displacement of adult structures into embryonic stages. On the other hand, if a gene slowed down the growth of a trait, that trait would appear later in the cycle and paedomorphosis or neoteny would occur, just the reverse of what the biogenetic law predicted. "Many undoubted cases of recapitulation," Huxley wagered, "will be found to owe their origin not to any mysterious phyletic law, but to embryological convenience" (1932, pp. 234-240). And third, Huxley appeared very conscious of the connection between differential growth rates and Darwinian evolution: "The existence of growth-gradients gives opportunity for mutation and selection to affect a number of parts in a correlated way, thus greatly simplifying the picture of the genetic and selective process at work" (1932, p. 222).

The first edition of *Embryology and Evolution* appeared in 1930, two years before Huxley's *Problems of Relative Growth*; it is obvious, however, that de Beer drew extensively upon Huxley's ideas which had been expressed in article-length versions. De Beer's thesis was simple. He spoke of a personal "simmering revolt" (1930, p.i) within, as he had become disillusioned with Haeckel. He was now setting out to demolish both the descriptive claims of recapitulation and the causal chain embedded in Haeckel's biogenetic law. Instead of finding past adult forms compressed into embryonic stages of present organisms, de Beer described eight different evolutionary patterns that in themselves had no causal significance but were to be explained in terms of heterochronies or those differential growth rates with which Huxley was so much preoccupied.

There is a second less obvious aspect to the work. De Beer was very conscious to establish in the minds of his readers the difference between the internal factors of heredity and the interplay between those factors and the external environment that produces each separate ontogenetic cycle. These views appear to be derived to a minor extent from Child's work on polarities in eggs and very pronouncedly from Goodrich's comparative anatomy. On the basis of this distinction, between the factors of inheritance and the processes of development, de Beer rejected the inheritance of acquired characteristics and dismissed the causal implication of the biogenetic law—that phylogeny caused ontogeny. De Beer was attuned to the increasing awareness on the part of the biological community of the important difference between inheritance and development, between the germ-plasm and the soma, between the genotype and the phenotype. Some of us (including Mayr and Churchill) have emphasized that this distinction seen from a number of different directions was fundamental to the formation of modern biology; as with most basic concepts in science, these terms themselves are not as important as the eyes of the beholder. De Beer looked on this distinction, as Goodrich and Child (and Lankester before them) had done, not as a geneticist or cytologist but as an embryologist. A statement de Beer borrowed from Goodrich shows completely their mutual perspective: "An organism," Goodrich wrote, "is moulded as the result of the interaction between the conditions or stimuli which make up its environment and the factors of inheritance. *No single part* is completely acquired or due to inheritance alone" (de Beer, 1930, p. 15).

Sensible enough as far as this statement goes. But what about the germinal constitution? What about those genes, bound to the chromosomes in every nucleus, aren't they the result of "inheritance alone"? Aren't they legitimate "parts"?

De Beer would probably have agreed if confronted with the question, and he might even have rejoined that the phrase "no single part" should not be taken so literally, a verbal quibble. But this "part," I would insist, is exactly what is at stake with the modern distinction between genotype and phenotype. Watch what happens when de Beer talks about the gene a chapter later as he turns his discussion to differential growth.

At the outset of this chapter, de Beer speaks in a wholly familiar fashion about discrete Mendelian factors or genes associated with the chromosomes. Yet when he turns to Goldschmidt's work on sex determination in moths, de Beer slides into curious expressions; he speaks of stronger and weaker genes; he writes that "genes . . . can alter [a] structure by varying the rate at which they work," and that "genes produce their different effects by working at different speeds" (1930, p. 25). That these are not just awkward phrases of the moment is indicated when de Beer returns to the same subject in his summarizing statements. He writes that "Evolution is brought about by acquisition of qualitative novelties, and by the production of novel situations by quantitative alteration of the rate of action of the internal factors" (1930, p. 108). What had started out as "hard," discrete Mendelian factors that interacted with the environment seems to have dissolved into "soft," variable factors with a range of action dependent on the environment.

In contrast, in his *Problems of Relative Growth*, Huxley chooses his words with greater care. He writes of "rate-genes" and "genes which determine that rate of the developmental process" (1932, p. 229).

The difference between rate-genes and genes that change their rate is very significant. To quantify rate-genes, one counts the frequency in a transmission phenomenon; to quantify genes that change their rate, one measures the qualitative differences in a developmental phenomenon. The first expression implies a conceptual readiness to think about population genetics and to fit it into the broader issue of the evolution of natural populations; the second implies a readiness to focus on the descriptive relationship between ontogeny and phylogeny. These two subtly different expressions stand as telltale signs of the contrast between Huxley and de Beer. Between 1934 and 1942, that is, the years between their combined text on experimental embryology and Huxley's *Modern Synthesis*, these two phrases signaled the diverging aspirations of the two Oxford-trained biologists.

Delivering the presidential address for the zoological section of the

British Association in 1936, Huxley outlined what was later to become the book of 1942 (1936, 1942). In the very first sentence he spoke of a "synthesis" of many isolated disciplines, which in his mind included paleontology, systematics, natural history, and biogeography. The most interesting comments in the paper centered on the recent developments in genetics; Huxley was fully attuned to thinking in terms of gene complexes including the complexities of modifiers, recombinations, and mutations. He drew on Haldane and Fisher for the elucidation of this concept and contrasted it with the older view of Mendelian or unit characters. When he turned to the origin of species, he appeared very much at home discussing the temporal and geographic movements of populations, and he spoke of such populations diverging, converging, and interacting in complex patterns. He placed his earlier examination of differential growth rates within the context of a selection of rate-genes. Some of his details may have been inaccurate even for that day, but his perspective is completely that of the evolutionary synthesis. I see Huxley's book of 1942 as little more than an updating and enrichment in details of his statement of 1936—even to the last curious section on evolutionary progress. Huxley seemed to make the transition from problems of relative growth to the modern synthesis with complete ease.

De Beer's second edition of his classic statement against the biogenetic law appeared in 1940, exactly a decade after the first edition. It remained virtually unchanged, even to the statements about gene rates. One minor change is worth noting: de Beer changed the title from *Embryology and Evolution* to *Embryos and Ancestors*. Had a sales-conscious editor conjured up a jazzier title? Or had de Beer sensed that his subject had contracted in scope and importance?

Less problematic is de Beer's paper of 1938. It raises another bibliographic confusion, for its title was also "Embryology and Evolution" and it appeared in a commemorative volume honoring Goodrich, in turn edited by de Beer. De Beer's message had not changed significantly since 1930. Once again he attacks the descriptive accuracy of recapitulation; again, he resorts to the same expressions about the "activity rate of controlling genes" (p. 60). The paper's organization brings out other aspects of de Beer's approach. There is no question but that he was abreast of some of the most recent work in genetics; after all, both Ford and Haldane contributed to the same volume. Curiously, however, de Beer uses Fisher's analysis of dominance in poultry, *Drosophila* genetics, the concept of a gene-complex, and experimental embryology solely to undermine both a genetic and a phenotypic criterion of homology. As he concluded, "The interesting paradox remains that, while continuity of homologous structures implies affinity between organisms in phylogeny, it does not necessarily imply similarity of genetic factors or of ontogenetic processes in the production of homologous structures (pp. 64-71). In the same spirit he used the most recent work of the Spemann school to discredit anew the germ-layer theory (pp. 71-75). Both his demonstrations militated once again against the public menace, the "biogenetic law." St. George was not satisfied in slaying the dragon but was driven on to dismember its moribund parts.

De Beer was every bit an embryologist; he saw the whole field ranging from Goodrich's functional homologies and Garstang's adaptations of larval forms to Huxley's relative growth rates, Spemann's organizers, and even Needham's biochemistry advancing as a unified front against morphological and descriptive explanations of form. When genetics could be brought into the fray, all the better. Embryology for de Beer was a cause-directed field; it explained the mechanics and the physiology and the chemistry of form. On the other hand, the study of phylogeny for de Beer was a historical and descriptive endeavor that produced lineages. Haeckel's great mistake had been trying to connect these two very different domains of science. That biology possessed an irreconcilable dual nature was brought home in de Beer's final paragraph: "A living organism must be studied from two distinct aspects. One of these is the causal-analytic aspect which is so fruitfully applicable to ontogeny. The other is the historical descriptive aspect which is unravelling lines of phylogeny with ever-increasing precision. Each of these aspects may make suggestions concerning the possible significance of events seen under the other, but does not explain or translate them into simpler terms" (pp. 76-77).

De Beer invoked Huxley's name to add weight to this distinction. But whereas de Beer took it literally and enforced it in his own studies of the twenties and thirties, Huxley did not. When Ernst Mayr recently mentioned, as an afterthought, that "Huxley wasn't really an embryologist," I thought this exclusion was utterly unfair; I now find it curiously accurate. Huxley was an eclectic of extraordinary range. His own research was as diverse as ethology and physiology. He collaborated with Haldane, Ford, de Beer, and even H. G. Wells. He also had a remarkable ability to draw the details of a field together into a summary statement, as we find with *Problems of Relative Growth* and *The Elements of Experimental Embryology* as well as with *The Modern Synthesis*. Despite his professions to de Beer, I am convinced that biology did not possess two aspects for Huxley. Whereas the embryologist de Beer had become sensitized through Haeckel's failure to avoid connecting phylogeny and ontogeny in a causal manner, the eclectic biologist Huxley quickly saw that population genetics and the study of natural populations filled the very same causal role vacated by the much discredited biogenetic law.

My historical thesis then is simply this: the very success of causally directed embryological research steered embryologists away from a fresh appraisal of the causal connection between phylogeny and ontogeny.

References

Baxter, A. L. 1974. Edmund Beecher Wilson and the problem of development: from the germ layer theory to the chromosome theory of inheritance. Doctoral dissertation, Yale University.

de Beer, G. R. 1930. *Embryology and evolution*. Oxford: Oxford University Press. (The second edition, which otherwise contained only minor changes, was retitled *Embryos and ancestors*. Oxford: Oxford University Press, 1940.)

———— 1938. Embryology and evolution. In *Evolution: essays presented to E. S. Goodrich*, ed. G. R. de Beer. Oxford: Oxford University Press.

Churchill, F. B. 1974. William Johannsen and the genotype concept. *Journal of the History of Biology* 7:5-30.

Garstang, W. 1928. The origin and evolution of larval forms. *Report of the British Association for the Advancement of Science*. Glasgow.

Goette, A. W. 1875. *Die Entwicklungsgeschichte der Unke (Bombinator igneus) als Grundlage einer vergleichenden Morphologie der Wirbelthiere*. Leipzig.

Gould, S. J. 1977. *Ontogeny and phylogeny*. Cambridge, Massachusetts: Harvard University Press.

Hertwig, O. 1906. Ueber die Stellung der vergleichenden Entwickelungslehre zur vergleichenden Anatomie, zur Systematik und Descendenztheorie. (Das biogenetische Grundgesetz, Palingenese und Cenogenese.) In *Handbuch der vergleichenden und experimentellen Entwickelungslehre der Wirbeltiere*, ed. O. Hertwig, vol. 3, pt. 3. Jena: Fischer, pp. 149-180.

His, W. 1874. *Unsere Körperform und das physiologische Problem ihrer Entstehung, Briefe an einen befreundeten Naturforscher*. Leipzig: Vogel.

Huxley, J. S. 1932. *Problems of relative growth*. London: Methuen.

———— 1936. Natural selection and evolutionary progress. *Reports to the British Association for the Advancement of Science* 106:81-100.

———— 1942. *Evolution: the modern synthesis*. London: Allen and Unwin.

———— and G. R. de Beer. 1923. Studies in differentiation. IV. Resorption and differential inhibition in Obelia and Campanularia. *Quarterly Journal of Microscopical Science* 67:473-495.

———— and G. R. de Beer. 1934. *The elements of experimental embryology*. Cambridge: Cambridge University Press.

MacBride, E. 1914. *Invertebrata*, vol. 1 of *Text-book of embryology*, ed. W. Heape. London: Macmillan.

———— 1917. Recapitulation as a proof of the inheritance of acquired characteristics. *Scientia* 22:425-434.

Maienschein, J. 1978. Cell lineage, ancestral reminiscence, and the biogenetic law. *Journal of the History of Biology* 11:129-158.

Mangold, O. 1923. Transplantationsversuche zur Frage der Spezifität und der Bildung der Keimblätter. *Archiv für mikroskopische Anatomie und Entwicklungsmechanik* 100:198-301.

Mayr, E. 1973. The recent historiography of genetics. *Journal of the History of Biology* 6:125-154.

Nature. 1972. Sir Gavin de Beer. 239:179-180.

Needham, J. 1930. The biochemical aspect of the recapitulation theory. *Biological Reviews* 5:142-158.

———— 1975. Huxley remembered. *Nature* 254:2-3.

Oppenheimer, J. 1940. The non-specificity of the germ-layers. In *Essays in the history of embryology and biology*, ed. J. Oppenheimer. Cambridge, Massachusetts: MIT Press, pp. 256-294.

Ospovat, D. 1976. The influence of Karl Ernst von Baer's embryology, 1828-1859: a reappraisal in light of Richard Owen's and William B. Carpenter's "Palaeontological application of 'Von Baer's law.' " *Journal of the History of Biology* 9:1-28.

Rauber, A. 1880. *Formbildung und Formstörung in der Entwicklung von Wirbelthieren.* Leipzig.

Russell, E. S. 1916. *Form and function, a contribution to the history of animal morphology.* London: John Murray.

Sedgwick, A. 1909. The influence of Darwin on the study of animal embryology. In *Darwin and modern science*, ed. A. C. Seward. Cambridge: Cambridge University Press, pp. 171-184.

Shumway, W. 1932. The recapitulation theory. *Quarterly Review of Biology* 7:93-99.

Wilson, E. B. 1905. The problem of development. *Science* 21:281-294.

4 Systematics

The Role of Systematics
in the Evolutionary Synthesis

Ernst Mayr

Describing the role of systematics in the evolutionary synthesis is difficult because of the enormous diversity in the views of the systematists.
On the one hand were those taxonomists, and they were by far the majority, on whom the publication of the *Origin* in 1859 and the subsequent
battles had had no visible impact, particularly those working in poorly
known groups of animals and plants. The spirit of their publications is
not visibly different from that of Linnaeus a hundred or more years earlier. They were essentialists, concerned with descriptions and names. On
the other hand, when we consider Darwin's most enthusiastic supporters,
the most committed adherents of natural selection, we discover that
nearly all of them were naturalists-systematists.

We have no history yet of the manifold impact of Darwinism on all
aspects of taxonomy except for an excellent treatment of this subject in
ornithology (Stresemann, 1975). Recent historians have focused on the
role of genetics to the extent that they universally overlooked what an
important conceptual revolution systematic biology experienced in the
post-Darwinian period, reaching its culmination in the new systematics.
These developments, mostly prior to 1900, were quite independent of the
developments in genetics during the first three decades of the twentieth
century; in fact, they preceded the birth of population genetics by many
decades.

The leading systematists between 1910 and 1935 were so unwilling to
combine their own findings with the new findings of genetics because
they had adopted a series of traditional beliefs in the nature of variation
and inheritance that seemed consistent with their own observations and,
in their opinion, were better able to explain evolution than the views of

the Mendelians. They were in fact the same beliefs as those held by Darwin.

In my interpretation I place considerable stress on the tenuousness of the line of communication between genetics and taxonomy. The representatives of both areas were little interested in what the others were doing and did not read the literature of the other field. We cannot ask why the systematists did not keep up better with the advances in genetics; we will never understand the nature of the argument between the Mendelians and the naturalists if we do not accept the situation as it was. The situation was, of course, symmetrical, as is evident from D. S. Jordan's complaint of the total neglect of the systematists' speciation literature by the Mendelians.

Misconceptions

Reconstructing and understanding the nature of the arguments between the Mendelians and the naturalists is impossible in terms of our contemporary understanding of variation and inheritance. Unfortunately, one recent historical treatment of genetics does not mention the de Vriesian concept of mutation. It is necessary to go back to the writings of Bateson, de Vries, and Johannsen to understand the arguments of the presynthesis period. What happened as a result of the rediscovery of Mendel's rules was that the "formerly universally accepted opinion [first expressed by Darwin] that species evolve gradually from one another was abandoned by nearly all workers as a result of the publications of Johannsen and de Vries, and replaced by the theory of saltational changes" (Rensch, 1929, p. 118).

The Nature of Variation

The most important dogma of the period, classical already in Darwin's day, was that there are two kinds of variation, discontinuous variation (sports or mutations) and gradual (continuous, individual, or fluctuating) variation. The early Mendelians and the naturalists agreed on the existence of these two kinds of variation but differed in their interpretation of the evolutionary significance of mutations versus continuous individual variation. Both kinds of variation were repeatedly reaffirmed in the ensuing decades by naturalists as well as geneticists. Sumner, after years of crossing individuals from different wild populations of *Peromyscus*, concluded: "Taken at face value, the evidence shows that we have to do here with two different types of variation and two different types of heredity" (1918, p. 447).

The pioneering work of Nilsson-Ehle, Baur, East, and later the Morgan school showing that these kinds of variation are only the extremes of a continuous spectrum was largely ignored by the nongeneticists. Bateson and de Vries had claimed that discontinuous variation results from mutation and that inferior mutants are eliminated by selection, while successful mutations are the true material of speciation and evolution. The systematists did not deny the occurrence of mutations, even in natural populations. Sumner found them in his *Peromyscus* populations: "The clearcut . . . Mendelian segregation, in respect to these mutant characters, is in striking contrast to the complete lack of segregation . . . in respect to the subspecific characters" (1918, p. 440). Two ornithologists, Chapman (1923, 1928) and Stresemann (1926), agreed that "mutations" are common in birds. Indeed, Stresemann devoted more than a score of papers to the description of such mutations and postulated that the occurrence of striking differences among geographic races of the same species is the result of mutation. But neither he nor Chapman ascribed all variation or speciation to these more or less exceptional mutations. Chapman was quite explicit: "The presence or absence of a pectoral band, vertical streak or superciliary line [mutations described by him] does not materially affect a species' chances of success or failure, I also believe that natural selection has played no part in their development" (1923, p. 274). He further stated that it is not a question of *either* mutation *or* the direct action of the environment because "it seems to me that both may be operative" (1923, p. 243). When discussing gradually varying characters, such as "variation in size and color . . . [it] has convinced me of the profound influence exerted by observable environmental factors (chiefly climatic) on the species." It is rather evident that Chapman was a neo-Lamarckian. How much the interpretation of mutations among naturalists differed from that of a modern evolutionist may be further illustrated. The famous herpetologist Noble thought that mutation could lead to sympatric speciation: "Species may arise within the range of an ancestral stock by physiological isolation. For example, female toads are attracted towards the male by his voice; if a change in the voice of the male should occur (as a mutational, a genetic difference), he probably would not attract a female, but if he should happen to seize a female, their offspring could only interbreed, for the voice of the offspring of the mutant would not be attractive to females of the original stock" (Osborn, 1927, p. 16; the quotation reveals how widespread the belief in homozygous, pure lines was at that time).

Most taxonomists, even though they accepted the existence of mutations, differed quite drastically from the Mendelians in their interpretation of the evolutionary potential of mutations. The interpretation of variation adopted by most naturalists was that discontinuous variation, as exemplified by mutations, is of no evolutionary significance. What is important in evolution, they felt, is gradual, continuous, adaptive variation.

In the period of early Mendelism from about 1901 to 1905, the two interpretations differed in just about every single detail. The geneticists gradually changed their interpretation rather drastically; they also adopted additional, at that time seemingly rather sophisticated, concepts such as pleiotropy (multiple effects of single genes) and modifying genes or polygeny (effect of multiple genes on a single character). Unfortunately, the systematists largely ignored these changes of interpretation, even after they had been incorporated in the standard textbooks of genetics. Both Osborn (1927) and Rensch (1929) make a distinction between evolution by mutation—that is, by discontinuous saltations—and evolution by speciation—that is, by gradual, continuous, genetic changes. Both Osborn and Rensch, as well as various authors who make a similar sharp distinction between saltational mutations and gradual adaptive variation, were apparently unaware of the transfer by Morgan of the term mutation from the de Vriesian phenomenon of saltations to the genetic component of gradual individual variation (Allen, 1968).

As a result, the systematists continued to believe that these two kinds of variation control two very different evolutionary processes. This belief was articulated in Osborn's well-known statement (1927). Rensch adopted Osborn's concept of speciation and was therefore convinced that in order to refute evolution "by mutation," all one needed to do was to demonstrate gradual geographic variation and the formation of polytypic species. To prove this point Rensch devoted a major part of his book to showing how widespread polytypic species (*Rassenkreise*) are in the better known groups of animals, and also how widespread character gradients (later called "clines" by Huxley) are in most widely distributed species.

The attempt to interpret gradual geographic variation as the product of mutations, subsequently directionally organized by natural selection, was rejected by Rensch because he could find no "steps" in the smooth clines, an observation that was in striking contrast to the phenotypic discontinuity of mutations occurring in natural populations. Like Sumner (1918, p. 448), Rensch found it implausible to explain in terms of unit factors the "insensible gradations that occur throughout considerable ranges of territory."

Even though Rensch, like all naturalists, believed in selection, he thought at that time the power of selection had ultimate limits. For two reasons he felt that selection could not be invoked to explain character gradients. The characters involved, like differences in the shade of brown in the cap of a subspecies of chickadees, could not be of selective significance and furthermore the differences between adjacent populations were considered to be too slight to be affected by selection. Even Goldschmidt (1918, p. 48) who believed in the genetic basis of continuous variation, found it difficult to explain the parallel variation in mimic and model and the parallel variation of sympatric species in terms of selection. Sumner (1918, p. 299) regarded the fact that in the zone where two subspecies met there was no segregation of the subspecific characters but gliding intergradation as evidence for the absence of Mendelian factors.

Contributions Made by Naturalists-Systematists

The fifty to seventy-five years preceding the synthesis were characterized by much ferment in systematics. An increasing number of taxonomists rejected the static Linnaean species concept, indeed essentialistic taxonomy as a whole. This rejection eventually led to the development of the new systematics, the most distinctive aspect of which was the study of populations (the reason it is also referred to as "population systematics"). I have described its characteristic features elsewhere (Mayr, 1969, pp. 51-53). These changes made systematics far more receptive to the new developments in genetics than was essentialistic systematics. In order to highlight some of the major conceptual contributions of systematics to the new synthesis, I present a short discussion here of some of the concepts of the new systematics.

Population Thinking

Systematists had been leading in the break with essentialism and in the introduction of population thinking. Long before Darwin some pointed out that no two individuals of a natural species are identical and that a "series"—that is, a more or less extensive population sample—must be collected to understand the variation of species. By the 1860s and 1870s students of birds, mammals, and snails quite customarily collected such series. They were surpassed, however, by the extraordinary statistical analyses of herring populations made by Heincke (1898) and continued by Duncker, Averinzev, and Schnakenbeck (1931).

By 1900 the populational analysis of species was routine for students of mammals, birds, fishes, snails, and some other groups. This approach spread from systematics to genetics as the time of publication of the first population genetics studies indicates. Sumner (1915-1932) based his approach on that of Osgood (1909); Schmidt (1918, 1920) on that of Heincke and his followers; and Chetverikov on that of earlier Russian taxonomists (Menzbier, Semenov-Tianshansky). Goldschmidt reports that his work on the genetics of Lepidopteran populations was stimulated by his "constant contact with a progressive group of taxonomists [at the R. Bavarian Museum in Munich]" (1940, p. 32).

Not all taxonomists adopted or understood the populational approach. Crampton (1932), originally an experimental zoologist, dealt with his *Partula* collections in a rather typological manner despite his elaborate statistical analyses, as did Kinsey (1936) in his treatment of the cynipid wasps.

The populational interpretation of species necessitated an entirely new solution for the problem of speciation. De Vries's attempt to explain the origin of new species by saltations made no sense when species were conceived as populations. Furthermore, the populational approach stimulated the rise of population genetics, which originated in the work of Chetverikov (1926) and the mathematical population geneticists.

The Immense Variability of Natural Populations

Typological thinking had been widespread, if not prevailing, among experimental geneticists, who made every effort to eliminate all variables in the hope of achieving pure types. Johannsen's pure lines represented this type of thinking as did, in a milder form, T. H. Morgan's discussions of the "wild type." Only a few, such as Castle, East, and Baur, stressed variation and individuality. Most naturalists, by contrast, stressed that variability is a normal attribute of populations and that what characterizes populations is indeed the kind and amount of variability. Natural selection is meaningless if there is no variation on which it can work.

In addition to visible variation there is a great deal of concealed variation, a phenomenon to which Darwin gave special attention in his investigations of reversion and latency. Sumner (1918) produced a number of such reversions (homozygous recessives) when inbreeding *Peromyscus* stocks. No one appreciated better the enormous amount of variability of wild populations than Chetverikov (1926) in his work on *Drosophila*. Not surprisingly, only geneticists who had a background in systematics (like Chetverikov and Dobzhansky) studied intrapopulational and geographic variation. Generally, biologists appreciated this variation only after the publication of Dobzhansky's book (1937, pp. 49-60).

The Gradualness of Evolution

Bateson (1894) and de Vries (1900-1902) had rebelled against Darwin's thesis that evolution is gradual and had succeeded in making many converts (greatly impressing T. H. Morgan, for instance). In reaction, the naturalists-systematists between 1910 and 1930 redoubled their efforts to demonstrate by studying geographic variation that Darwin had been right and that evolution in nature is indeed gradual. Sumner (1918) showed that each locality has a different population and that each character of dimension or color varies independently of the others.

Rensch (1929, p. 81) went one step further and showed that all the characters normally used by taxonomists to discriminate between species may be geographically variable within species. He also demonstrated that extreme geographic races are often morphologically more different from each other than are good sympatric species.

As long as ranges are contiguous there is no evidence of discontinuity in this geographic variation. Even where subspecies are recognized in such cases, they are connected by intermediate or intergrading populations. Character gradients, later designated as clines by Huxley (1939), are found in all geographically variable species, further evidence of the gradualness of variation.

The dividing line in the discussion of gradualness versus saltationism was not between straight taxonomists and geneticists but rather between naturalists and most experimentalists. A number of naturalists like Sumner and Schmidt were also breeding populations and an author like Goldschmidt derived his major ideas in this area from his friends at the Munich Museum. Sumner, Schmidt, and Goldschmidt, therefore, thought about populations and their variation like those naturalists who did not do any breeding. Their thinking was quite different from that of Johannsen or Morgan, whose background was in experimental laboratories. Geneticists with connections to the breeders, like Castle, East, Emerson, and Baur, occupied a somewhat intermediate position. Many authors were inconsistent in their writings.

The Genetic Nature of Gradual Evolution
Some Mendelians continued to belittle the evolutionary importance of continuous variation until far into the second decade of the nineteenth century, as documented by Bateson's well-known statement (1913, p. 248). He still insisted on saltational speciation: "We have direct perception that new forms of life may arise sporadically, and that they differ from their progenitors quite sufficiently to pass for species." Unsupported assertions like this brought the systematists up in arms. The situation was made worse by the often repeated claims of the Mendelians that they alone were qualified to investigate species and speciation: "It is impossible for the systematist with the means at his disposal to form a judgment of value [on species status] in any given case. Their business is purely that of the cataloguer, and beyond that they can not go. They will serve science best by giving names freely" (Bateson, 1913). This statement was taken as a deliberate insult by the more perceptive systema-

tists, who were actually far ahead of Bateson in their evolutionary thinking. Poulton (1909, pp. 258-280) has well documented to what extent the Mendelians (such as Punnett and Johannsen) stressed discontinuous variation or even considered continuous variation as nongenetic.

This attitude of the Mendelians, now largely forgotten, must be kept in mind when we discover what strenuous efforts the naturalists made in the ensuing twenty years to prove the genetic nature of continuous and particularly of geographic variation and its evolutionary importance. As long as adaptive geographic variation, such as documented by Rensch (1929) in his Climatic Rules, was considered, in the sense of Johannsen, as a nongenetic response of the phenotype, evolution by mutation pressure was defensible. As soon as the genetic nature of gradual geographic variation was established, its control by natural selection became inescapable. To be sure, the advocates of soft inheritance also favored the genetic nature of this variation because they believed that phenotypic adaptation could be converted into a genotypically based adaptation. However, the changeover from this type of gradualism to the neo-Darwinian position (as shown by the later attitudes of Sumner, Rensch, and Mayr) was easier than the changeover from saltationism to adaptive geographic variation controlled by natural selection. Neither Bateson nor de Vries succeeded in making this transformation. The breeding work of Sumner, Schmidt, and Goldschmidt, the extensive review by Rensch (1929, pp. 90-97), and the accelerating discovery of small mutations in various genetic laboratories, led eventually to an abandonment of the saltationist thesis, even though it was once more revived by Goldschmidt (1940) and Schindewolf (1936).

Geographic Speciation
To an essentialist, no other speciation is possible than the origin of an individual representing a new essence, a new type. This concept of speciation prevailed to the end of the eighteenth century and indeed until 1859. The most perceptive naturalists, however, made rather casual proposals of geographic speciation, beginning with von Buch (1825). Later, Rossmaessler (1836), Darwin (in his Notebooks), and Wagner (1841) also supported the idea that a peripheral population of a species might turn into a new species during a period of isolation. Darwin himself, unfortunately, later became confused when he discovered all sorts of difficult situations in plants and when he began to use the term variety not only for populations but also for aberrant individuals (Sulloway, 1979). Yet other authors in the 1860s, 1870s, and 1880s produced more and more evidence in favor of a widespread occurrence of geographic speciation. This mode of speciation had the tremendous advantage of reconciling

two opposing viewpoints because it permitted a gradual evolutionary change of the isolated populations and yet resulted ultimately in sharply separated discontinuous species.

The theory of geographic speciation was set back, however, when Weismann (1872), de Vries (1889), and Bateson (1894) revived saltational theories explaining the production of new species without geographic isolation. De Vries's mutation theory based on *Oenothera* drastically reinforced the popularity of saltational theories. D. S. Jordan (1905) deplored the fact that the theory of geographic speciation, even though it "is accepted as almost self-evident by every competent student of species or of the geographical distribution of species . . . has been almost universally ignored . . . in the literature of evolution of the present day," as a result of the ideas of the Mendelians.

The saltationists were not totally wrong, for in plants polyploidy can indeed lead to the instantaneous saltational origin of new species. However, this is presumably the minority process even in plants; in animals, it is virtually unknown. Not surprisingly, the naturalists maintained their faith in geographic speciation, the claims of the mutationists notwithstanding. Osborn (1927) stated that it was the prevailing theory of speciation among the staff of the American Museum of Natural History. Goldschmidt in 1918 supported geographic speciation but abandoned it later (1933, 1940). In 1939 Rensch gave a long list of authors who still denied the occurrence of geographic speciation. Indeed, neither Morgan's evolutionary writings nor those of Muller prior to 1937 mention geographic speciation.

It was not until the 1930s that the vigorous promotion of the theory of geographic speciation by Rensch (1929, 1933), Stresemann (1936), and others changed general opinion. Dobzhansky's 1937 book was particularly important. He continued the tradition of Russian naturalists in upholding geographic speciation, indeed taking it completely for granted. Goldschmidt's 1940 book was a temporary setback; this continuing resistance induced me to devote twenty-four pages of my 1942 book to a series of proofs of geographic speciation. By permitting slow evolution, the gradual acquisition of isolating mechanisms, and the entering of new niches or adaptive zones, geographic speciation is ideally suited for an application of the neo-Darwinian interpretation of evolution.

The Adaptive Nature of Observed Variation
Naturalists have always tended to consider the variation of species as adaptive. Yet in most cases their belief was based on wrong premises. In the days of natural theology variation was held adaptive because it was part of the design of this world. Lamarckians expected it to be adaptive

as evidence of the capacity of organisms to respond to the environment. Some early Mendelians claimed that gradual evolution was often non-genetic and that it was nonadaptive in the cases where it was genetic. Bateson (1913, p. 131) summarized his views in this statement: "A broad survey of the facts shows beyond question that it is impossible to recon-cile the mode of distribution of local forms with any belief that they are on the whole adaptational." He added, "I incline far more to agree with Gulick [than with J. A. Allen] who, after years of study of the local vari-ation of the *Achatinellidae* came to the conclusion that it was useless to expect that such local differentiation can be referred to adaptation in any sense" (p. 133). Because Bateson minimized the role of natural selection as compared to mutation pressure, he had a difficult time in reconciling gradual adaptive variation and mutation.

The naturalists, with a few exceptions, continued to accumulate evi-dence that favored the adaptive nature of geographic variation. A re-markable pioneering contribution to this field was Gloger's early work (1833) on the geographic variation of birds. Literally scores of similar studies were summarized by Rensch (1929) and Mayr (1942, pp. 85-99). Rensch's survey of the Climatic Rules, many of which he was the first to establish, was particularly impressive. Two American groups made major contributions in this area—Grinnell and his school in California on birds, and Sumner and Dice and their school on *Peromyscus* and other mammals. Dobzhansky (1937) wholeheartedly endorsed the find-ings of Rensch and others on the adaptive nature of geographic variation and in fact contributed new evidence himself in his studies on the geo-graphic variation of coccinellid beetles.

Not all naturalists agreed. Gulick (1873-1905) ascribed the local diver-sity of Hawaiian snails entirely to random variation. Crampton (1932) essentially adopted the same viewpoint in his work on the *Partula* snails of the Society Islands. Kinsey (1936) emphatically denied any adapta-tional significance in the geographic variation of cynipid wasps. Many authors, quoted by Robson and Richards (1936), denied not only the adaptive significance of species characters but also of geographic varia-tion. These dissenting voices were, however, very much in the minority. Gradual adaptive variation was clearly in conflict with the mutationist thesis of Bateson and de Vries.

Belief in the Importance of Natural Selection
The theory of natural selection was rather unpopular right up to the syn-thesis. The systematists who worked with natural populations were the strongest supporters of natural selection. It was almost universally re-jected by embryologists, physiologists, and most other experimental

biologists. The situation worsened after the rediscovery of the Mendelian rules in 1900. Neither Bateson nor de Vries nor Johannsen had much use for natural selection except as a process for "the elimination of the unadapted" (Crampton, 1932, p. 188). Crampton thought only two alternatives could explain phenotypic variation; either it was caused by mutation pressure or by induction by the environment. Because he quite rightly rejected the latter, mutation pressure was the only conceivable evolutionary factor (pp. 191-194). Not only the Mendelians but also the biometricians (Pearl, 1917) and a considerable portion of the camp of the taxonomists (Robson, 1928, pp. 212-222) held natural selection in low esteem.

The evolutionists who were the most enthusiastic supporters of natural selection in the years after 1859—Wallace, Bates, Hooker, Müller, Poulton, and K. Jordan—were all systematists-naturalists. Most, if not all of them, also believed in soft inheritance. Weismann, largely a laboratory zoologist, seems to be an exception but he had been an ardent student of butterflies since his boyhood and belonged far more to the camp of the naturalists than to that of the experimentalists. Huxley, by training an anatomist and medical man, was never happy with natural selection. How little of a naturalist he was is evident when one compares his *Rattlesnake* journals with those of Darwin on the *Beagle*.

The tradition of a simultaneous belief in soft inheritance and natural selection continued from Darwin until the 1920s and 1930s. Plate, who wrote an important work on the selection theory (with a fourth edition in 1913) strongly believed in the importance of natural selection but also in an inheritance of acquired characters. I believe that most of my own university teachers in the 1920s accepted the same combination. The history of neo-Lamarckism and neo-Darwinism cannot be understood without recognizing the wide distribution of this seemingly contradictory combination.

Those Mendelians who minimized natural selection and attributed most of evolution to mutation pressure had to stress the random nature of most evolution. They were particularly delighted at the findings of Gulick and Crampton, who denied the existence of adaptive variation in snails. Bateson concluded, "The extreme irregularity, for example, of the local combination of types of shells in *Helix*, *Partula*, and *Achatinella* makes it impossible to regard them as local adaptations" (1913, p. 43).

The naturalists, in contradistinction, made every effort to show correlations between variation and the environment, partly to demonstrate soft inheritance. At the same time, nearly all of them also stressed the importance of natural selection, far more so than any of the experimental geneticists of the period. It was only slowly that the efficacy of selection was recognized by the geneticists in the post-Mendelian period, as a result of the work of Castle, Lutz (1911), Payne, MacDowell, East, and of many animal and plant breeders. The naturalists were slow in recognizing, prior to the work of the mathematical geneticists, the importance of even very small (1 to 5 percent) selective advantages.

Macroevolutionary Phenomena and Gradual Evolution
The most determined opponents of the Darwinian theory were unquestionably the paleontologists. They furnished the leading neo-Lamarckians, the leading orthogenesists, and some of the leading saltationists. Because higher taxa in fossils cannot be crossed with each other, the material of paleontology seemed to be singularly remote from any possibility of genetic analysis. The gap between the findings of genetics laboratories and the findings of paleontologists seemed to be unbridgeable because of the difficulties of a genetic interpretation of macroevolution. This gap, in turn, delayed the synthesis until the division was closed by zoologists and paleontologists (Rensch and Simpson), with a minor contribution by Huxley (1942) and Mayr (1942). Their work did not prove that macroevolution is governed by variation and selection, but simply showed that macroevolutionary phenomena can be explained in terms of the neo-Darwinian theory (see chapters 6, 9, and biographical essays).

References

Allen, G. E. 1968. Thomas Hunt Morgan and the problem of natural selection. *Journal of the History of Biology* 1:130.

Bateson, W. 1894. *Materials for the study of variation.* New York: Macmillan.
――― 1913. *Problems of genetics.* New Haven: Yale University Press.

Buch, L. von. 1825. *Physicalische Beschreibung der canarischen Inseln.* Berlin: Königliche Akademie der Wissenschaften, pp. 132-133.

Chapman, F. M. 1923. Mutation among birds in the genus *Buarremon. Bulletin of the American Museum of Natural History* 43:243-278.
――― 1928. Mutation in *Capito auratus. American Museum Novitates,* no. 335.

Chetverikov, S. S. 1926. On certain aspects of the evolutionary process from the standpoint of modern genetics (Russian). *Zhurnal Eksperimental'noi Biologii* A2:3-54 (English translation, 1961, *Proceedings of the American Philosophical Society* 105:167-195).

Crampton, H. E. 1932. Studies on the variation, distribution, and evolution of the genus *Partula. Carnegie Institution of Washington Publication* 410:1-335.

de Vries, H. 1889. *Intracelluläre Pangenesis.* Jena: Fischer.

———— 1906. *Species and varieties, their origin by mutation.* Chicago: Open Court.

Dobzhansky, Th. 1937. *Genetics and the origin of species,* 1st ed. New York: Columbia University Press.

Gloger, C. L. 1833. *Das Abändern der Vögel durch Einfluss des Klimas.* Breslau.

Goldschmidt, R. 1918. A preliminary report on some genetic experiments concerning evolution. *American Naturalist* 52:28-50.

———— 1933. Some aspects of evolution. *Science* 78:539-547.

———— 1940. *The material basis of evolution.* New Haven: Yale University Press.

Heincke, E. 1898. Naturgeschichte des Herings. *Abhandlungen des deutschen Seefischereivereins* II:1-223.

Huxley, J. 1939. Clines: an auxiliary method in taxonomy. *Bijdragen tot de dierkunde* 27:491-520.

———— 1942. *Evolution, the modern synthesis.* London: Allen and Unwin.

Jordan, D. S. 1905. The origin of species through isolation. *Science* 22:545-562.

Kinsey, A. C. 1936. The origin of higher categories in *Cynips. Indiana University Publications,* Science Series 4:1-577.

Lutz, F. E. 1911. Experiments with *Drosophila ampelophila* concerning evolution. *Carnegie Institution of Washington Publication* 143.

Mayr, E. 1942. *Systematics and the origin of species.* New York: Columbia University Press.

———— 1969. *Principles of systematic zoology.* New York: McGraw-Hill.

Osborn, H. F. 1927. Speciation and mutation. *American Naturalist* 61:5-42.

Osgood, W. H. 1909. Revision of the mice of the American genus *Peromyscus. North American Fauna* 28:285.

Pearl, R. 1917. The selection problem. *American Naturalist* 51:65-91.

Plate, L. 1913. *Selektionsprinzip und Probleme der Artbildung,* 4th ed. Leipzig and Berlin: Engelmann.

Poulton, E. B. 1909. Fifty years of Darwinism. In *Fifty years of Darwinism,* ed. Anon. [T. C. Chamberlin]. New York: Holt, pp. 8-56.

Rensch, B. 1929. *Das Prinzip geographischer Rassenkreise und das Problem der Artbildung.* Berlin: Borntraeger.

———— 1933. Zoologische Systematik und Artbildungsproblem. *Verhandlungen der Deutschen Zoologischen Gesellschaft* 1933:19-38.

———— 1939. Typen der Artbildung. *Biological Reviews* 14:180-222.

Robson, G. C. 1928. *The species problem.* Edinburgh: Oliver and Boyd.

———— and O. W. Richards. 1936. *The variation of animals in nature.* London: Longmans, Green.

Rossmaessler, E. A. 1836. *Iconographie der Land- und Süsswasser-Mollusken.* Dresden and Leipzig.

Schindewolf, O. H. 1936. *Paläontologie, Entwicklungslehre, und Genetik.* Berlin: Borntraeger.

Schmidt, J. 1918. Racial studies in fishes. I. Statistical investigations with *Zoarces viviparus* L. *Journal of Genetics* 7:105-118.

———— 1920. Experimental investigations of *Zoarces viviparus. Compte rendu des travaux du Laboratoire de Carlsberg* 14 (9).

Schnakenbeck, W. 1931. Zum Rassenproblem bei den Fischen. *Zeitschrift für Morphologie und Ökologie der Tiere* 21:409-566.

Stresemann, E. 1926. Übersicht über die Mutationsstudien I-XXIV und ihre wichtigsten Ergebnisse. *Journal für Ornithologie* 74:377-385.

—— 1936. Zur Frage der Artbildung in der Gattung Geospiza. *Orgaan der Club van Nederlandsche Vogelkundigen* 9:13-21.

—— 1975. *Ornithology: from Aristotle to the present.* Cambridge, Massachusetts: Harvard University Press.

Sulloway, F. J. 1979. Geographic isolation in Darwin's thinking: the vicissitudes of a crucial idea. *Studies in the History of Biology* 3:23-65.

Sumner, F. B. 1918. Continuous and discontinuous variation and their inheritance in *Peromyscus. American Naturalist* 52:177-208.

—— 1932. Genetic, distributional, and evolutionary studies of the subspecies of deer mice (*Peromyscus*). *Genetica* 9:1-106.

Wagner, M. 1841. *Reisen in der Regentschaft Algier.* Leipzig.

Weismann, A. 1872. *Ueber den Einfluss der Isolierung auf die Artbildung.* Leipzig: Wilhelm Engelmann.

5 Botany

Impressive in Darwin's writings, including *On the Origin of Species, Animals and Plants under Domestication, Fertilization of Orchids,* and *Cross- and Self-fertilization in the Vegetable Kingdom,* is the great contribution that plants made to his thinking. Botany, however, does not seem to have been equally important in the evolutionary synthesis. In the 1930s and 1940s no botanist published a book comparable in impact to the books of Dobzhansky, Huxley, Mayr, Rensch, Simpson, or other architects of the synthesis. In this chapter G. L. Stebbins reviews the situation in botany prior to the synthesis, describes the contributions made by botany, and compares them with the contributions made by other biological disciplines.

To nonbotanists, botany appeared to face two serious handicaps with which the zoologists did not have to cope. First, as Turesson (1922) pointed out, the museum (herbarium) worker and the field worker in botany were more widely separated than their counterparts in zoology. Most herbarium workers were collectors in an almost Linnaean tradition; genecology (the analysis of natural populations) was quite generally considered an independent botanical discipline. No such schism had developed in zoology, where it had long been routine for taxonomists to carry out population analyses. Second, genetic systems in plants tend to be a good deal more complicated than those of such animal groups as birds, butterflies, snails, or insects with the help of which the synthesis was achieved. Such phenomena as polyploidy, the frequent introgressive hybridization of sympatric species, various forms of apomixis, aspects of cytoplasmic inheritance, and great plasticity of the phenotype are much rarer in animals than in plants. These phenomena had to be fully understood before the synthesis could be achieved.

The numerous complexities of plant species prevented unanimity in the adoption of a uniform species concept. Some authors continued to adhere to the Linnaean (morphological) species concept; others adopted Turesson's ecospecies and cenospecies. Even the most progressive authors defined species strictly in terms of the sterility barrier and recognized the existence of other isolating mechanisms (such as pollinators) only much later.

In zoology it had been standard practice since the middle of the nineteenth century to collect population samples or "series," as the museum people called them. This tradition virtually did not exist in botany. In larger plants, the so-called duplicates were usually all taken from the same plant. I think three botanists were primarily responsible for the rebellion against this tradition and promoted the concept of "mass collections": W. B. Turrill in England, Norman Fassett in Wisconsin, and Edgar Anderson. Anderson probably did more than any other plant scientist to spread the population concept among botanists by demonstrating with his students how plant populations could be studied in nature.

We have no history yet of the respective contributions made by the various centers of evolutionary plant research in different parts of the world. There was an unusually broad base: several institutions in the USSR; Sweden (Gustafson, Müntzing, Turesson, Langlet); Denmark (Winge, Jens Clausen); Germany (Baur); England (Darlington, Turrill); United States (East, and his students Sax, Anderson, and Rick; Babcock, Blakeslee, R. E. Clausen, Cleland, Epling). For a more thorough survey, see Stebbins, 1979.

More than anything else, it was Stebbins' book, *Variation and Evolution in Plants* (1950), that brought botany into the synthesis. It had the same impact in botany as Dobzhansky's book in population genetics, integrating the widely scattered literature of plant evolution and providing abundant suggestions for further research. It was far more than a compilation; perhaps its most original chapter (chapter 5) was devoted to genetic systems and their role in evolution. Few later works dealing with the evolutionary systematics of plants have not been very deeply affected by Stebbins' work.   E.M.

References

Stebbins, G. L. 1950. *Variation and evolution in plants.* New York: Columbia University Press.

———— 1979. Fifty years of plant evolution. In *Topics in plant population biology,* ed. O. Solbrig and others. New York: Columbia University Press.

Turesson, G. 1922. The genotypical response of the plant species to the habitat. *Hereditas* 3:211-350.

Botany and the Synthetic Theory of Evolution

G. Ledyard Stebbins

Botany made three kinds of contributions to the synthetic theory: first, its discovery of the facts that were necessary before the synthesis could be built; second, plant science's actual syntheses; and third, the reception of these syntheses by both other plant scientists and the biological community in general.

Genetic Discoveries Made by Botanists

With the exception of the fact that genes lie on chromosomes, which was established by Morgan and his coworkers, every other significant fact about genetics needed for the synthetic theory was worked out from research on higher plants. Genetics began with Mendel's research on peas, which was followed by the concept of mutation, developed by de Vries on the basis of research in the evening primrose, *Oenothera*. The distinction between genotype and phenotype was developed by Johannsen from his research on the garden bean. Bateson demonstrated linkage, using genes of peas. Multiple-factor inheritance was first recognized by Nilsson-Ehle on the basis of seed colors in wheat, followed by corolla size in *Nicotiana* by East. Later came research on the chromosomal basis of hybrid sterility: translocations first recognized by Belling from research on several plant species, inversions worked out first by McClintock on corn, and genetic systems worked out by Darlington. Polyploidy is, of course, a phenomenon found chiefly in plants. How did plant scientists apply this knowledge to the problem of species and evolution in plants?

From the standpoint of the synthetic theory, Erwin Baur, the great plant geneticist in Germany of the 1920s, was a tragic case. For twenty years before his death, he explored not only the genetics of the garden snapdragon, *Antirrhinum majus*, but also species relationships among its relatives. The section *Antirrhinastrum* of the genus *Antirrhinum* is endemic to southern Europe, particularly Spain. This exploration included studies of the variation in and spatial isolation of natural populations, both large and small; hybridization between representatives of different populations recognized by taxonomists as species; and the growing of extensive F_1 and F_2 generations from these hybrids. He estimated the number of gene differences needed to separate two plant species. He found that hybrids between most of these species are fertile. Nevertheless, Mather demonstrated reproductive isolation in one case by testing the selective discrimination of the bees that pollinate them. In nature, all different populations are allopatric; two populations with different characteristics never grow together. Baur recorded intermediate degrees of reproductive isolation that are like those found in related entities recently recognized as semispecies. In the only paper he published on this work, he realized that he was just beginning and outlined an extensive program of future research. This 46-page paper appeared in 1932; Baur died of a heart attack the following year, at the age of fifty-eight. If he had lived, he would probably be recognized now as one of the fathers of the synthetic theory of evolution in plants. After his death his associates continued to work on *Antirrhinum*, but did only formal genetics, mutagenesis, and similar research.

Between 1922 and 1931, Turesson published a series of studies in which he attacked the three basic problems that a plant scientist must solve before starting the synthesis. The first, already started by Johannsen, is the significance of phenotypic versus genotypic variation. This problem is more serious in plants than in animals because in plants the germ plasm is not separate from somatic tissues, as Weismann showed for animals. New germ cells are differentiated every year from embryonic or meristematic cells. Until the molecular revolution, which demonstrated that DNA replication is independent of the environment, there was no theoretical reason for denying the inheritance of acquired modification. For example, it was easy to say that the frost hardening of fruit trees changed the cells, so that in the next year the buds containing those changed cells could produce hardened offspring that would be more resistant to frost. However, horticulturists had tried this experiment repeatedly and, except for charlatans like Lysenko, nobody claimed success. Without the necessary theoretical background, botanists had to make the most careful experiments possible to demonstrate that there are genetic differences within species. In addition, a compact, highly integrated insect such as *Drosophila* does not tolerate nearly as much phenotypic variation as is found in plants. Therefore, when comparing two plants—a dwarf on a high mountain or a very large one in a more lush environment—one never can say from simple observations how much of their difference is based on heredity and how much is based on environment. The distinction can be made only on the basis of carefully controlled experiments. Jens Clausen and his successors did such experiments very precisely in a controlled environment and with sufficient surveillance so as to avoid contamination.

Before 1922 the most widely known experiments were those of Bonnier in France, who maintained that he had transformed species by environmental manipulation. Bonnier transplanted various well-known European species and claimed that when a lowland species was brought to the Alps it was converted into an alpine species, but others could easily see that there was contamination. Clements, the eminent ecologist, did exactly the same thing in Colorado. When I looked at his material later, I knew it was very susceptible to contamination from the local plants. Turesson then established the fact that there are genetically different races in any widespread species of plant known in Scandinavia and that any observable phenotypic difference has a genetic component and an environmental component such that when two plants from different localities are brought together in the same controlled garden they do not look quite as different as they did in nature but they still look different. Clausen's data show that the amount of difference between a lowland and a highland race is governed about half by the difference in genotype and half by the effects of environment. Moreover, Turesson showed that if a highland race is growing in a lower altitude, the plants do not become progressively different over time. The change that is imposed on them by the first initial environment is maintained, which successfully refutes Bonnier's claim that one species or race can be transformed into another by environmental manipulation.

Unfortunately, Turesson was mistaken about the pattern of racial variation within the species. As Mayr has commented, Turesson was definitely a typologist. He insisted that each species he studied was divided into recognizably different ecotypes and argued that in intermediate localities there are mixtures of ecotypes. Other people who saw his work, however, had difficulty recognizing them because his conclusions were based on questionable sampling techniques.

Another Scandinavian, Langlet, reached completely different conclusions. After a start as a cytogeneticist he went to the Royal Forestry Institute in Stockholm and studied racial variation in *Pinus sylvestris*. He developed several large arboretums in which he grew pines side by side from all parts of Sweden. He measured their rate of growth and such physiological characters as the water content of the dormant leaves, characters of great importance to foresters. Langlet told me at the International Congress of Genetics in 1948 that the genetic nature of populations of pines in Sweden accurately reflects the variations in the Swedish climate. Chiefly on the basis of water content of leaves, he had a climatic map of Sweden reflected by the racial diversities of Scotch pine. Therefore, where the gradients were continuous he had continuous clines, and where the gradients were abrupt, as between central Sweden and Lapland, he had sudden change. Langlet's careful, more complete survey showed that Turesson was incorrect.

Third, although Turesson knew about reproductive isolation barriers between species, he did very little to explore them. On the basis of what he knew and what he had read, he established a hierarchy of species classification that included the ecotype, the ecospecies, the cenospecies, and the comparium, a concept first developed by Danser. Ecospecies are populations separated by imperfect barriers of reproductive isolation. Cenospecies are clusters of ecospecies separated by complete barriers of isolation. Comparia are groups of cenospecies between which F_1 hybrids can be obtained, but these are completely sterile. This hierarchy is fine in principle, but unfortunately in plants it does not lead to a more objective solution of the species problem than the morphological method because the degrees of imperfection are quantitative. The distinction between ecotype and ecospecies must be based upon a quantitative estimate of the degree of fertility of the F_1 hybrid between two populations.

Next came the research of Clausen, Keck, and Hiesey. In the 1920s Harvey M. Hall was associated with Frederic Clements on his transplant project. Clements was a Lamarckian. Because Hall could not agree with him, he asked the Carnegie Institution for money to set up a similar series of stations in California, over which Hall would have complete control. After Hall's death in 1932, Spoehr, then head of the Carnegie Laboratory of Plant Biology, asked Ernest Babcock for advice about completing Hall's experiment. Babcock suggested writing to Turesson, who in turn recommended hiring Jens Clausen, who had done brilliant work on the Danish violets. Clausen went to California, saw the stations, and realized that they were just what he wanted. He also decided that he needed a team of botanists trained in different disciplines, choosing Keck, who had just received his degree in taxonomy under Jepson at Berkeley, and Hiesey, a graduate student in physiology at the University of California, Berkeley.

Studying chiefly the widespread species, *Potentilla glandulosa*, they showed that Turesson's Swedish experiments were equally valid in California. In addition, they showed that every morphological character that separates ecotypes in nature segregates either according to a multiple-factor pattern with a very large number of gene loci, or a three- or four-locus pattern. They did not get a single sample of simple Mendelian inheritance, although they used such large numbers as 500 to 700 plants. They then attacked the problem of reproductive isolation, using as material the California tarweeds (Compositae, tribe Madiinae). Here they failed, even though they obtained extremely important data. Unlike Turesson, Clausen was reluctant to publish until every detail was understood. As a result, after about fifteen years of work, Clausen, Keck, and Hiesey still had no publications on tarweeds except very summary annual reports to the Carnegie Institution. During World War Two, they thought that they would help our country by breeding better forage grasses. Their research on speciation in tarweeds was neglected and unpublished. The only publication of their work on species is a short book published by Clausen in 1951 as a result of his lectures at Cornell, based on data obtained between 1934 and 1941.

Achillea, on which they also worked, is very similar to *Potentilla glandulosa*. Clausen found clines that he would not recognize because he saw discontinuities in them. He recognized two species because one had 36 chromosomes and the other 54. However, because these plants are perennial, pentaploid hybrids between them with 45 chromosomes, although largely sterile, could still produce considerable amounts of seed. Ehrendorfer and I explored the boundary between tetraploids and hexaploids in the Sierra foothills. We saw definite indications of intergradation. We took clones to Hiesey, but his team never studied them. Morphologically, the tetraploids and the hexaploids are difficult to distinguish; they appear to be good subspecies. Despite the lowered fertility of the hybrids, gene exchange probably occurs. Consequently, I doubt that *Achillea* provided any insight on the species problem.

Because *Potentilla glandulosa* is a single polytypic species, the Madiinae (tarweeds) enabled the Carnegie group to tackle the species problem. The tarweeds are a subtribe of the sunflower tribe (Heliantheae), nearly all endemic to California. Most are annual, and many are quite common weeds with a variety of chromosome numbers. They are very good for experimental work. Nearly all of their chromosome numbers were already known. A very large number of crosses were attempted and the successful crosses were carefully recorded. Botanists who visited the laboratory saw their data on hybrid fertility, chromosome behavior, and many other characteristics. Among other things, they had an excellent example of sibling species, the genus *Holocarpha*. They found that the single species recognized in Jepson's manual was three morphological species, within at least two of which exist a whole series of populations that either would not cross or made completely sterile hybrids if they did cross. The actual number of sibling species in the genus was never worked out. In the same tribe, they found polytypic species, just as they had in *Achillea* and *Potentilla*. For instance, *Layia platyglossa*, the common tidy tips, has distinct maritime ecotypes and different subspecies in northern and southern California. One example that I used in *Processes of Organic Evolution* consists of six species of *Layia* that represent by far the best example in plants of Jordan's law (that is, the two most closely related species are in ecologically similar but allopatric areas). They made all possible hybrids between these six species and obtained careful data on their vigor and fertility. I still think that their demonstration is one of the best either in animals or plants for the geographic theory of speciation. It didn't appear until 1951, but it was worked out in the 1930s. Similarly, they had a fine example of a carefully analyzed natural allopolyploid. The tragedy, I think, was that this didn't come out in a series of papers during the 1930s and early 1940s.

In the genus *Zauchneria*, Clausen, Keck, and Hiesey had a fine example of the phenomenon Dobzhansky called "hybrid breakdown." They recognized three diploid species and several subspecies within one variable tetraploid. Each of the three diploids could form fertile F_1 hybrids when crossed, but they obtained very few healthy plants from the F_2.

English botanists working at about the same time did not discover any general principles concerning the problems of speciation in addition to those discovered by the Scandinavians and Clausen (a transplanted Scandinavian). Gregor's work on *Plantago* was important. He recognized topoclines and ecoclines, but along the same lines. He was also a very good friend of Clausen and they exchanged data all the time. Clausen used some of Gregor's stations when he went to work on his grasses, and they were very much the same in their philosophy. Salisbury worked out a case of semispecies in *Quercus*, the two well-known English oaks, which in some parts of England apparently behave like perfectly separate species and in others intergrade completely. Turrill did similar research on *Silene*.

In Russia, Sinskaia for the first time clearly distinguished between ecotypes or races that are geographic in nature and edaphic ecotypes, races that are sympatric but occupy a different ecological niche, an expected situation in a sedentary plant. Other Russian work in this period (see Stebbins, 1950) included the study of races of *Camelina sativa*, the *linicola* group, which Zinger showed were adapted to various climates, and methods of culture in flax fields. *Camelina* is one of the best examples of adaptive radiation by natural selection, because the relationship to the flax plant produces character differences that are really different from the usual climatic effects. The biotic relationship is very clear, according to Zinger's description.

Polyploidy is also an important basis for speciation found chiefly in plants. Autopolyploids were first recognized by Lutz and Gates, who demonstrated in 1907 and 1909 that one of de Vries's mutant *Oenothera* (*gigas*) had 28 chromosomes rather than the usual 14. Between 1909 and 1922, polyploid series were recognized in several plant genera, but their genetic nature was not understood. Polyploidy as a form of speciation came onto the scene with the research on *Primula kewensis*. This hybrid (*P. floribunda* x *verticillata*) appeared first at Kew and was completely sterile. Then an offshoot from a clonal division acquired large leaves, robust growth, and high fertility. This remarkable change was first studied by Farmer and Digby, who greatly confused the issue. Eventually Darlington provided correct chromosome drawings of this allopolyploid, which were published by Newton and Pellew (1929, pp. 405-466). The final report was by Upcott (1939, pp. 79-100).

Meanwhile, Winge published a theoretical paper (1917) that postulated exactly the kind of change which could convert a sterile hybrid into a fertile allopolyploid. So an anomalous situation existed during World War One. The English had the allopolyploid and a Danish geneticist had the theory but they couldn't put the two together because the war prevented communication between the two nations. Winge wrote a very interesting and comprehensive paper bearing the naive title, "The Chromosomes: Their Numbers and General Importance," in 1917. He clearly showed what would happen if one doubled the chromosome number of a hybrid that forms only univalents. He was preceded by Rosenberg's work on the allopolyploid *Drosera*, which we now call *anglica*. He crossed this species with one of its diploid ancestors, and established what we call the *Drosera* scheme of chromosome pairing: 10 bivalents and 10 univalents. Using this scheme, Winge argued that in a case involving two species with chromosomes that could not pair, thus producing a sterile hybrid, the chromosomes of the doubled hybrid should pair without difficulty and breed true. But Winge produced no examples. The first proof of Winge's theory was provided in 1925 when Roy Clausen, working with T. H. Goodspeed on *Nicotiana*, produced the amphiploid hybrid *Nicotiana tabacum* x *Nicotiana glutinosa*. They called it *digluta* and argued that it was the experimental verification of Winge's hypothesis.

In the history of polyploidy, Kihara was the first person to analyze a whole genus, to distinguish between allopolyploidy and autopolyploidy. Kihara used the two terms for the first time and developed the genome concept for the species of *Aegilops* and *Triticum*. His work was followed quickly by similar studies of *Gossypium* by Skovsted, Harland, and Beasley; *Brassica* by Nagahuru U; I did a little with *Paeonia*. In addition, *Spartina townsendii* demonstrated that a polyploid can arise in nature as a result of both artificial introduction and changing the ecology, as in the harbor at Southampton. It seems now very clear that the spread of *S. townsendii* as an allopolyploid resulted from the introduction of the American *Spartina* as well as the enormous changes in Southampton Harbor and its surrounding area caused by industrialization. Another case of polyploidy made a great impression on me. In 1930 I read Müntzing's monograph on *Galeopsis*, in which he showed that a species, *G. tetrahit*, already known to Linnaeus, had originated by hybridization and chromosome doubling from two other species. Thus he showed conclusively that a common species in nature has been derived by recognizable processes from other species of the same genus. This special case of polyploidy was the first demonstration of the origin of a species. By 1940 the concept of a polyploid complex consisting of several diploid species that are well separated from each other and that contribute their collective variability to the superstructure of polyploids was well established with numerous examples, starting with Kihara. Polyploidy contributed to the synthesis by showing to plant geneticists as well as the more progressive taxonomists that the origin of some species can be analyzed and duplicated in the laboratory and in the garden.

The positive contributions that botanists made to our understanding of plant evolution and speciation laid the foundation for the coming synthesis. However, some plant scientists developed theories and promoted concepts that delayed the synthesis. Botanists were not alone; even Morgan (1932), who expressed a definitely antiselectionist point of view, and Bateson delayed the synthesis.

De Vries's theory that new species result from single mutations—a theory based on *Oenothera*—was quite a roadblock for botanists, as was Johannsen's claim that two kinds of variation exist and that all that selection did was to separate pure lines. I used to tell my students that Johannsen was a Dane looking at the Danish landscape with its nice neat gardens full of pure lines of peas and beans. His world included no conception of the confusion that exists in a tropical rain forest or in the waste lots of the subtropics, where weeds are coming in and the entire ecosystem is disturbed. He thought that his world was the natural world. We know now that these confusing worlds are much more natural and productive of evolution.

Speciation and Evolution in Plants

Attempts to synthesize all this information to develop a consistent theory of speciation and evolution in plants were done poorly by botanists. One effort was a very short paper in 1939 by Clausen, Keck, and Hiesey, simply categorizing the ecospecies, the cenospecies, and other terms used by Turesson and referring to their examples, including the unpublished work on Madiinae. No book, before mine appeared in 1950, dealt with plants. However, a series of fairly careful monographs of individual plant genera in which these principles were taken into account was published. Unfortunately, the *Aegilops* work never appeared as a monograph. The research on *Gossypium* did not appear as a monograph until

Hutchinson and Stephens (1947). Baur's research on *Antirrhinum* appeared as a preliminary paper in 1932.

One of the pioneering works of the 1920s and 1930s was a long treatment of the grass family, *Gramineae*, by N. P. Avdulov in 1931. Avdulov was in Levitzky's laboratory in Leningrad and therefore in Vavilov's Institute. (Vavilov's work on cultivated plants yielded nothing of theoretical importance for the synthetic theory of evolution.) Avdulov discussed relationships between genera and tribes in a broad synopsis. He obtained representatives of all of the tribes of this complex family and showed that characters other than the usual morphological ones can reveal new relationships. The new characters were provided by histology, particularly of the leaves but also of the glumes. He studied the structure of the seeds, the size of the embryo in relation to endosperm, the structure of the seedlings and the methods of germination, the size of the chromosomes and the fine structure, as far as they could recognize it, of the nuclei. On that basis he realigned the whole grass family, recognizing three main groups. The first group contained bamboos, canes, reeds, and a few other genera, all of which are generally regarded as primitive. These grasses have small chromosomes, chiefly the basic chromosome number $N = 12$, and a great deal of polyploidy. A second group included the familiar grasses of the Northern Hemisphere plus such familiar cereals as barley, wheat, and rye with large chromosomes, $N = 7$, and seedlings with very long slender first leaves and a large amount of endosperm relative to the size of the embryo. These genera are temperate in distribution. A third group had smaller grains; larger embryo relative to endosperm; broad, flat first leaves; distinctive histological characters; small chromosomes; and a tropical distribution.

How could he explain the relationship between climatic distribution and chromosome size? It bothered him very much, I think, that there was a distinct correlation. The grasses of cool regions have bigger chromosomes than those of warm regions. Because there is no reason why they should have more genes, he asked, why should this be? Although he did not follow up that problem, his discovery was a notable synthetic contribution.

A second important work of this period is the monograph on the genus *Crepis* by E. B. Babcock. *Crepis* is a large genus of the dandelion tribe of the Compositae family with about 150 species, principally perennials and annuals of the Old World, having low chromosome numbers: $N = 6$, $N = 5$, $N = 4$, $N = 3$, but with polyploids and in particular polyploid groups in the New World. Babcock was a major pioneer in using the synthetic method for evolution above the species level who also contributed important data on variations within species. For instance, he and Marion Cave worked out rather carefully a polytypic species *Crepis foetida* and showed that one entity which previously had been put in a separate genus—*Rodigia commutata*—was actually just a race of *C. foetida*. The supposed generic character was the presence of palea on the receptacle. He demoted that character in another species when one of his students showed that paleae could appear on the receptacle of *C. capillaris* by a simple mutational difference. Marta Sherman (now Marta Walters) showed that a related entity which seemed on morphological grounds to have the same relationship to the subspecies that they had to each other, actually had a different chromosome number and made completely sterile hybrids with the *C. foetida* group. In that case, the origin of a 4-chromosome type from a 5-chromosome type was explained on the basis of reciprocal translocations, followed by the loss of a centromere. In similar research, Tobgy demonstrated that a 3-paired species, *C. fuliginosa*, was derived from a 4-paired species, *C. neglecta*, by a series of translocations plus deletion of a pair of centromeres.

Babcock's monograph also reviewed a series of studies by his students delimiting species on the basis of reproductive isolation barriers that were principally hybrid sterility resulting from chromosome repatterning, but that also involved some cases of hybrid inviability. For instance, one paper that was featured fairly prominently in the first edition of Dobzhansky's *Genetics and the Origin of Species* was Hollingshead's essay showing that *C. tectorum* has a gene which has no effect within *tectorum* but which when put into the hybrid with *C. capillaris* makes the hybrid die at the cotyledon stage. In other words, he demonstrated a specific lethal gene that could bring about reproductive isolation by the death of hybrids.

Babcock's *Crepis* monograph appeared in 1947, but the Hollingshead work and the Babcock and Cave work were already being done during the 1930s. After collecting all the material for the monograph, about 1942, Babcock spent about four years reading paleontology and other subjects and making his synthesis. He found just the reverse situation from that of Avdulov on grasses, with respect to chromosome size. He showed that in several lines of the genus *Crepis*, reduction in plant size, reduction in length of life-cycle from perennial to annual, specialization of the achenes, and particularly the appearance of beaks and of protective phyllaries surrounding the group of achenes, could be associated with a reduction in chromosome size. We know now through studies in similar genera that this reduction is actually a reduction in the amount of the DNA and that a similar phenomenon occurs in fishes, possibly because of elimination of extra copies of highly repetitive DNA sequences. Avdulov and Babcock described a situation that, when thoroughly understood, will be an important addition to our understanding of how DNA and chromatin affect adaptive characters in plants.

The part of *Crepis* that dealt with polyploidy and apomixis had already been worked out in *Hieracium* and *Taraxacum* by Gustafsson and several others. Babcock and I published a monograph of the American species of *Crepis* in 1938, which finally became incorporated in the 1947 monograph. We showed that among those species which had apomixis, very careful search revealed diploids with all the morphological and ecological characters found among the whole series of tetraploids. Some of those diploids were extremely rare, but it was possible to show that nothing new was added in the way of morphology by the hybridization and polyploidy that produced this enormous polyploid complex. We also showed that there are two groups forming quite different systems based on different ecological relationships in the American species of *Crepis* with the chromosome number $X = 11$. One contains the polyploid complex and the other consists of a single complex polytypic species, *C. runcinata*. The difference in the ecology of these two groups is highly significant. The apomicts and their very isolated sexual ancestors are plants of mountain areas; the diploids of rocky places—often in rock slides or continually shifting habitats. The *C. runcinata* polytypic species inhabits the borders of streams and meadows—that is, the linear, continuous habitats that probably have been linear and continuous since *Crepis* came to North America. This habitat has not allowed the geographic isolation that built up the parents of the apomictic group.

In general, the botanists of the 1930s and 1940s did not accept this work unless they were genetically trained. Most herbarium taxonomists were very skeptical, except for Camp among a few others. Camp and Gilly produced another classification of the different kinds of species. British workers also produced some of these classifications. However, the major concern of the taxonomists was unresolved: shall we change our concepts of species to fit in with the new discoveries, or shall we say there are the taxonomists' species, which are the species we like, and there are ecospecies, cenospecies, demes, and other entities that can be disregarded, and we won't try to put them together?

Turesson's dictum—that ecotypes can be recognized only when grown in a garden—caused repeated difficulties. Many botanists asked: How can I do that with my genus of tropical forest trees growing in the Amazon? These botanists recognized the theoretical value of Turesson's work but, because it could not be put into practice, they felt it was irrelevant to their research. This problem is now being solved by indirect methods.

Botanists also resisted the idea that the differences that are adaptive in subspecies are of the same nature as the differences among species of the same genus and among genera and among families. Many taxonomists argued that differences among races are largely in leaves and stems and other superficial characters; differences among species are largely in reproductive parts and are not adaptive.

Such ideas were held by leading American botanists, including Jepson, Maguire, Mason, Constance, and Cronquist. They were the ones who taught plant taxonomy at American universities. I suspect they adopted this attitude because, in contrast to animal or at least vertebrate taxonomists, plant taxonomists had little interest in natural history. Although they went into the field to collect specimens, they paid little attention to plants as living organisms.

Another pioneer was Edgar Anderson, who recognized the importance of secondary hybridization more clearly than anyone else. He made two great contributions. He demonstrated by experiments that by crossing two morphologically very different species and then back-crossing the hybrid with one of its parents, a few offspring will result even when the hybrid is very sterile. The first generation of back-crossed individuals are so much like the recurrent parent that the genes of the nonrecurrent parents cannot be recognized except by special methods. This discovery said at once that the taxonomist, if he recognizes hybrids at all, can recognize only the F_1. Anderson had some very good stories; for example, he anticipated the existence of introgression in loco weeds in Colorado even though he claimed to have never seen the nonrecurrent species. Soon after, he found them. I have been to that same locality and that nonrecurrent species is so easy to see I can't imagine that he could have missed it.

The reason that no botanist in the 1930s and 1940s did for botany what Simpson and Rensch did for zoology is perhaps that the only people who thought above the species level were the morphologists and the anatomists. They were dominated by people like Arber, Bower, and Bailey, who regarded adaptation and selection as teleology and were strongly opposed to truly Darwinian concepts. None of these botanists ever studied plants in nature; it was not fashionable. Perhaps the fact that animals have obvious behavior made them more attractive than plants to biologists who wanted to spend long hours in the field. Both Arber and Bailey emphasized descriptive morphology because the relation between structure and function is much harder to establish in plants than in animals. And Bailey never, even though he might have answers in his wood structures, speculated or even experimented on the differences in function between the wood of an advanced angiosperm and the wood of a pine. After all, pines and oaks grow side by side, one with a very primitive wood and one with a very advanced wood. Both are highly successful.

The Bailey group saw no benefit in trying to explain such phenomena. Their hypotheses did not deal with the processes that produced change, but rather with the descriptive morphology of the types that were supposed to have been present. It was pure phylogeny, based upon supposed morphological homologies. Admittedly, the findings of population genetics are not directly applicable at the macroevolutionary level. Yet it is legitimate to relate evolutionary changes in morphology to the environment. Simpson and Rensch related morphological changes to changes in the environment in terms of changing selection pressures in the animal world, precisely what the plant morphologists had failed to do in the plant world.

References

Avdulov, N. P. 1931. Karyo-systematische Untersuchung der Familie Gramineen. *Bulletin of Applied Botany (St. Petersburg)* 44.

Babcock, E. B. 1947. The genus *Crepis*, I and II. *University of California Publications, Botany* 21 and 22.

Baur, E. 1932. Artumgrenzung und Artbildung in der Gattung *Antirrhinum*, Sektion Antirrhinastrum. *Zeitschrift für induktive Abstammungs- und Vererbungslehre* 63:256-302.

Clausen, J. 1951. *Stages in the evolution of plant species*. Ithaca: Cornell University Press.

———, D. D. Keck, and M. W. Hiesey. 1939. The concept of species based on experiment. *American Journal of Botany* 26:103-106.

Gregor, J. W. 1939. Experimental taxonomy. IV. Population differentiation in North American and European sea plantains allied to *Plantago maritima* L. *New Phytologist* 38:293-322.

Hutchinson, J. B., and S. G. Stephens. 1947. *The evolution of* Gossypium. London and New York: Oxford University Press.

Langlet, O. 1936. Studien über die physiologische Variabilität der Kiefer und deren Zusammenhang mit dem Klima. Beiträge zur Kenntnis der Öcotypen von *Pinus silvestris* L. *Meddelelser fra Statens Skogsförsöksanstalt* 29: 219-470.

Müntzing, A. 1930. Outlines to a genetic monograph of *Galeopsis*. *Hereditas* 13:185-341.

Newton, W. C. F., and C. Pellew. 1929. *Primula kewensis* and its derivatives. *Journal of Genetics* 20:405-467.

Sinskaia, E. N., and A. A. Beztuzheva. 1931. The forms of *Camelina sativa* in connection with climate, flax, and man. *Bulletin of Applied Botany, Genetics, and Plant Breeding (Leningrad)* 25:98-200.

Stebbins, G. L. 1950. *Variation and evolution in plants*. New York: Columbia University Press. (Contains many valuable earlier references.)

Turesson, G. 1922. The genotypical response of the plant species to the habitat.

Hereditas 3:211-350.

Upcott, M. 1939. The nature of polyploidy in *Primula kewensis. Journal of Genetics* 39:79-100.

Winge, O. 1917. The chromosomes. Their numbers and general importance. *Compte rendu des travaux du Laboratoire de Carlsberg* 13:131-275.

Zinger, H. B. 1909. On the species of *Camelina* and *Spergularia* occurring as weeds in sowing the flax and their origin. *Travaux du Musée botanique de l'Academie impériale des sciences de St.-Petersbourg* 6:1-303. (In Russian.)

6 Paleontology

George Gaylord Simpson was one of the most important architects of the synthesis. He engineered the marriage of paleontology with genetics and more broadly with the rest of evolutionary biology. Because Simpson could not attend either the May or the October 1974 workshop, for reasons of health or previous commitment, two of Simpson's friends and closest professional colleagues, Everett Olson and Bobb Schaeffer, agreed to attend and present their views on the role of paleontology in the evolutionary synthesis. (A complete transcription of their presentations, together with the ensuing discussion, is filed with the American Philosophical Society.) S. J. Gould, one of the current leaders of paleontology, also attended the conference; his analysis of paleontology is included in this chapter. See also the biographical essays for further detail on Simpson's life and views.   E.M.

G. G. Simpson, Paleontology, and the Modern Synthesis
Stephen Jay Gould

If science really progressed by empirical cataloguing and subsequent induction of theory from regularities in the catalogues, then paleontology should stand at the forefront as leader of the evolutionary sciences. For paleontology represents all we know directly about the actual course of life's history. Yet paleontology as an evolutionary discipline either languished in intellectual poverty or rushed up blind alleys with exuberance during most of its history since 1859.[1] Darwin himself viewed paleontol-

1. Reasons include: (1) Science does not develop its theoretical structure by induction from empirical catalogues. (2) The empirical catalogues of paleontology are compromised by systematic imperfections and biases in the fossil record (see Raup, 1976, on some preliminary quantitative attempts to remove these biases). (3) Most paleon-

ogy more as an embarrassment than as an aid to his theory. The *Origin's* primary chapter on paleontology is not a ringing manifesto on "evolution as proved from its direct evidence," but an extended apologia blaming imperfections in the fossil record for an apparent discordance between evolutionary expectations and the facts of paleontology. Fortunately, for those of us who labor in this profession, paleontology's status as an evolutionary discipline has been upgraded to respectability and far beyond during the past thirty years. If Cinderella has moved from the kitchen to the dance, we must thank Simpson for supplying the glass slipper.

Although paleontologists, with some notable exceptions (Agassiz's and Dawson's strong opposition, Owen's ambiguity), quickly embraced evolution, few accepted Darwin's theory of natural selection as its mechanism. Paleontologists have led a number of non-Darwinian movements. The powerful school of American neo-Lamarckism was spearheaded by two paleontologists, E. D. Cope and A. Hyatt. The finalist theory of Father Teilhard, for all its idiosyncracies (Teilhard de Chardin, 1959), belongs, in many respects, to a mainstream of paleontological tradition. When Dobzhansky heralded the modern synthesis in 1937, I do not think that a single prominent paleontologist had recorded himself as a consistent Darwinian (L. Dollo and W. D. Matthew, both then recently deceased, came closest; W. K. Gregory, still under Osborn's thumb, might have been a closet Darwinian). In part, this reluctance to embrace Darwinism arose from particular paleontological traditions and the nature of fossil evidence.[2] But, in large measure, it reflected little more than paleontology's usual inclination to follow biological orthodoxy, perhaps with a slight lag. For the mid-1930s was not so far removed from the era of Batesonian agnosticism among leading biologists: "Less and less was heard about evolution in genetical circles and now the topic is dropped.

---

tologists have been trained as geologists and have practiced their professions as biostratigraphers without much interest in or knowledge of evolutionary theory.

2. Particularly, the peculiar vantage point of gazing down from the present on lineages moving toward the modern biota. In this perspective, it is easy to fall into a bad habit of viewing evolution as intrinsically directed by some non-Darwinian, internal force. In addition, the literal appearance of stasis within species and abrupt replacement between them has bothered many paleontologists (including Darwin), who have equated the operation of natural selection with a prediction of imperceptibly gradual change. Paleontologists have generally invoked imperfection of the fossil record to explain away its episodic character; I accept the literal appearance of stasis and abrupt replacement as a correct geological expression of evolutionary expectations (Gould and Eldredge, 1977).

When students of other sciences ask us what is now currently believed about the origin of species we have no clear answer to give. Faith has given place to agnosticism" (Bateson, 1922, p. 57).

As the synthesis dawned, very few paleontologists took an active interest in evolutionary theory, and even fewer followed the progress of genetics. I know of only one pre-Simpsonian attempt to base the phenomena of paleontology explicitly upon genetical theory—the anti-Darwinian work of Schindewolf (1936) with its invocation of de Vriesian saltation as an explanation for discontinuities and episodic pulsations in the fossil record. Most paleontologists adopted one of two attitudes toward the evolutionary theory of geneticists and other "neontologists" (a word that we paleontologists invented to designate everyone else). Some, like Simpson's teacher Richard S. Lull at Yale, adopted an eclectic attitude and adduced paleontological support for a range of evolutionary phenomena often associated with contradictory theories—from selection to internal trends outside the control of selection (Lull, 1924). Others, like Henry Fairfield Osborn, searched the paleontological record for inductive generalizations and concluded that no neontological theory could encompass the regularities. He therefore argued that some special process must operate at scales of time longer than those available to geneticists for study—macroevolution requires a different theory associated with unknown mechanisms that cannot be observed within *Drosophila* bottles. In a typical article, he cites the following paleontological "facts" against the three major evolutionary theories of the 1920s (1925, pp. 9-11). Against Lamarckism: 1) "Mechanical adaptation" of small-brained, cold-blooded Mesozoic reptiles was as great, if not greater, as that of Cenozoic mammals ("against the Lamarckian hypothesis of the influence of animal intelligence upon evolution," in Osborn's view). 2) "Nerveless" plants evolve as rapidly and multifariously as animals ("against Lamarckism which involves the efforts, desires, and movements of animals"). 3) "In the horse family . . . mechanical evolution of the limbs and feet, which are rapidly improved and adapted by habit (ontogeny), is less rapid and less remarkable than the mechanical evolution of the teeth, organs which are entirely preformed by heredity and are destroyed by use and habit in ontogeny." Against Darwinism: 1) Lack of correlation between evolutionary rates and generation time. "During the 500,000 years of the Pleistocene period there was an intensely rapid evolution of the dental mechanism of the slow-breeding elephant and little or no evolution in the dental evolution of the fast-breeding rodents." 2) We can conceive no use for the incipient stages of adaptations in gradual evolutionary sequences (slight thickening of skull bones that presage the titanothere horn in its ancestors). Against "mutational or saltatory hypotheses of evolution": paleontological sequences display imperceptibly gradual transition. "Biomechanical evolution of the skeleton and teeth as observed in paleontology assumes and follows its firm and undeviating order."

Paleontology, according to Osborn, must therefore be studied at its own level. Regularities can be delimited with confidence—gradual transition, often across incipient stages of inadaptation, orthogenesis (Osborn, 1922), and a set of morphological rules, with fancy new names, for usual paths of transition. These regularities are inconsistent with all neontological theories and guarantee a discordance in process between microevolution and macroevolution. Macroevolution has its own modes of action, unfortunately mysterious and not likely to be uncovered; we must content ourselves with an empirical accounting of regularities. Osborn concluded:

> We are as remote from adequate explanation of the nature and causes of mechanical evolution of the hard parts of animals as we were when Aristotle first speculated on this subject . . . I think it is possible that we may never fathom all the causes of mechanical evolution or of the origin of new mechanical characters, but shall have to remain content with observing the modes of mechanical evolution, just as embryologists and geneticists are observing the modes of development, from the fertilized ovum to the mature individual, without in the least understanding either the cause or the nature of the process of development which goes on under their eyes every day (pp. 141-142).

This advocacy of an intrinsically separate theory of macroevolution, combined with such a pessimistic outlook on the possibility of ever encompassing it, scarcely encouraged hope for a synthesis of evolutionary theory—especially when the dean of American paleontology was the advocate.

But at least Osborn maintained an interest in evolutionary theory. Most invertebrate paleontologists (the vast majority of the profession) either professed no interest in evolution at all or else worked with outdated biological concepts. In 1941, after Simpson had largely completed *Tempo and Mode in Evolution*, Percy Raymond of Harvard University wrote the following assessment for the fiftieth anniversary volume of the Geological Society of America:

> Invertebrate paleontologists are constantly accumulating masses of information toward a study of evolution, but few of them have concerned themselves with the theories . . . The school of experimental biologists has so firmly convinced them that they can never prove anything that, whatever their beliefs, paleontologists have

generally chosen to remain silent. Probably most are Lamarckians of some shade; to the uncharitable critic it might even seem that many out-Lamarck Lamarck . . . Recently orthogenesis has become popular, particularly with the younger generation. This idea is natural for the paleontologist, whose lines of descent are necessarily straight (pp. 98-99).

If paleontology, in the late 1930s, was not quite a "great, dry, confused heap of stones" (Haeckel's designation of systematics in 1866), it was not bursting with dynamic ideas either.

Tempo and Mode

Intent and Content

For a biographical account of Simpson on the circumstances surrounding his decision to write *Tempo and Mode in Evolution* (1944), see Simpson's own account of these subjects in the biographical essays later in this book. I don't wish to overemphasize the "great man" or "great book" approach to history; it is always an oversimplification, and often a pernicious one. Other paleontologists emphasized small, random variation with direction imparted by selection leading toward adaptation (Wood, 1934, for example). Others advocated a continuity between microevolution and macroevolution and hoped actively for some kind of synthesis. But Simpson's book was as close to a "one-man show" as any movement I know in the history of paleontology. It was idiosyncratic, unique, and surprising, and it both annoyed traditionalists (or left them utterly confused) and inspired a generation of younger workers. After all, paleontology is a small profession and evolutionary theorists are a very small subset within it. Until a few years ago, the national meeting of the Society of Vertebrate Paleontology had no formal agenda. Everybody simply came and talked about what they were doing. In this context, one man can have enormous impact. Thus, I do not treat Simpson as a synecdoche for a movement; I am discussing the movement.[3]

In the introduction of *Tempo and Mode*, Simpson aptly summarized the mutual incomprehension and even covert hostility of the two fields he would attempt to reconcile:

3. Others, Rensch (1947) and Stebbins (1950) in particular, played important roles in stressing the continuity of Darwinian processes within populations and the history of life (see also the last chapter of Mayr, 1942). But none of these major figures were paleontologists and none shared Simpson's professional background with its firm tradition for separation between micro- and macroevolution and its general suspicion of any sort of theory.

The attempted synthesis of paleontology and genetics, an essential part of the present study, may be particularly surprising and possibly hazardous. Not long ago, paleontologists felt that a geneticist was a person who shut himself in a room, pulled down the shades, watched small flies disporting themselves in milk bottles, and thought that he was studying nature . . . On the other hand, the geneticists said that paleontology had no further contributions to make to biology, that its only point had been the completed demonstration of the truth of evolution, and that it was a subject too purely descriptive to merit the name "science." The paleontologist, they believed, is like a man who undertakes to study the principles of the internal combustion engine by standing on a street corner and watching the motor cars whiz by . . . It is not surprising that workers in the two fields viewed each other with distrust and sometimes with the scorn of ignorance (pp. xv-xvi).

Tempo and Mode, like so many seminal books, lies completely outside the traditions of its profession. To be sure, paleontologists had written copiously about "evolution"; but, in the profession, this word referred to the documentation of history, specifically to the establishment of phylogeny, not to a study of processes and mechanisms. (A. S. Romer, for example, continued to speak of "evolution" with this meaning throughout his life.) Paleontological works on evolution proceeded in descriptive and chronological order. If they attempted any closing statements on theoretical generalities, they tried to portray such conclusions as inductions in the enumerative mode from the facts of phylogeny—hence the various "laws"—Cope's, Williston's, Dollo's—of the classical literature (see Gould, 1970). Simpson turned this procedure around. Instead of an exhaustive tome in documentation, he wrote 217 pages of stimulating suggestions. He started from the principles of neontological Darwinism as he saw it emerging, finally triumphant after so many years in limbo, from a synthesis of population genetics (with its emphasis on small-scale Mendelian variation rather than de Vriesian macromutation) and the various subdisciplines of natural history—and particularly as expressed in Dobzhansky's pivotal work (1937). He then asked if major features of the fossil record could be rendered consistent with this modern version of Darwinism, and not in need of a special macroevolutionary theory. *Tempo and Mode* contains 36 figures, but only one portrays an animal—actually only the lower second molar and fourth premolar of the Eocene condylarth *Phenacodus*, cribbed from Osborn (fig. 9, p. 43). The rest are graphs, frequency distributions, and pictorial models. No paleontological innovation could have been more stunning than this.

This use of quantified information provided Simpson's second greatest departure from traditional paleontological practices. Paleontologists were dubious enough about any kind of quantification (Simpson and Roe, 1939, played a major part in dispelling this attitude). Such quantitative work as had been done (Trueman, 1922, for example) explored the empirical statistics of variation within populations and among successive samples of lineages (Brinkmann, 1929). But statistical characterization of populations remained within the traditional purview of paleontology, albeit with an important new twist—the empirical documentation of individual collections from the field. Simpson introduced a novel style of quantification by drawing models (often by analogy) from demography and population genetics and applying them to large-scale patterns of diversity in the history of life. Do all groups evolve at about the same rate, or do clams, for example, change more slowly than mammals (based on origination and extinction of taxa, analogized with birth and death of demographic models)? If we plot the distribution of evolutionary rates within major higher taxa, do they form a single curve (the horotelic distribution in Simpson's terms), or can we identify separate modes with either slower (bradytelic) or faster (tachytelic) characteristic rates? If these separate modes (in the technical sense) exist, what do they tell us about different modes (in the vernacular sense) of evolution? (Simpson inferred a tachytelic mode from discontinuities in the record, but argued that limitations of paleontological data would not permit its actual measurement. He did identify a bradytelic mode in several groups by finding a small clumping of "immortal" forms in the distribution of living genera, but not among extinct genera of the same higher taxon. He could provide no satisfactory evolutionary interpretation of this bradytelic mode, though he offered several suggestions. This problem remains intriguing and unsolved today.)

Revolutions in science do not succeed by mere verbiage; they must supply methods and suggestions for a reformulation of practice. Simpson's models for large-scale quantification did just this. The idea that one might use an empirical distribution of rates to make inferences about evolutionary processes is a fruitful and exciting concept and a great improvement over previous reliance upon speculative stories based on individual cases or their selective accumulation. It continues to serve today as a guiding method of paleobiology (for example, Bambach, 1977, on within-habitat diversity through time; Sepkoski, 1978, on stasis and disruption in ordinal marine diversity; Stanley, 1978, on characteristic durations for species; and Gould and others, 1977, on patterns in the morphology of clade-diversity diagrams compared with stochastic models).

Tempo and Mode may seem an unlikely title for such a revisionary work, but Simpson used it to express what he regarded as uniquely pale-ontological among the potential objects of evolutionary study. Neontol-ogists could study functional morphology, immediate adaptation, and community structure. Only paleontologists could study rates of evolu-tion (tempo). Quantitative distributions of rates could then be used to infer modes. If the panoply of modes, inferred from tempos, could be rendered consistent with the models of population genetics, then an evo-lutionary synthesis might be forged. Thus, Simpson designed his work. The first chapter is not a bombastic statement of first principles, but an exploration of various ways to identify and measure evolutionary tem-pos. Chapter 2, the longest in the book, takes up determinants of evolu-tion (variation, mutation, population size, length of generations, and natural selection). Can these factors alone encompass both the rates of chapter 1 and the phenomena of chapters to follow? In chapter 3, Simp-son discusses the continuum between microevolution (within species) and megaevolution (across discontinuities between higher taxa above the level of genera) and argues that the determinants of chapter 2 are suffi-cient at all levels. Chapter 4 then presents the model (and data) of horo-telic, tachytelic, and bradytelic distributions of rate both to show that quantification of tempos provides insight about modes and to argue that all modes are consistent with Darwinian principles. Chapter 5 treats the classical phenomena of non-Darwinian, internally directed evolution—inertia, trends, and momentum—and argues that all cases of apparently rectilinear evolution are consistent with control by selection leading to adaptation. Chapter 6 develops the model of adaptive zones and grids, now famous as a standard depiction for the evolution of higher taxa. Fi-nally, chapter 7 identifies modes of evolution (speciation, phyletic evolu-tion, and quantum evolution, as inferred from tempos), and explores their consistency with genetical explanations. The general argument thus moves from empirical studies of tempos, through standard phenomena (and some new ones) of macroevolution, and finally to the modes in-ferred from tempos. The consistency of all paleontological data with mechanisms of modern genetics, particularly with Darwinian models of evolution directed toward adaptation by natural selection, forms the primary theme and brings paleontology into the developing synthesis of evolutionary theory. Reflecting in 1976, Simpson spoke of his intentions in *Tempo and Mode* and of the book's impact: "It was a success at least to the extent that it did bring a new field of study and a new thesis into the development of the synthetic theory, and that its thesis has stood up well. That thesis, in briefest form, is that the history of life, as indicated by the available fossil record, is consistent with the evolutionary process of genetic mutation and variation, guided towards adaptation of populations by natural selection" (p. 5).

Developing the Argument

One cannot understand and assess Simpson's argument in *Tempo and Mode* without recognizing first its unusual character among styles of doing science. It is not a claim of proof or demonstration (though it does seek to eliminate some causal schemes); it is, instead, a consistency argument. It maintains that all the events of macroevolution, as seen in the fossil record, are consistent with the panoply of models devised by geneticists to explain microevolutionary change—no special causes or forces, operating only at high taxonomic levels or across vast stretches of time, are needed. Consistency is claimed with a range of genetic models (including Wrightian drift), not with a monistic version of causation by a single determinant. This pluralism becomes quite important at a crucial stage of the argument. His later move toward a more rigid selectionism (in the 1953 adumbration and expansion, *The Major Features of Evolution*) marks an important change both in Simpson's thought and in the history of the modern synthesis.

Nonetheless, Simpson certainly had strong candidates for predominant determinants and styles of evolution. First, he regarded natural selection as the primary controller of rate and direction, and adaptation to local environments as its result. He examined (chapter 2) all the determinants of evolution and chose selection as a primary control. Other potential factors either failed the test of data (lack of correlation between generation time and evolutionary rate, for example), did not limit the operation of selection (variation, he argued, is rarely inadequate), or worked in tandem with selection (differing styles in small and large populations, for example).

Second, he viewed this "creative" style of selection-toward-adaptation as occurring predominantly in the phyletic mode by the continuous (and usually gradual) transformation of populations in lineages, not primarily by splitting or fractionation. This view was common among paleontologists and differed strongly from Mayr's emphasis upon speciation (by splitting) as the source of evolutionary novelty. He denies (p. 24) that a selective sorting out among species can impart any direction to evolution, and advocates transforming selection *within* lineages as the creative style of evolution: "Natural selection . . . acts on both, but its action on intergroup variation can produce nothing new; it is purely an eliminating, not an originating, force. Despite its critics, the action of natural selection on intragroup (or interindividual) variation is essentially an originating force; it produces definitely new sorts of groups (populations), and the interbreeding group is the essential unit in evolution" (p. 31). He regards the process of speciation by splitting largely as a parceling out of existing variation, not as a way of incorporating evolutionary novelty:

> This sort of differentiation draws mainly on the store of preexisting variability in the population. The group variability is parceled out among subgroups, or a lesser group, pinched off from the main mass, carries with it only part of the general store of variability . . . Speciation, in this sense, is more likely to be a matter of changing proportions of alleles than of absolute genetic distinctions, although the latter also occur . . . The phenotypic differences involved in this mode of evolution are likely to be of a minor sort or degree. They are mostly shifting averages of color patterns and scale counts, small changes in sizes and proportions, and analogous modifications (p. 201).

But speciation, according to Simpson, is unimportant to the general direction of evolution both in effect and in representation among the data provided by fossils; for fully "nine-tenths of the pertinent data of paleontology fall into patterns in the phyletic mode" (p. 203; that is, the continuous transformation of lineages). The trends of phyletic evolution dominate the fossil record and produce directional, nonrandom, adaptive change: "It is within this mode that evolution tends to be most strictly adaptive. Hence the overgeneralization, e.g. of Osborn, that all evolution is adaptive. There is little or no random change. Adaptively unstable marginal variants do, indeed, appear but in the usual pattern they do not persist, presumably being either eliminated by selection or diluted or lost as phenotypes by back breeding." (p. 203). Thus, Simpson approaches the fossil record in *Tempo and Mode* with three primary ingredients for encompassing it—a premise that it might be rendered consistent with genetic models devised by neontologists, a preference for natural selection leading to adaptation as the primary cause and result of evolutionary change, and a belief that continuous transformation of populations in the phyletic mode represents the main source of directional change in the history of life.

At first, he is able to present a strong and positive argument. He quantifies evolutionary rates for different lineages and times and concludes that the copious variety of tempos cannot reflect "internal" forces operating outside the control of selection; the variation itself makes sense in adaptive terms (accelerated tempos after a passage to new adaptive zones, moderate rates within zones, higher rates in complex animals living in complex environments). In what still remains the paradigm for such studies (and still engenders debate; see Schopf and others, 1975; Stanley, 1978), he borrows survivorship analysis from demography and, working with the geological duration of genera, illustrates the higher evolutionary rates of mammals vs. clams. In a striking analogy to curves of birth and death schedules in *Drosophila*, he writes: "The *Drosophila* curve is based on life spans of individual flies and so is only analogous to the generic curves, not homologous; but the latter might be said to give a picture of a sort of evolutionary metabolism in the two groups [clams and mammals] concerned, much as the *Drosophila* curve portrays a sort of vital metabolism in the corresponding population" (p. 26). He attributes the differences in "metabolism" to different intensities of selection acting on different inherited *Baupläne* (constrained and simple vs. complex) in basically different environments (monotonous vs. heterogeneous and varying).

Simpson then arrives at his most difficult problem—to apply the consistency argument to those phenomena of macroevolution that seem, on the surface, least amenable to subsumption under the genetic models of neontology: undeviating trends that seem to operate outside the control of selection (momentum effects, orthogenesis, racial senescence); and the classic dilemma of paleontology since Cuvier's time—the absence of transitional forms between major alterations of *Bauplan* (the apparently sudden origin of new morphologies that encouraged saltationist theories among paleontologists like Schindewolf and geneticists like Goldschmidt). This will be the primary test of Simpson's vision: can microevolutionary processes be extrapolated to encompass all events in the history of life? (As I understand it, Simpson's thought can be epitomized by recognizing two themes: that he is framing a *consistency* argument for explaining macroevolution by *extrapolation* of processes operating within species and amenable to genetic analysis by neontologists): "The most important difference of opinion, at present, is between those who believe that discontinuity arises by intensification or combination of the differentiating processes already effective within a potentially or really continuous population and those who maintain that some essentially different factors are involved. This is related to the old but still vital problem of micro-evolution as opposed to macro-evolution . . . If the two proved to be basically different, the innumerable studies of micro-evolution would become relatively unimportant and would have minor value to the study of evolution as a whole" (p. 97).

The first macroevolutionary challenge provides little difficulty. Simpson shows that standard evidence for internally directed, undeviating evolution is either false (the phylogeny of horses is an adaptively oriented bush, not a preconceived ladder to one toe and hypsodont teeth), or subject to an alternative, Darwinian interpretation ("overcoiling" in *Gryphaea* might only eliminate senile, postreproductive individuals). The individual arguments are not always correct (*Gryphaea*, in fact, did not overcoil at all—Gould, 1972), but the general point is persuasive and has not been seriously challenged since.

The second issue is a more thorny problem indeed. "Why," in Darwin's words, "do we not find interminable varieties, connecting together all the extinct and existing forms of life by the finest graduated steps?" (Nonpaleontologists often do not realize how notoriously uninformative the fossil record is in documenting morphological transitions between higher taxa of fundamentally different body plan.) Darwin regarded this lack of intermediate forms as the greatest stumbling block to the acceptance of natural selection. He interpreted this absence as the artifact of an imperfect fossil record and declared: "He who rejects these views on the nature of the geological record will rightly reject my whole theory" (1859, p. 342). Cuvier (1817, p. 115) had read the discontinuities as a guarantee that evolution could not occur at all. (Indeed, the same argument is a mainstay of the fundamentalist credo today; see Gish, 1978, for example.) In the late 1930s, it stood as the major support behind anti-Darwinian saltationist views for a distinction in kind between micro- and macroevolution.

Simpson relied in part on the classical argument of an imperfect fossil record, but he concluded that such an outstanding regularity could not be entirely artificial. Yet he recognized that, as a real event, he could not express it by gradualistic Darwinian selection in the ordinary phyletic mode that he favored so strongly. Thus, he made his one major departure from control by selection-toward-adaptation and, in his most striking and original contribution, framed the hypothesis of quantum evolution. He was clearly quite pleased with this formulation, for he ended the book with a 12-page defense of quantum evolution, and called it "perhaps the most important outcome of this investigation, but also the most controversial and hypothetical" (p. 206). Faced with the prospect of abandoning strict selection in the gradual, phyletic mode, he framed a hypothesis that stuck rigidly to his more important goal—to render macroevolution by genetical models operating within species and amenable to study by neontologists. Thus, he focused on the one major phenomenon in the literature of population genetics that granted control of direction to a phenomenon other than selection (mutation pressure he rejected as unsupported; saltation by macromutation he rejected as contrary to the spirit of *population* genetics)—Sewall Wright's genetic drift.

He envisaged the major transition as occurring within small populations (where drift might be effective and preservation in the fossil record virtually inconceivable). He chose the phrase "quantum evolution" because he envisioned the process as an "all-or-none reaction" (p. 199) propelling a small population across an "inadaptive phase" from one stable adaptive peak to another. Since selection could not initiate this departure from the ancestral peak, he called upon drift to carry the population into an unstable intermediary position, where it must either die, retreat, or be drawn rapidly by selection to a new stable position—"the relatively rapid shift of a biotic population in disequilibrium to an equilibrium distinctly unlike an ancestral condition" (p. 206). Simpson felt that, with quantum evolution, he had carried his consistency argument to completion by showing that the genetical models of neontology could encompass the most resistant and mysterious of all evolutionary events. Quantum evolution, he wrote, "is believed to be the dominant and most essential process in the origin of taxonomic units of relatively high rank, such as families, orders, and classes. It is believed to include circumstances that explain the mystery that hovers over the origins of such major groups" (p. 206). Simpson could therefore conclude: "The materials for evolution and the factors inducing and directing it are also believed to be the same at all levels and to differ in mega-evolution only in combination and in intensity" (p. 124).

Simpson's emphasis on quantum evolution underscores a final feature of his original system—its pluralistic view of evolutionary mechanisms. He wished to render macroevolution as the potential result of microevolutionary processes, not to rely dogmatically upon any single process. Although he favored selection-toward-adaptation as a primary (and dominating) theme, he explicitly denied that all evolution is adaptive and under selective control. He concludes: "The aspects of tempo and mode that have now been discussed give little support to the extreme dictum that all evolution is primarily adaptive. Selection is a truly creative force and not solely negative in action. It is one of the crucial determinants of evolution, although under special circumstances it may be ineffective, and the rise of characters indifferent or even opposed to selection is explicable and does not contradict this usually decisive influence" (p. 180).

The Major Features of Evolution (1953): The Synthesis Hardens

When pressured for a new edition of *Tempo and Mode*, Simpson realized that too much had happened in the intervening ten years to permit a reissue or even a simple revision. The field that he pioneered had stabilized and flourished: "It was [in the late 1930s] to me a new and exciting idea to try to apply population genetics to interpretation of the fossil record and conversely to check the broader validity of genetical theory and to extend its field by means of the fossil record. That idea is now a common-place" (1953, p. ix). Thus, Simpson followed the outline of *Tempo and Mode*, but wrote a new book more than double the length of its ancestor — *The Major Features of Evolution* (1953). It differs from *Tempo and Mode* in refinement and extension of technique and in multiplication of examples; it also displays some subtle but important shifts in theoretical emphasis and content. These shifts mirror some general trends in the modern synthesis, as its theory won adherents, gained prestige, and (unfortunately in some respects) hardened. In particular, increasingly exclusive reliance on selection-toward-adaptation (for Simpson, in the gradual, phyletic mode), coupled with a greater willingness to reject al-ternatives more firmly than the evidence warranted, marks both Simp-son's new book and the growing confidence of the synthetic theory in general.

Major Features still abounds with the pluralistic statements that made its predecessor such an expansive and suggestive book. Simpson recog-nizes, in a striking visual picture, that complex evolutionary events can-not be dissected into separate effects with exclusive reliance upon one or a few members of the set: "It is impossible to pile up words in layers, and linear sequence requires taking up one aspect of the whole after an-other and using different names for the aspects chosen. The only correc-tive is continual reminder that the analysis is artificial, while hoping that the parts can eventually be gathered mentally into the natural whole" (p. 141). Although he relies so heavily on selection oriented toward adapta-tion in the phyletic mode, he is careful to remind us that other factors are important and, in some cases, decisive: "The conviction that evolution is usually and mainly oriented by adaptation involving selection does not exclude the necessity for considering some other factors" (p. 266). "Abso-lutely or relatively inadaptive phases occur and all organisms develop nonadaptive and inadaptive characteristics, but over-all patterns of evo-lution are predominantly adaptive and adaptation has been seen to be the usual orienting relationship even in minor details of the pattern" (p. 199).

Nonetheless, *Major Features*, in numerous statements both subtle and explicit, and in general tenor as well, takes a much harder, much less gen-erous, much more uncompromising line than its predecessor on the domination of evolutionary pattern by selection-toward-adaptation in the phyletic mode. Though he chides others (quite properly) for assum-ing that structures are inadaptive because they cannot imagine a use for them,[4] he often constructs adaptive scenarios, in the speculative mode

4. "Human judgment is notoriously fallible and perhaps seldom more so than in facile decisions that a character has no adaptive significance because we do not know the use of it" (p. 166).

and on the opposite (and equally invalid) assumption that prominent features must have some immediate use. Although he warns against such story-telling in discussing the ear tufts of some squirrels (p. 170), he pursues this tactic in many other places himself—in explaining the ammonite suture (p. 167), in maintaining that the advantage of "damp, dark caves" as lairs outweighed the high degree of bone disease that such a domicile imposed on cave bears (p. 288), in arguing that virtually all the complexities of the labyrinthodont skull are either primary adaptations to environment or secondary adaptations to the primary effects (pp. 278-280).

Moreover, in places, Simpson's text verges on the impatience of incipient dogmatism. Of Schindewolf's support for macromutationism, he grumbles: "The possibility has deserved a hearing but now—more than sixty-five years since De Vries began work on *Oenothera* and fifty since he elaborated the theory (1901), and after due consideration of the later arguments in the light of later fact and theory—it seems that the old discussion might well come to an end" (p. 111). (In this case, I do not disagree with Simpson's rejection of the extreme view that new taxa may arise by saltation in a single step by a single large mutation. But the effect of his abrupt dismissal extended beyond the specific view he attacked and cast a pall of disrepute on all defenses of discontinuous change in macroevolution—a concept that I, for one, find indispensable and had to rediscover for myself after a professional upbringing in the Simpsonian tradition.) In another passage, he draws a sweeping conclusion from the demonstration that very small selection pressures can lead to adaptation: "This same evidence also completely rules out any alternative theories, such as those of the mutationist school, that consider adaptation as the result of chance or random processes alone" (p. 146). (Again, I may not disagree, but the demonstration of a correlation in some or even many cases does not prove its ubiquity.) Later, he dismisses the venerable argument (still bothersome, in some cases, in my opinion) that incipient stages of useful structures have no evident function themselves: "This long seemed an extremely forceful argument, but now it can be dismissed with little serious discussion. If a trend is advantageous at any point, even its earliest stages have *some* advantage; thus if an animal butts others with its head, as titanotheres surely did, the slightest thickening as presage of later horns already reduced danger of fractures by however small an amount" (p. 270).

But the most dramatic difference between the two books lies in his demotion to insignificance of the concept that was once his delight and greatest pride—quantum evolution. It had embodied the pluralism of his original approach—reliance on a *range* of genetical models. For he had advocated genetic drift to propel very small populations off adaptive peaks into an ultimately untenable inadaptive phase. And he had explicitly christened quantum evolution as a mode different in *kind*, not only in rate, from phyletic transformation within lineages. But now, as the adaptationist program of the synthesis hardened, Simpson decided that genetic drift could not trigger any major evolutionary event: "Genetic drift is certainly not involved in all or in most origins of higher categories, even of very high categories such as classes or phyla" (p. 355). (Wright, when reviewing *Tempo and Mode* in 1945, doubted the creative power of drift at the high taxonomic levels advocated in quantum evolution. But Wright praised the concept of quantum evolution and the idea of rapid transition through inadaptive phases. He merely tried to encompass it under his shifting balance theory with small demes in occasional contact and with interdemic selection among them.)

In *Major Features* quantum evolution merits only four pages in an enlarged final chapter on modes of evolution. More importantly, it has now become what Simpson explicitly denied before—merely a name for phyletic evolution when it proceeds at its most rapid rates, a style of evolution differing only in degree from the leisurely, gradual transformation of populations. Quantum evolution, he now writes, "is not a different sort of evolution from phyletic evolution, or even a distinctly different element of the total phylogenetic pattern. It is a special, more or less extreme and limiting case of phyletic evolution" (p. 389). Quantum evolution is listed as one among the four styles of phyletic evolution—and all four are characterized by "the continuous maintenance of adaptation" (p. 385). The bold hypothesis of an absolutely inadaptive phase has been replaced by the semantic notion of a relatively inadaptive phase (an intermediary stage inferior in design to either ancestral or descendant *Bauplan*). But *relative* inadaptation is no threat to the adaptationist paradigm, for it matters little that an intermediate is not so well suited for its environment as its ancestor was for a different habitat (since they are not in competition). And it matters even less that the intermediate is not so well designed as its descendant will be (for there is even less opportunity for competition here!). In short, relatively inadaptive populations are fully adaptive to their own environments in their own time—and quantum evolution moves comfortably under the umbrella of the adaptationist program. Simpson even suggests that quantum evolution may be more rigidly controlled by selection than other modes of evolution (though he still invokes inadaptation for the initial trigger): "Indeed the relatively rapid change in such a shift is more rigidly adaptive than are slower phases of phyletic change, for the direction and the rate of change result from strong selection pressure once the threshold is crossed" (p. 391).

Epilogue

In rereading Simpson's two great books, I was reminded forcefully of the transforming influence that he exerted on my profession. I am a child of the first generation tutored by his works. So many of the major concepts that I learned as eternal verities of my calling date only to 1944 when I was quite alive (though a bit young to appreciate them). The revitalization of paleontology, which is now enjoying its greatest flowering since the halcyon days of the early nineteenth century when many of Europe's greatest scientists became paleontologists, began with Simpson's synthesis and his development of operational research strategies for furthering it. Biological theory is the lifeblood of excitement in paleontology; otherwise we become a technical service industry to biostratigraphy, a congeries of experts on different taxa from different times.

Simpson's synthesis was the very step that evolutionary paleontology required in his time. As such, it is one of the great achievements in the history of my profession. Simpson found a field wallowing in the certainty that its empirical catalogues held the key to a separate kind of evolutionary theory, one that could not be apprehended in neontological research. But paleontologists had found no way to state what this "special something" was and they often retreated to mystery or terminological obfuscation—for induction from empirical catalogues is rarely the path to major theoretical breakthrough. This deadlock not only kept paleontology in the doldrums, but also prevented evolutionary theory from reaching any satisfactory state—for how can anyone be satisfied with a theory that explains the colors of *Biston* and the bands of *Cepaea*, but cannot encompass the rise of reptiles or the evolution of horses? As Simpson wrote with what I take to be a warranted chauvinism: Neontologists "may reveal what happens to a hundred rats in the course of ten years under fixed and simple conditions, but not what happened to a billion rats in the course of ten million years under the fluctuating conditions of earth history. Obviously, the latter problem is more important" (1944, p. xvii).

Simpson achieved this synthesis with a consistency argument based on extrapolating the processes operating within populations to encompass all the phenomena of macroevolution. It worked, but it was only a consistency argument, not a proof. The success of this argument, in my opinion more than any other achievement of the modern synthesis, established the vision of a truly unified evolutionary theory.

Although Simpson's argument did so much for my profession, and although it represented what had to be done at its moment in history, the argument also contained a disturbing dilemma for its later development as a ruling paradigm in paleontology. For if macroevolution is no more than a set of results produced entirely by processes operating within species, then what role can paleontology play in evolutionary theory beyond the mere documentation of how processes discovered elsewhere work in the long run? Simpson's synthesis unified paleontology with evolutionary theory, but at a high price indeed—at the price of admitting that no fundamental theory can arise from the study of major events and patterns in the history of life. Why be a paleontologist if all fundamental theory must arise elsewhere, if predominantly passive application must be our highest evolutionary role?

I freely confess to a partisan stance (Eldredge and Gould, 1972; Gould and Eldredge, 1977), but I do believe that a resolution of this dilemma is in sight. As a first stage, the older paleontologists searched for an independent macroevolutionary theory, but they could not formulate it and they posed its undeveloped necessity against the evolutionary synthesis. In a second stage, Simpson bought unity at the price of independence. In a third stage (and I say "a" rather than "the," for there will surely be a fourth and a fifth if the profession is worth anything), we seek a new style of independence. We do not advance some special theory for long times and large transitions, fundamentally opposed to the processes of microevolution. Rather, we maintain that nature is organized hierarchically and that no smooth continuum leads across levels. We may attain a unified theory of process, but the processes work differently at different levels and we cannot extrapolate from one level to encompass all events at the next. I believe, in fact, that a process Simpson identified as relatively unimportant in macroevolution—speciation by splitting—guarantees that macroevolution must be studied at its own level. For while Simpson denied that speciation could be a creative force (although Wright, in his 1945 review, vigorously defended it), selection among species—not an extrapolation of changes in gene frequencies within populations—may be the motor of macroevolutionary trends (see Eldredge and Gould, 1972, and Stanley, 1975). If macroevolution is, as I believe, mainly a story of the differential success of certain kinds of species and, if most species change little in the phyletic mode during the course of their existence (Gould and Eldredge, 1977), then microevolutionary change within populations is not the stuff (by extrapolation) of major transformations. Simpson's own bias for gradual change in the phyletic mode allowed him to advance the notion of extrapolation and may have subtly directed his thinking away from this style of independence for his profession. But then no one was more aware than Simpson of the role that such biases can play, for he wrote with characteristic honesty: "The science of systematics has long been affected by profound philosophical preconceptions, which have been all the more influential for being usually covert, even subconscious" (1953, p. 340).

Can I pay any higher tribute to a man than to state that his work both established a profession and sowed the seeds for its own revision? If Simpson had reached final truth, he either would have been a priest or would have chosen a dull profession. The history of life cannot be a dull profession.

References

Bambach, R. K. 1977. Species richness in marine benthic habitats through the Phanerozoic. *Paleobiology* 3:152-167.

Bateson, W. 1922. Evolutionary faith and modern doubts. *Science* 55:55-61.

Brinkmann, R. 1929. Statistisch-biostratigraphische Untersuchungen an mitteljurassischen Ammoniten über Artbegriff und Stammesentwicklung. *Abhandlungen der Königlichen Gesellschaft der Wissenschaften zu Göttingen*, Math-Phys. Kl., 8 (3).

Cuvier, G. 1817. *Essay on the theory of the earth*, trans. R. Jameson. Edinburgh: Blackwood.

Darwin, C. 1859. *On the origin of species*. London: John Murray.

Dobzhansky, Th. 1937. *Genetics and the origin of species*. New York: Columbia University Press.

Eldredge, N., and S. J. Gould. 1972. Punctuated equilibria: an alternative to phyletic gradualism. In *Models in Paleobiology*, ed. T. J. M. Schopf. San Francisco: Freeman, Cooper, pp. 82-115.

Gish, D. T. 1978. *Evolution—the fossils say no!* San Diego: Creation Life Publishers.

Gould, S. J. 1970. Dollo on Dollo's law: irreversibility and the status of evolutionary laws. *Journal of the History of Biology* 3(2):189-212.

——— 1972. Allometric fallacies and the evolution of *Gryphaea*: a new interpretation based on White's criterion of geometric similarity." In *Evolutionary biology*, ed. Th. Dobzhansky and others. New York: Appleton-Century-Crofts, vol. 6, pp. 91-118.

——— and N. Eldredge. 1977. Punctuated equilibria: the tempo and mode of evolution reconsidered. *Paleobiology* 3(2):115-151.

———, D. M. Raup, J. J. Sepkoski, Jr., T. J. M. Schopf, and D. S. Simberloff. 1977. The shape of evolution: a comparison of real and random clades. *Paleobiology* 3(1):23-40.

Lull, R. S. 1924. *Organic evolution*. New York: Macmillan.

Mayr, E. 1942. *Systematics and the origin of species*. New York: Columbia University Press.

Osborn, H. F. 1922. Orthogenesis as observed from paleontological evidence beginning in the year 1889. *American Naturalist* 56:134-143.

———— 1925. The origin of species as revealed by vertebrate paleontology. *Nature*, June 13 and June 20.

Raup, D. M. 1976. Species diversity in the Phanerozoic: an interpretation. *Paleobiology* 2:289-297.

Raymond, P. 1941. Invertebrate paleontology. In *Geology 1888-1938*. Washington, D.C.: Geological Society of America, 50th anniversary volume, pp. 73-103.

Rensch, B. 1947. *Neuere Probleme der Abstammungslehre*. Stuttgart: Enke.

Schindewolf, O. H. 1936. *Paläontologie, Entwicklungslehre und Genetik*. Berlin: Bornträger.

Schopf, T. J. M., D. M. Raup, S. J. Gould, and D. S. Simberloff. 1975. Genomic vs. morphological rates of evolution: influence of morphologic complexity. *Paleobiology* 1:63-70.

Sepkoski, J. J. 1978. A kinetic model of Phanerozoic taxonomic diversity. I. Analysis of marine orders. *Paleobiology* 4:223-251.

Simpson, G. G. 1944. *Tempo and mode in evolution*. New York: Columbia University Press.

———— 1953. *The major features of evolution*. New York: Columbia University Press.

———— 1976. The compleat paleontologist? *Annual Review of Earth Planetary Science* 4:1-13.

———— and A. Roe. 1939. *Quantitative zoology*. New York: McGraw-Hill.

Stanley, S. M. 1975. A theory of evolution above the species level. *Proceedings of the National Academy of Sciences* 72:646-650.

———— 1978. Chronospecies' longevities, the origin of genera, and the punctuational model of evolution. *Paleobiology* 4:26-40.

Stebbins, G. L. 1950. *Variation and evolution in plants*. New York: Columbia University Press.

Teilhard de Chardin, P. 1959. *The phenomenon of man*. New York: Harper and Brothers.

Trueman, A. E. 1922. The use of *Gryphaea* in the correlation of the Lower Lias. *Geological Magazine* 59:256-268.

Wood, H. E. 1934. Revision of the Hyrachyidae. *Bulletin of the American Museum of Natural History* 67:181-295.

Wright, S. 1945. *Tempo and mode in evolution*: a critical review. *Ecology* 26:415-419.

7 Morphology

No field of biology, except for biogeography, supplied Darwin with more evidence in favor of evolution than morphology. Gegenbaur, T. H. Huxley, and other morphologists were the most enthusiastic supporters of Darwin's theory of "common descent" immediately after 1859. The result was an extraordinary flourishing of comparative anatomy that continued right into the period of the evolutionary synthesis. In the conference discussions on paleontology, reference was made quite rightly to the great interest in evolution displayed by Goodrich, Gregory, Romer, and by other students of macroevolution who were perhaps as much (if not more) comparative anatomists as they were paleontologists. Under these circumstances morphology might have played as dominant a role in the evolutionary synthesis as it had in leading the original evolutionary theory to victory. Did this indeed happen, and if not, why not?

We had hoped that Bernhard Rensch of Münster would lead the discussion on the role of morphology in the synthesis, but illness prevented his attendance. In his absence William Coleman and Michael Ghiselin kindly agreed to fill in for him. They supply evidence in this chapter that comparative anatomy more or less stagnated after the promising start in Darwin's lifetime. By 1959 and later, when a new way of thinking made itself felt (Davis, Bock, and Wahlert), the evolutionary synthesis was already an accomplished fact. In the USSR, as shown in Mark Adams's account in this chapter, the synthesis was not yet apparent in Severtsov's work (1931), but Schmalhausen had already fully endorsed it (in the 1940s and 1950s).

Morphology made no concrete contribution to the synthesis, but rather the reverse: the synthesis had an impact on the field of morphology. The synthesis spread quickly after 1940, but the developments in morphology demonstrate that there was not a simultaneous conflagration in all branches of biology and in all countries. This early lag in morphology is noteworthy in view of the growing importance of contemporary evolutionary morphology.   E.M.

Morphology in the Evolutionary Synthesis

William Coleman

The role of morphology in the formation of the evolutionary synthesis was negligible. Comparative anatomy in the United States by 1930 had been largely assimilated by premedical education. It certainly has never recovered. This role has been a disaster in many ways. We have discovered that man is an animal. Because man is an animal, and a vertebrate as well, the study of vertebrate morphology has seemed the best way to prepare for medical school. On the undergraduate level, in the hands of a capable instructor, comparative anatomy might truly be an introduction to the substance and issues of biology (such was my own fortunate experience with a splendid teacher, L. E. DeLanney); with lesser talent, these courses can rapidly degenerate to a tedious and perilous hurdle on the route to medical school admission. In most cases by 1970 the course, and thus the subject, simply disappeared from the curriculum, destroyed by its own uninspired routine and by competing new glories on the molecular level.

Within the discipline, the focus of work since the mid-nineteenth century has been to study vertebrate morphology in terms of evolutionary change. Yet, this focus has been on the bare fact of descent and not at all on the process or processes governing descent. The question of mechanisms of change has been given a nod of approval by morphologists but very little analysis. Whereas the synthesis, whatever it be and however we characterize it, does seem to attend particularly to the process of evolutionary change in terms of mechanisms, morphologists of the twentieth century have, by contrast, largely attended to re-creating the history of life in the past. The fundamental fact emphasized throughout twentieth-century morphology is homology. As Hyman stated in 1922 (and continued to repeat in her later editions): "The whole aim of comparative anatomy is to discover what structures are homologous" (p. 3). These words are printed in black block letters that stand out from the page. The first truth to strike the reader is that Hyman never retreats from the primacy of homology.

Romer (1949) says no less. In his preface, which remained unchanged over the years and through the various editions, he points out the paramount question. How could blind mutation possibly produce the many and complex patterns of resemblance that we witness? He offers a few words about adaptation. He employs the word genetics. He then, probably with relief, drops the entire subject. Romer fails to develop possible interrelationships between genetics and morphology; nor does he include anything pertaining to organ functions other than precise description of parts that might indeed be active. There is no hint, at least to my mind, in *The Vertebrate Body* of the long-term evolutionary processes involved. I say this not as a critic of Romer, although the number of critics may well be large today. Romer's was, however strange the admission may seem, one of the most exciting works I used as a textbook.

Goodrich also emphasizes homology. In his enormous collection of studies he states that morphology must concern itself with "divergent phylogenetic lines." He asks: Why should we be concerned with phylogenetic lines? We do so because phylogenetic lines, as recorded by vertebrate structure and development, play "an important part in the elucidation of the evolutionary process" (1930, p. v). Although his words sound promising, all he means by "process" is that organisms change in time. Thus, we are merely returned to the same phylogenetic lines. Goodrich offers no discussion throughout his descriptive studies about how organic structures might change in time and makes no reference to mechanisms that might bring about such changes. It seems to me that it is, in essence, an obsession with the question of the transforming process which defines the modern synthesis. A morphologist, sensitive to the problem of change of structure and function, would translate this concern into terms of adaptation. Goodrich scarcely uses the word.

De Beer's *Vertebrate Zoology* (1928) is suggestive from two points of view. It offers de Beer's own traditional statement of objectives and Julian Huxley's introduction that announces why we shall find a new zoology in this volume. De Beer states that vertebrates present a remarkably homogeneous group and, being homogeneous, constitute a group peculiarly suitable for effecting comparisons among its members. This claim should be weighed against what Davis has argued for many years: the interest of vertebrates resides in their differences and not their homogeneity (1964, pp. 11-13, 322-327).

De Beer announces that from comparison we can move toward generalization, the latter being the central purpose of our science. What are his generalizations? The first is that homology is the whole art of morphology. He cites thyroid structure as a splendid instance by which evolutionary history can be re-created. He asks if this is not a clever trick of the anatomist. Function here is no guide to phylogeny; we must rule out any consideration of function. Huxley acknowledges de Beer's effort. He tells us, however, that morphology has far too long been self-contained. It is time that morphology moved out of its narrow constraints; it is time that it made connections with biology. Nonetheless, if you follow in the body of the volume how de Beer's discussion develops Huxley's claims you find that this book could have been written, had the factual material been available, by Gegenbaur or virtually any other late nineteenth-century morphologist. A new program is announced but the conceptual structure and execution persist unchanged.

The second generalization by de Beer in this 1928 volume is that morphology permits us to study convergence in the evolutionary process. Similarities can be understood not as the result of descent but of similar adaptive courses (paddlelike limbs, for example). Once again, however, de Beer, like his teacher Goodrich, attends only to the adaptive product, the organs employed in comparisons, and ignores the process by which adaptation takes place. Huxley's introduction declared that genetics is changing our whole view of the science of morphology. We must reformulate our very conception of the science, but de Beer does not develop, even modestly, the implications of the programmatic introduction to his book. Huxley's notions, it must be admitted, would have been very difficult to develop in the context of comparative anatomy. He tells us that the object and idea of homology have been revolutionized because identical but independent mutation of genes has occurred, which is what *Drosophila* research is all about. Therefore, identical but independent mutations are obviously similar forms with no real connection in descent. Evolutionary history may well be ignored—but the recovery of that history was really what defined the contemporary pursuit of morphology. The pertinence of genetics to morphology was not at all obvious to students of the latter.

This singular point of view is the evolutionary synthesis as stated in 1928 and again in 1932 by morphologists. This argument survives, for example, in the concise, but sophisticated *Einführung in die vergleichende Morphologie der Wirbeltiere* by Portmann (1948). Portmann tells us that species and genera (that is, the basic systematic categories) are no longer clear units for discussion. They themselves are the problems of the science. He uses a term that catches our eye; he tells us that henceforth morphology has to deal with populations. Population and populations, he reports, are subject to genetic analysis. Three categories of genetic change are important: gene mutation, which Morgan and Chetverikov dealt with; chromosomal mutation (such as inversions and translocations); and genome mutation, referring to polyploidy. Portmann then proposes that analysis of gene mutation seems to be the most promising course for the study of evolution, but only for microevolution. Yet, at this point, he launches into a discussion of the distinction between microevolution and macroevolution and concludes that macroevolution, which is a problem that the paleontologists have defined, can only be resolved by as yet unknown explanatory factors. Alas, he here simply gives up the effort to unite genetics and morphology and turns to developing a book of descriptive morphology. Incidentally, in his bibliography the titles he cites as most germane to the modern study of evolution are by Mayr (1942), Dobzhansky (1937), Cuénot (1932), and Guyénot (1930),—a curious combination, but perhaps less so to a Swiss open to simultaneous influence from the contrasting domains of French and German-American-English evolutionary biology.

Friedrich Maurer, a morphologist who occupied a very distinguished chair in comparative anatomy at Heidelberg, previously held by Gegenbaur, explored the relationship of genetics to evolution in a brief essay (1929). He saw that genetics was of interest because it dealt with discrete variants and considered the problem of how such variation might be summed. Nevertheless, to Maurer it was clear that variation dealt only with superficial characteristics. It touched nothing of decisive import to the organism. He ultimately concluded that experimental genetics is not true to nature; the essence of an organism lies in its *Grundplan* or *Bauplan*. To understand genetics is, consequently, simply irrelevant to the morphologist. One could not, in short, move from an individual variant or even a sum thereof to significant change of the organism itself.

Nearly all these works and others on comparative anatomy emphasized the traditional approach to comparative morphology. Some authors felt obliged to make apologetic statements in response to what they sensed to be a new approach or at least an approach that might one day come to be. They did not, however, seek to develop systematically the implications of the new approach and did not appropriate suggestions made by others.

Why did anatomical opinion develop in this manner? Why was there disinterest on the part of morphologists in the evolutionary process? The systematic propaganda spread in the early twentieth century, above all in the United States, regarding the distinction between experimental and descriptive biology was important. The experimentalists felt that the future belonged to them. The term description meant not simply a descriptive statement but was used in a pejorative and condemnatory sense. Anyone who did descriptive work had thereby opted out of making any causal statement regarding organisms.

Against whom was this statement directed? At the beginning of the century I feel the critics were addressing the statement to an imaginary generation that had put forward any number of speculative hereditary schemes. Delage's vast book of 1903 offers some 700 pages of description and seeming analysis of the phenomena of inheritance. Yet the patterns described had only the slightest foundation in empirical fact. The experimentalist critics were ruling out speculation in this domain. More important, however, the traditional morphologist's objective was being attacked. That objective had assumed that the task of morphology was not the study of the evolutionary process; the true goal of evolutionary investigation was to re-create the history of life. Comparative anatomy—above all, vertebrate comparative anatomy—became the great instrument of this art.

The basically nonevolutionary interpretation of morphology rose again toward 1900. It appears to have been a national style in natural science, affecting above all study of plant and animal form. This view prospered in Germany; the style was marked by the reappearance of a radically idealistic morphology. This position was represented by such anatomists as Naef and Lubosch. The history of comparative anatomy that Lubosch wrote for the first volume of Bolk's *Handbuch der vergleichenden Anatomie der Wirbeltiere* seems to the pragmatic eye to be a never-never land of morphology (1931). During these years, furthermore, a tremendous interest in Goethe as a morphologist reawakened; Troll's important edition (1932) of Goethe's morphological writings appeared at this time (see also Arber, 1950).

I think this situation arose around 1900 because of developments in comparative anatomy in the immediate post-Darwinian years. Gegenbaur, the leader of comparative anatomy, converted to the Darwinian position during the 1860s and not only articulated the general goal of the science but also outlined the kinds of studies that comparative anatomists should pursue. In fact, his publications of the 1860s and 1870s are little different from anatomy textbooks of the recent past (for example, Romer). Gegenbaur tells us that homologies are the essential purpose of the study; comparative anatomy is fundamentally the study of formal relationships. He emphasized beyond all else that comparative anatomy really must begin with the premise of resemblance, although he let it be inferred that resemblance was also an unresolved and interesting problem. Actually he began with the assumption of resemblance, only then tracing slight variations. What was not studied were real divergences from similarities. With this attitude, he virtually ruled out the whole question of adaptation. It was in Gegenbaur's hands, and in those of others who worked after him, that comparative anatomy became the primary tool for what seemed to be the confirmation of the Darwinian position with regard to the fact of descent (Coleman, 1976).

More important is the fact that evolutionary thinking triumphed in biology at a time when the older style in comparative anatomy was falling into disrepute, primarily because of earlier idealistic speculation in Germany. Just as comparative anatomy made a great contribution to evolutionary thinking, evolutionary thinking as the new dispensation, the new generalization, confirmed the validity and proved the utility of comparative anatomy. Thus the two worked hand in hand to the end of the century. It seems to me that this legitimation of comparative anatomy by evolutionary doctrine in the post-Darwinian years has continued into the present in such works as those by Hyman, Goodrich, and Romer.

We may conclude from this account, I believe, that general anatomical emphasis changed in the 1930s. Before the modern synthesis, emphasis had been placed on the fact of descent, apart from any interest in the causal factors of descent, above all in the problems of adaptation. General characteristics of groups were emphasized, not the diversity of populations within them. As a consequence it seems to me easily understood why comparative anatomy during the 1920s and 1930s failed to contribute to the evolutionary synthesis. It was not until long after the synthesis that a new school of evolutionary morphology became established whose members did ask questions about the evolutionary process. The objectives of the new movement are clearly stated by Davis (1960, 1964), Bock and Wahlert (1965), and Böker (1935).

References

Arber, A. 1950. *The natural philosophy of plant form.* Cambridge: University Press.

Bock, W. J., and G. von Wahlert. 1965. Adaption and form-function complex. *Evolution* 19:269-299. An earlier but isolated call for functional anatomy was issued by Böker (see below).

Böker, H. 1935. *Einführung in die vergleichende biologische Anatomie der Wirbeltiere.* Jena: Fischer.

Coleman, W. 1976. Morphology between type concept and descent theory. *Journal of the History of Medicine* 31:149-175.

Cuénot, L. 1932. *Tardigrades,* vol. 24 of *Faune de France.* Paris: P. P. Lechevalier.

Davis, D. D. 1960. The proper goal of comparative anatomy. *Proceedings of the centenary and bicentenary congress of biology, Singapore, December 2-9, 1958,* ed. R. D. Purchon. Singapore: University of Malaya Press, pp. 44-50.

———— 1964. The giant panda, a morphological study of evolutionary mechanisms. *Fieldiana. Zoology Memoirs* 3.

de Beer, G. 1928. *Vertebrate zoology.* New York: Macmillan. (2d ed., 1932.)

Delage, Y. 1903. *L'hérédité et les grands problèmes de la biologie générale,* 2d ed. Paris: Schleicher.

Dobzhansky, Th. 1937. *Genetics and the origin of species,* 1st ed. New York: Columbia University Press.

Goodrich, E. S. 1930. *Studies on the structure and development of vertebrates.* London: Macmillan.

Guyénot, E. 1930. *La variation et l'évolution.* Paris: Doin.

Huxley, J. S. 1932. Introduction. In *Vertebrate zoology,* by G. de Beer. New York: Macmillan, pp. xi-xii.

Hyman, L. H. 1922. *A laboratory manual for comparative vertebrate anatomy.* Chicago: University of Chicago Press.

Lubosch, W. 1931. Geschichte der vergleichenden Anatomie. In *Handbuch der vergleichenden Anatomie der Wirbeltiere,* ed. L. Bolk. Berlin and Vienna: Urban and Schwartzenberg, vol. 1, pp. 3-76.

Maurer, F. 1929. Der gegenwärtige Stand der Lehre vom Darwinismus. *Deutsche medizinische Wochenschrift* 55:1433-1435.

Mayr, E. 1942. *Systematics and the origin of species.* New York: Columbia University Press.

Portmann, A. 1948. *Einführung in die vergleichende Morphologie der Wirbeltiere.* Basel: Benno Schwabe, pp. 13-31.

Romer, A. S. 1949. *The vertebrate body.* Philadelphia: Saunders.

Troll, W., ed. 1932. *Goethes morphologische Schriften.* Jena: Diederich.

The Failure of Morphology to Assimilate Darwinism

Michael T. Ghiselin

Mayr (1959) stressed the importance of ways of thinking as determining success and failure in scientific endeavor. An outstanding example is how typological (essentialistic) thought habits were most inappropriate to dealing with natural selection. This insight was most useful in my own efforts to understand what led Darwin to his epochal discovery (Ghiselin, 1969). Another important point insisted on by Mayr is that biology has to use modes of reasoning that differ from those appropriate to physical sciences. Here I argue along the same lines. A sort of ideology about what constitutes a proper scientific view of things strongly influenced the thinking of many biologists, pre-Darwinian and post-Darwinian alike. Biology had to be modeled upon chemistry and physics. Biologists often looked on the world from a theological point of view, but even so they commonly held that a naturalistic explanation could be found for such phenomena as adaptation and the natural system. However, the kinds of explanations they anticipated were the sort that were current in the physical sciences. Natural selection differed so much that scientists with the traditional point of view could neither discover nor cope with it.

Some historians may object to this hypothesis. For example, Winsor (1976) in a detailed and scholarly study of nineteenth-century taxonomy concludes that the science developed in response to empirical discoveries and was little influenced by theory. She does, to be sure, point out that physics was the ideal science. Naturalists expected to find what she calls an "order of life," but entertained no expectations as to what might be the cause of this order. Granted, evidence for metascientific influences is slim in published works and even in correspondence. But such negative evidence is not compelling. Scientists rarely write about such matters, although they occasionally do, and oblique references are not hard to find. As a rule they merely presuppose their fundamental assumptions. Only when something seems amiss do the metascientific issues come to the surface. Criticisms of Darwinism by Owen (1868), Mivart (1871), and others are manifest examples of efforts to salvage traditional metaphysical views. Furthermore, the ideologies in question are rarely inculcated systematically. Rather, they are taken for granted in the educational process and form tacit assumptions in research. It is not difficult to show how metaphysical and other metascientific influences have had considerable impact on morphology.

To many it has seemed enigmatic that morphology contributed virtually nothing to the synthetic theory of evolution. Beyond the accumulation of more phylogenetic data and the elucidation of long-term trends, it seems to have existed in another world. One might conjecture that there was something about morphologists that led them to dislike selection theory. Although morphology has perhaps attracted a disproportionately large number of antiselectionists (such as D'Arcy Thompson), many have been ardent advocates of selection. And there have been geneticists on both sides of the fence as well. Bateson, Morgan, and Goldschmidt—all, incidentally, originally trained as morphologists— opposed neo-Darwinism within genetics. Furthermore, the failure of morphology to participate to any great extent in the synthetic theory is paralleled by Darwin's synthesis. The theory stimulated new developments in morphology, but was not evoked by it. Morphology has contributed so little primarily because it has had so little to contribute. It is a descriptive science of form, and only when conjoined with other disciplines does it tell us anything about causes. But once a causal mechanism has been accepted, it can provide a valuable service. Nonetheless, for this very reason, morphology tends to be the sort of discipline that will follow, rather than lead, in the development of evolutionary theory. Some basic philosophical issues surrounding the mechanics of innovation and the history of science in general will clarify the reasons for this role (see Ghiselin, 1969, 1971, 1974a).

My research has been strongly influenced by the work of Kuhn (1970), and I accept the notion of a "paradigm" (in the sense of a "research tradition") but my outlook differs in some important respects. My views derive from the thesis that the evolution of thought and the evolution of organic beings are more particular instances of the same general phenomenon. In the first place, science has to be evolutionary, not revolutionary. Just as organisms never evolve by saltation, "paradigms must have predecessors." Novelty comes from without, by various forms of analogical transfer from some other field of experience (much as preexisting organs assume new functions). Perhaps most important, I reject Kuhn's assumption that paradigms should be analyzed at the level of the society. After all, only organisms can think. Hence I stress personality and individual experience, while not denying that the environment profoundly influences intellectual and organic evolution alike.

Closer to home, Mayr (1972) and I (especially Ghiselin, 1971) have both published analyses of the Darwinian revolution in relation to paradigm theory. So far as biology proper goes, it is hard to find two advocates of the synthetic theory and new systematics who concur in so much. We only part company on certain metascientific issues. Mayr's analysis focuses on taxonomy. I, on the other hand, maintain that the Darwinian revolution was basically a matter of ecology, not systematics (Ghiselin, 1974a). I view the later history of biology as a continued struggle to come to grips with a new conception of the natural economy. Indeed, economics and biology seem destined to become a single branch of knowledge (Ghiselin, 1974b, 1978). Mayr further maintains that biology differs from the physical sciences enough that separate philosophies are needed. Granted, there are differences between living and nonliving systems, but a good philosophy of science should apply to all branches of knowledge. Finally, Mayr suggests that before the Darwinian revolution a number of paradigms coexisted. There seems to me nothing a priori objectionable to this hypothesis. Nonetheless, the apparently contradictory paradigms can also be treated as aspects of a single, more general one. What was overthrown was so comprehensive in scope that it is hard to delimit it. I am tempted to say that the old paradigm was almost coextensive with Western thought. But that gives the wrong impression. Many pillars were torn out (teleology, for instance) and large portions of the edifice came tumbling down. But much of the foundation remained intact, and the ruins were remodeled and built up into a different structure.

We will never appreciate the history of twentieth-century evolutionary biology, especially morphology, until we go back to its earlier stages of historical development. In the half century before *On the Origin of Species* was published, there was a standard way of doing natural history. We need not decide whether this amounted to a Kuhnian paradigm. There was, at least, a set of common assumptions and procedures. The differences should be neither overestimated nor underestimated. They had to do with certain important details: unity of type, the significance of function, and the possibility of evolution—while not considered from a single point of view—remained, however, variations on one general theme. The search for order was modeled on physics and chemistry and incorporated a number of significant tacit assumptions. Research gave rise to a taxonomic system that was believed to reveal the plan of nature. It was presupposed that the taxonomic groups (taxa) were classes analogous to classes of minerals. Properties found in such groups were to be explained in terms of laws of nature—again like those of physics and chemistry.

In those days, laws of nature, as scientists viewed such matters, could do a lot more than we now are willing to admit. For us a law merely states that if *A* happens, *B* must happen. For our predecessors they were also fiats enunciated by God. As this power was infinite, there would be no trouble in God's providing whatever might be needed by organisms. It mattered little what theological superstructure might support this general position. There seemed, furthermore, to be no difficulty in embracing a teleological view of the universe: laws could account for that as well.

Systematics was an integral part of the mainstream of biological thought. The same laws of nature were known to govern the course of embryological development as governed the distribution of forms in the taxonomic hierarchy. Nobody could say precisely why at the time, but the growth of crystals could be analogized with that of organisms, and it was obvious that the structure of all three kingdoms of nature would have to be explained in the same way. Even such "evolutionists" as Lamarck propounded "laws" implying that change is necessary and progressive. So long as one persists in looking on organisms from this point of view, there will be no compelling reason to become an evolutionist or to search for a mechanism producing change, and there will be no hint where such a mechanism might be found.

Hence the fact that the Darwinian revolution did not arise out of morphology and taxonomy hardly poses an enigma. It arose out of biogeography and ecology. The discoverer was a geologist with extensive knowledge of natural history. He developed his theory of natural selection by overthrowing the traditional conception of the natural economy. The crucial insight, evoked by reading a work on classical economics, was realizing the consequences of reproductive competition between members of the same species. Morphology, like all branches of knowledge, only gradually assimilated natural selection. The new paradigm differed so radically from what was expected that biologists would take generations to come to grips with it.

A vast gulf separates the metaphysics of Darwinism from that which preceded it. Historians and biologists alike have tended to stress factitious similarities while overlooking fundamental differences. Yet we should reexamine the ontological status of the basic units of taxonomy. Before Darwin, taxa were classes of organisms, and their properties had to be explained in terms of laws of nature. Subsequently, taxa became what I have termed "chunks, so to speak, of the genealogical nexus." It follows deductively that taxa are, in the logical sense, individuals rather than classes, and that organisms are parts, not members, of taxa (Ghiselin, 1966, 1974b). Some very important consequences may be derived from this inference, some less obvious than others. Clearly, any causal explanation would have to invoke all sorts of contingencies in addition to laws. Furthermore, laws of evolution would have to be laws for classes of taxa, not for any individual taxon (just as in physics there are laws for classes of planets, but none for Saturn). Lately Hull (1975, 1976) has elaborated on this point from a philosophical perspective.

Throughout the history of Darwinism, controversy has hinged on the old versus the new ontology. Natural selection was objectionable because it did not fit in with the traditional notions as to what a law ought to be. Sir John Herschel dismissed it as a "law of higgledy-piggledy." Yet Darwin himself pointed out that selection is not a force, but rather a mechanism like erosion in geology. No law states that any feature of the landscape must erode, but the erosional processes do account for the various land forms. Such explanations require a mixture of laws and historical contingencies, such as what kind of rock has chanced to be deposited in a given place. However, nineteenth-century thinkers wanted laws of evolution that would allow them to dispense with contingencies. Herbert Spencer sought in vain for a law of necessary progress, and analogies with physical sciences pervade his works. From Peirce to the present day, philosophers have labored incessantly to salvage the old ontology.

For biologists seeking to find laws of evolution, one way out was orthogenesis. Once environmental contingencies had been ruled out, there remained something intrinsic to the organism itself. Phylogeny was analogous to ontogeny, which in turn was analogous to crystallization. Hence biology could be explained in terms of molecular forces and their laws. Throughout the history of evolutionary biology, orthogenesis has been the main alternative to natural selection, not only for anatomists such as Owen, but also for geneticists like Morgan. Orthogenesis no doubt was somewhat encouraged by the successes of experimental embryology toward the end of the nineteenth century. Lamarckians, too, sought a progressive force internal to the organism itself, but were obviously seeking another way out. The crystal analogy, though long virtually forgotten, played a very significant role in nineteenth- and early twentieth-century biology (see Maulitz, 1971). It was invoked in a very influential paper by Reil (1796). It played a very important role in the cell theory of

Schwann (1839). Hylozoists such as Jordan (1842) and Haeckel (1904 and elsewhere) argued that because crystals grow they must possess at least a vegetative soul and perhaps more.

Granted, however, that evolution has occurred, and accepting even tentatively natural selection as its mechanism, morphologists were in an admirable position to reinterpret their science along historical lines. It has repeatedly been asserted that they did nothing more—that they continued to do their work in the old manner, and merely superimposed a historical terminology on a static formal comparison. Although this notion is wrong, there is enough truth in it to make it superficially attractive. Vestiges of it can be found even in the writings of Dobzhansky and Simpson, and to some it is still dogma. The new paradigm was so radically different that its assimilation into morphology took considerable time and is still incomplete. New techniques developed gradually, while work went on more or less as it always had. Nonetheless the classification systems of such pre-Darwinians as Owen and Lamarck and those of Darwin and Haeckel are obviously different. For example, the later work abandoned a single scale, from higher to lower, for a divergent arrangement, so that the ladder was replaced by the tree. The dominating principle ceased to be development and became diversification, or what we now call "adaptive radiation," an ecological notion.

From the outset it was apparent that laws of nature could be derived from the theory of natural selection, which could test hypothesized genealogical relationships. As there can be no teleology in modern evolutionary biology, vestigial structures had to be explained in the light of the past, not the future. To Dohrn (1875) it was apparent that organs do not arise out of nowhere, and he sought out likely precursors in organs that change function. His principle of *Funktionswechsel* was followed by similar ones. These include canons of evidence for the direction of change, an important development because in nonevolutionary morphology series of forms can be read just as well in one direction as another.

In general the main role of morphology in evolutionary studies has been to confirm and illustrate the findings of other branches of biology. The vogue for evolution initiated by Darwin and Wallace evoked an immense amount of work on comparative anatomy. If evolution had occurred, then the evident discontinuities in the series of organized beings could perhaps be filled in by new research. Adult morphology could focus on the inadequately examined organisms, especially ones thought primitive or transitional. Embryological structure could be examined for transitory appearance of organs formerly apparent in adults. Paleontologists worked in the same spirit, using new fossil materials. The detailed documentation of the history of life has continued, despite the fact that evolutionary morphology succeeded almost to the point of rendering itself superfluous. After many gaps had been filled in, all but the diehards would be convinced that evolution has occurred. The subject passed into a period of "puzzle-solving," in a Kuhnian sense. But it was not the sort of normal science that led to a crisis or a new paradigm. Rather, it continued to provide ever more detail on evolutionary history. It has persistently attracted good minds, partly because the puzzles are fascinating, but partly also because greater understanding of evolutionary mechanisms revitalizes the subject from without.

The ability of morphology to provide evidence concerning the mechanisms of evolution was limited by the fact that its data could be explained by more than one hypothesis. Alternatives to natural selection were argued on a morphological basis, but selectionists were able to provide their own—at least equally convincing—interpretations. If the controversies were often indecisive, the selectionist position was at least encouraged through its ability to withstand criticism. Morphological considerations, for example, have been significant in controversies about supposedly nonadaptive or maladaptive features. Darwin—who had some predecessors on this point—realized that the laws of growth might lead to adverse or neutral side effects in basically adaptive changes. The issue continues to be debated. On the other hand, many efforts to find nonadaptive and maladaptive features have been based on fallacious reasoning. Bizarre organisms, living and fossil, were claimed to exemplify racial senescence or the like on the basis that their features are hard to explain. But the imagination has triumphed over error. Here morphology made a positive contribution to the synthetic theory.

Perhaps the main contribution of morphology to the synthetic theory was the documentation and explanation of macroevolutionary trends, rules, and laws (see Rensch, 1960). It often happens that evolution tends to proceed in one direction. Such trends could be explained in terms of internal or external influences or both. When Darwinism went out of fashion, advocates of orthogenesis and Lamarckism seized on such directionality as indicative of an inwardly determined necessity to change in one way. The rebuttal was provided by a number of evolutionary morphologists, including Sewertzoff and Simpson. Change has occurred in more than one direction, it has been reversed, and what has happened depended on environmental contingencies. Unidirectional evolution has turned out to be orthoselective, not orthogenetic. It is curious, but hardly unexpected, that the study of allometric growth inspired by the antiselectionist Thompson was successfully incorporated into the mainstream of the synthetic theory by Huxley, Rensch, and others.

Evolutionary biologists have always had a serious problem in distinguishing between laws and other kinds of generalizations. Particularly notorious is the relationship between embryological development and evolutionary history. Ernst Haeckel tried to recast the well-known parallelism between systematics and development in terms of the "biogenetic law"—individual development (ontogeny) recapitulates the stages of racial history (phylogeny). Matters are not so simple, and it is widely maintained that Haeckel was wrong on this issue and that the preevolutionists were right. Actually, the preevolutionists were wrong. They thought that the same laws govern systematics and ontogeny; they do not, and it was Darwin who provided the correct explanation: laws of heredity and contingencies of history. Von Baer (1828) actually said that the more general properties appear before the particular ones. This statement was intended as a law, and it is false. Milk, for instance, is secreted long after generic and specific characters appear in mammals. Haeckel said that ontogeny is a recapitulation of phylogeny conditioned by laws of heredity and adaptation. The qualification accounts for the exceptions, but only at the cost of predictive power and hence utility.

To say that there is an evolutionary trend or tendency is to enunciate a mere historical generalization. There is nothing necessary about such statements. Evolutionary laws, however, predict what will happen to species whenever certain conditions are met. As such conditions are difficult for biologists to know, the laws of evolution are often formulated in weak statistical terms or as nonpredictive "rules." Such rules have often been lumped together with laws and with generalizations about what happens to have occurred in evolutionary history. Much of the work of evolutionists has been directed toward finding out the causal basis for rules and distinguishing laws from historical generalizations. A good example of a nonlaw is Cope's rule, which says that organisms always evolve toward larger size. Rensch attempted to turn it into a law by showing advantages to increased body size. Stanley (1973), on the other hand, has argued that the trend toward increased size is a consequence of the fact that the original members of a taxon are small. Actually, we know of groups and environmental circumstances in which size reduction rather than increase is the rule (Ghiselin, 1974a). When the conditions can be adequately specified, we will have a legitimate law.

Evolutionary biology does indeed have many valid laws of nature. Under ordinary circumstances, for example, there can be no speciation without an extrinsic isolating mechanism (see Mayr, 1963). When the synthetic theory emerged, a host of evolutionary rules and laws were put forth and documented, especially by Rensch. In general, however, they were weakly formulated, they were not highly predictive, and the exceptions were inadequately accounted for. Many of the explanations were little more than plausible guesswork, and little effort at serious critical testing was done. Such lack of rigor seems not so undesirable as one might think. It should be regarded as the natural consequence of the mechanics of innovation and the particular historical situation. The synthetic theory was then in what I have called the "illustrative" phase (Ghiselin, 1971), and stringent criticism would follow later.

As the synthetic theory was elaborated, a new evolutionary morphology did emerge. On the whole, however, this was a late development, one that belongs to the "exploratory" phase of intellectual evolution. Once the theory had developed, it could be used in solving puzzles, such as traditional problems in phylogeny. But the time lag was considerable, and even in the 1950s and 1960s there was ample opportunity to join in the mopping-up operation against saltationism and to develop new techniques. Indeed, we are still struggling with many problems, such as how to deal with parallel evolution in phylogenetic inference and classification.

In his analysis of the synthetic theory (see chapter 4), Mayr rightly points out that the architects of the theory made heroic efforts to engage in interdisciplinary transfer. If, for example, they were geneticists, they looked to implications of genetics for speciation and population structure. Other disciplines incorporated genetics quite deliberately and actively. It did not just happen. The architects differed from the hod-carriers insofar as the latter continued to work as puzzle-solvers in the Kuhnian sense. This pattern of interdisciplinary transfer is anything but unique: witness Darwin's debt to Malthus. Furthermore, it was not a matter of a group of scientists finding inadequacies in their paradigm. Rather, individual thinkers with innovative mentalities reached out for something new. They saw that an opportunity was at hand and took advantage of it.

What determines that someone will have an innovative mentality and engage in interdisciplinary transfer? Valid explanations are many. They include hereditary makeup, family background, education, the reward system, and previous research experience. In many fields, training militates against certain kinds of transfer. The transfer that does occur under such circumstances tends to be relatively shallow. A morphologist can be productive without having to know much about genetics or ecology; hence this discipline will tend to be a closed system in relation to those fields. Morphologists can deal with adaptive significance with little or even no understanding of natural selection.

Transfer tends to flow upward in the integrational hierarchy. An anatomist is expected to know more about physiology than ecology, tending to put ecological anatomy into the hands of populational biologists who do not know enough about organisms to treat the subject with great sophistication. The taxonomic division of labor restricts "horizontal" transfer. Innovations in methodology and even important generalizations or useful facts will tend to be known only to experts on a given taxon. Because so much has been done on vertebrates, and because they are so familiar, this group enjoys the action. Even though the phylum Mollusca is larger and more diverse than Vertebrata, its comparative anatomy is less developed and its findings are less well known. Furthermore the division of labor makes it hard to generalize on a broad scale.

One morphologist who did help develop the synthetic theory is Aleksi Nikolaevitch Sewertzoff (1866-1936) of the University of Moscow (see especially Sewertzoff, 1916, 1931). A more or less traditional comparative anatomist early in his career, he claims to have adopted the Russian approach of searching for morphological laws of evolution that developed around 1908. His works show that he immersed himself in the literature on evolutionary theory from an early period. On the other hand his anatomical publications until well into the 1920s show little evidence of his ultimate contribution. He worked on the details of lower vertebrate phylogeny, using embryology to find vestigial structures. Gradually he incorporated his two distinctive contributions into his narrative of vertebrate history. First, drawing on his broad knowledge of fish embryology he developed a sophisticated analysis of the various modes of evolution through changes in developmental processes. His analysis included a magisterial treatment of the ontogeny-phylogeny issue, which justified Haeckel but accounted for the many exceptions to the biogenetic law. Second, his interpretations took on an increasingly functional and ecological character. He showed how changes in function, intensification of activity, shifts in habitats, and habits could be used to relate structure to the environment. To him this relationship implied a selective rather than "endogenetic" evolutionary mechanism. These themes were not entirely new, but he developed them at considerable length. The advantage of his dynamic approach over the static ones of such idealistic morphologists as Naef (1931) are obvious. Sewertzoff's views were further elaborated by some of the architects of the synthetic theory, especially Schmalhausen and Rensch, but curiously not Simpson or Huxley. Dobzhansky (personal communication) and other Russian evolutionists obviously knew his work quite well. Thus he fits the pattern here suggested for synthetic theorists, in that he helped synthesize his field with others, especially embryology, physiology, and ecology.

Edwin S. Goodrich (1869-1946) of Oxford, as Linacre Professor of Zoology and Comparative Anatomy, influenced such figures as Ford (see

Hardy, 1946). He was undoubtedly one of the greatest morphologists of his time, yet neither his work nor the morphological research of his students contributed much if anything to the synthetic theory. As Churchill points out (chapter 3), he contented himself with working out details of homology and drawing an ever more accurate phylogenetic tree. He seems to have epitomized Kuhn's notion of a puzzle-solver. His published works show that he certainly could have participated in the formulation of the synthetic theory had something—perhaps lack of imagination—not held him back.

His popular book on evolution, *Living Organisms* (1924), shows that Goodrich possessed an excellent and up-to-date understanding of evolutionary biology and was an orthodox Darwinian. Yet it is virtually the only place where he deals at length with general principles, and he hardly says a word about morphology. Nowhere does Goodrich discuss either the principles or the implications of evolutionary anatomy, save in passing remarks. As with so many authors, he presupposes much, and only through indirect evidence can we infer how he did his work. It is particularly clear from his obituary of Lankester (Goodrich, 1931) that he continued to do research in the tradition that was worked out by Darwin's supporters late in the nineteenth century. And clearly he had mastered both the techniques and the results. This we can infer from scattered comments in his major works (1895, 1909, 1930, 1945). In addition to the pre-Darwinian criteria of homology, his methods included the principles of functional continuity (1909, p. 74; 1930, p. 474) and the need for precursors (1930, pp. v, 594, 596). Goodrich was anything but an idealistic morphologist. He repeatedly comments upon the difficulties of the subject, such as reversion and convergence (1909, pp. 65, 117, 368). His methodology seems not to have evolved much over fifty years.

Many of his views might well have influenced the synthetic theory, but probably did not. For one thing, Goodrich was adamantly opposed to classification by grades (1909, p. 92), even opposing garbage-can assemblages for ancestral stocks. He even went so far as to assert that because the class Reptilia is polyphyletic, the group is artificial (1930, p. 575). And the reason is clear enough: gradal systems obscure divergences and introduce an element of anthropomorphism. It is well known that Huxley and Simpson advocated gradal and polyphyletic taxa. Goodrich was in an admirable position to champion the cause of genealogical taxonomy, but failed to do so, and open discussion of such matters was long delayed.

One paper alone deals with the mechanisms of evolutionary change (Goodrich, 1913). He demonstrated that limbs may change their morphological connections from one vertebra to another, a reasonable extrapolation from the fact well known to experienced comparative anatomists that the principle of constancy of connections does not always hold true. The architects of the synthetic theory tended to emphasize changes in developmental timing and in body proportions when they discussed structural change. Modifications in position and other modes of reorganization could have received more attention. Morphologists such as Goodrich might have driven home such points; but for whatever reason such contributions were most exceptional.

If morphology was not contributing to the synthetic theory when so many other fields were, what then were the morphologists seeking to accomplish? A full answer would require an examination, in detail, of many materials. As a comparative anatomist who has read a fair sample of the literature, I can offer some impressions that seem consistent with the facts. During the first half of this century, many morphologists continued to work in the traditions of the previous century: Darwinian phylogenetics, idealistic morphology, and the like. This research led to the accumulation of much valuable data, particularly useful for systematics. But on the whole it tended to stick just another twig on the phylogenetic tree. Numerous anatomists and morphologists ceased to have evolution as their primary focus of interest. A shift of emphasis occurred toward elucidating the adaptive significance of features and discovering how they work. The resulting "functional morphology," even when comparative, tended to ask physiological questions, and cultivated physical-science thought habits to the detriment of biological ones. Others, especially those interested in embryology, sought to explain form in terms of purely mechanical considerations. Such an attitude was widespread among biologists, including Morgan. The element of historical contingency would be eliminated, chance would depart from the universe, and the physical-science ideology would emerge triumphant. Of course, such hopes were in vain, for there is no contradiction. Mechanical stresses and the like have their place in determining organic form but only as constraining influences. (Much as the strength of materials helps to explain the conformation of a Gothic cathedral, but hardly accounts for everything.) The legitimate explanation for the structure of organisms has to be sought not in physics, but in economics (Ghiselin, 1978). This point was driven home in discussion by Hamburger, who remarked how hard he found it to conceive of natural selection acting on the inside of an organism. One could hardly ask for a more metaphysical question. When one asks how it is that what goes on in the marketplace can influence what happens on the assembly line, the answer is obvious.

References

Baer, K. E. von. 1828. *Über Entwickelungsgeschichte der Thiere. Beobachtung und Reflexion*, vol. 1. Königsberg: Bornträger.

Dohrn, A. 1875. *Der Ursprung der Wirbelthiere und das Prinzip des Funktionswechsels. Genealogische Skizzen*.

Ghiselin, M. T. 1966. On psychologism in the logic of taxonomic controversies. *Systematic Zoology* 15:207-215.

―――― 1969. *The triumph of the Darwinian method*. Berkeley: University of California Press.

―――― 1971. The individual in the Darwinian Revolution. *New Literary History* 3:113-134.

―――― 1974a. *The economy of nature and the evolution of sex*. Berkeley: University of California Press.

―――― 1974b. A radical solution to the species problem. *Systematic Zoology* 23:536-544.

―――― 1978. The economy of the body. *American Economic Review* 68:233-237.

Goodrich, E. S. 1895. On the coelom, genital ducts, and nephridia. *Quarterly Journal of Microscopical Science* 37:477-510.

―――― 1909. Vertebrata Craniata (first fascicle: cyclostomes and fishes). In *A treatise on zoology*, ed. R. Lankester. London: Adam and Charles Black, pt. 9.

―――― 1913. Metameric segmentation and homology. *Quarterly Journal of Microscopical Science* 59:227-248.

―――― 1924. *Living organisms: an account of their origin & evolution*. Oxford: Clarendon Press.

―――― 1930. *Studies on the structure and development of vertebrates*. London: Macmillan.

―――― 1931. The scientific work of Edwin Ray Lankester. *Quarterly Journal of Microscopical Science* 74:365-381.

―――― 1945. The study of nephridia and genital ducts since 1895. *Quarterly Journal of Microscopical Science* 86:113-392.

Haeckel, E. 1904. *The wonders of life*. New York: Harper & Brothers.

Hardy, A. C. 1946. Edwin Stephen Goodrich. *Quarterly Journal of Microscopical Science* 87:317-355.

Hull, D. L. 1975. Central subjects and historical narratives. *History and Theory* 14:253-274.

―――― 1976. Are species really individuals? *Systematic Zoology* 25:174-191.

Jordan, H. 1842. Der Wiederersatz verstümmelter Kristalle als Beitrag zur näheren Kenntnis dieser Individuen und zu ihrer Vergleichung mit denen der organischen Natur. *Archiv für Anatomie und Physiologie* 35:46-56.

Kuhn, T. S. 1970. *The structure of scientific revolutions*, 2d ed. Chicago: University of Chicago Press.

Maulitz, R. C. 1971. Schwann's way: cells and crystals. *Journal of the History of Medicine and Allied Sciences* 26:422-437.

Mayr, E. 1959. Darwin and evolutionary theory in biology. In *Evolution and anthropology: a centennial appraisal,* ed. B. J. Meggers. Washington: Anthropological Society of Washington, pp. 1-10.

—— 1963. *Animal species and evolution.* Cambridge, Massachusetts: Harvard University Press.

—— 1972. The nature of the Darwinian revolution. *Science* 176:981-989.

Mivart, St. G. 1871. *On the genesis of species.* London: Macmillan.

Naef, A. 1931. Phylogenie der Tiere. *Handbuch der Vererbungswissenschaft* 3:1-200.

Owen, R. 1868. *On the anatomy of vertebrates. Vol. 3. Mammals.* London: Longmans, Green.

Reil, J. C. 1796. Von der Lebenskraft. *Archiv für Physiologie* 1:8-162.

Rensch, B. 1960. *Evolution above the species level.* New York: Columbia University Press.

Schwann, T. 1839. *Mikroskopische Untersuchungen über die Uebereinstimmung in der Struktur und dem Wachstum der Thiere und Pflanzen.* Berlin: Sanders.

Sewertzoff, A. N. 1916. Etudes sur l'évolution des Vertébrés inférieurs. I. Morphologie du squelette et de la musculature de la tête des Cyclostomes. *Archives russes d'anatomie, d'histologie et d'embryologie* 1:1-104.

—— 1931. *Morphologische Gesetzmässigkeiten der Evolution.* Jena: Fischer.

Stanley, S. M. 1973. An explanation for Cope's rule. *Evolution* 27:1-26.

Winsor, M. P. 1976. *Starfish, jellyfish, and the order of life: issues in nineteenth-century science.* New Haven: Yale University Press.

Severtsov and Schmalhausen:
Russian Morphology and the Evolutionary Synthesis
Mark B. Adams

The evolutionary synthesis had both an intellectual and a disciplinary aspect: we associate it not only with a theoretical integration of Darwinism and genetics, but also with a growing consensus among biologists of different disciplines that they could comfortably share, and operate within, a common theoretical framework.[1] In general terms, this common framework consisted in the view that evolution took place by natural selection acting on populations whose heritable variation was gov-

1. I should like to express my appreciation to S. R. Mikulinsky, Director of the USSR Academy of Sciences Institute of the History of Science and Technology, and B. V. Levshin, Director of the Archives of the USSR Academy of Sciences, for helping me to see some of the materials in the Severtsov archive in Moscow; and to A. A. Malinovsky and V. V. Babkov for giving me original editions of several of the books of Severtsov and Schmalhausen.

erned by combinations of genes: new alleles were added to a population through mutation and migration, and those combinations conferring advantage were selectively favored, resulting in biological changes in the population over time. The development of population genetics in the 1920s and 1930s provided theoretical and experimental demonstration that such processes could and did occur in natural populations. But the consensus went beyond this general agreement: it included the view that such a process was sufficient to account for evolutionary phenomena and that other proposed theories of evolution—saltatory mutations, Lamarckian factors, autogenetic tendencies—played no role and could safely be rejected.

However for many biologists, the solution to the *microevolutionary* question—provided by the development of population genetics—did not automatically entail a solution to the *macroevolutionary* question. Investigators may have demonstrated that allele frequencies in natural populations change over time, but this was hardly the same thing as demonstrating that such processes were sufficient to explain, for instance, the origin of vertebrates—a process presumably occurring over long stretches of time in the distant past and hardly subject to direct experimental confirmation. Biologists who agreed with Iurii Filipchenko (1882-1930), Russia's leading geneticist, that "the characteristics of higher systematic categories are in principle different from those of lower taxonomic units" were likely to conclude, with him, that "we do not have the right to apply mechanically what is established for the origin of lower units to the origin of higher ones" (Filipchenko, 1929, p. 256). And what of the tantalizing relationship between ontogeny and phylogeny, one of the central problems of late nineteenth-century morphology?

In this context the expansion of the evolutionary synthesis to systematics, paleontology, and morphology is especially intriguing, for these were the disciplines most intimately concerned with the macroevolutionary problem. Although those who founded population genetics in the 1920s shared the conviction that macroevolutionary changes were the result of microevolutionary processes, only Chetverikov—a butterfly taxonomist by training—attempted to address the special concerns of these macroevolutionary disciplines, and he had limited success (Adams, 1970). We associate the expansion of the evolutionary synthesis to these disciplines with works published well after population genetics was a going concern: for systematics, the works of Huxley (1940, 1942) and Mayr (1942); for paleontology, the works of Simpson (1944, 1953); for morphology, the works of Rensch (1947) and Schmalhausen (1946b, 1949).

In evaluating the situation in morphology on the eve of the synthesis,

Alexei Nikolaevich Severtsov is a particularly appropriate example for several reasons.[2] His active career spans the relevant period. He was born in 1866, just seven years after the publication of the *Origin*, and died in 1936 in the midst of the synthesis. From his first paper in 1891 through his last work written just before his death and published posthumously in 1939, Severtsov published 79 scientific works, of which 39 were in Russian, 32 in German, 7 in English, and 1 in French (see Severtsov, 1949, pp. 10-13). Although his early papers (1891-1910) were mainly technical studies of the *Entwicklungsgeschichte* of various organisms and organ groups (nerves, muscles, skeleton), including studies of metamerism, in 1912 he published the first of a series of major theoretical statements on the evolutionary process, *Studies on the Theory of Evolution: Individual Development and Evolution* (reprinted in 1921). Two years later, in 1914, he published his second major tract on a totally different subject: *Current Tasks of Evolutionary Theory* dealt with the origin of the vertebrates in connection with issues of progress, regress, and correlative variability. In 1925 his third major book appeared, *Principal Directions of the Evolutionary Process*; originally subtitled *Progress, Regress, and Adaptation*, by its second edition in 1934 it had gone from 83 to 146 pages and gained a new subtitle—*The Morphobiological Theory of Evolution*. In 1928 he published the 83-page essay "Directions of Evolution" —his only theoretical work in English. His major work in a Western language, however, was his 1931 text *Morphologische Gesetzmässigkeiten der Evolution*, a 372-page classic that embodied his most complete exposition of evolution. The last six years of his life were mainly devoted to preparing a Russian edition of this book, but he had expanded and reworked it so thoroughly that by the time of its posthumous publication in 1939, *Morphological Laws of Evolution* was over 600 pages and included two new introductory chapters on the history of evolutionary studies and neo-Lamarckism. Thus, Severtsov was most prolific in his exposition of the morphological approach to evolutionary theory at precisely the time when the evolutionary synthesis was gaining momentum.

Severtsov is also important because he was without doubt the best and most influential Russian morphologist of his generation. While serving as a professor at Dorpat (1898-1902), Kiev (1902-1910), and Moscow (1910-

2. Throughout this essay I have used a simplified standard system of transliteration for Russian names, except in those cases where a person is well known in the West by a different spelling. Severtsov's name also appears variously in Western sources with different spellings: Sewertzoff, Sewertzov, Severtzov, Severzov. All translations from Russian and German sources that appear in quotation are my own. In all quotations the italics are those of the original text.

1930), he trained a large school of followers. With some justice Soviet historians refer to him as the father of Russian evolutionary morphology: it would be difficult to find a major Soviet animal morphologist today who was not trained by Severtsov or one of his students. In 1930 he organized the Laboratory of Evolutionary Morphology of the USSR Academy of Sciences; in 1934 it became the Institute of Evolutionary Morphology and Paleontology, and was named the Severtsov Institute in 1936 while he was still alive, in honor of his seventieth birthday. Despite its various subsequent reorganizations and name changes, to this day it bears his name (it is now the A. N. Severtsov Institute of Evolutionary Animal Morphology and Ecology). Thus, Severtsov was responsible for setting into place the primary Soviet research institution in animal morphology, and current Soviet research in the field continues to bear his mark.

Throughout his career Severtsov was a strict Darwinian. He not only accepted evolution, but also held that natural selection was sufficient to account for the facts of evolutionary history. Considering his background, this belief is hardly surprising. Severtsov was the only son of the Russian explorer and naturalist N. A. Severtsov (1827-1885) who was one of the earliest and most outspoken of Darwin's supporters in Russia. Following the death of his father, the younger Severtsov studied and earned his degrees at Moscow University in the Department of Comparative Anatomy under the direction of M. A. Menzbir (1855-1935), a close friend of his father who had inherited his enormous collection of birds and was one of the most outstanding ornithologists of the last century. More important, throughout his career, Menzbir was an energetic defender of Darwin's natural selection theory, publishing many polemical articles defending it from attacks by creationists, experimental biologists, vitalists, orthogeneticists, geneticists, and others (see, for example, Menzbir, 1927).

When Severtsov was publishing his major theoretical works on evolutionary theory in the 1920s and 1930s, population genetics was being developed by Chetverikov and his group under the aegis of N. K. Kol'tsov, who had begun his career as a morphologist before turning to experimental biology. Although Severtsov and Kol'tsov had their differences, both intellectual and political (Severtsov had won his post at Moscow University only because both Menzbir and Kol'tsov had left in protest after the political takeover of the university by Kasso, the tsarist minister of education), they were friends from their student days in Menzbir's department, and worked in Moscow throughout most of their careers, keeping close professional contact. Thus Severtsov had the opportunity to become aware of Chetverikov's work and more generally of the developments in population genetics.

Finally, Severtsov intellectually influenced two of the evolutionary theorists most involved in relating the synthesis to macroevolutionary and morphological questions: Rensch and Schmalhausen. Indeed, Ivan Ivanovich Schmalhausen (1884-1963) studied under Severtsov, first at Kiev (1906-1910), and then at Moscow University (1910-1917), where he was appointed a senior lecturer in 1912. His own professional career showed a remarkable parallel to that of his teacher: like Severtsov, he became a professor first at Dorpat (1917-1920), then at Kiev (1920-1937), and finally at Moscow University, where he became the head of the Department of Darwinism (1937-1948). In 1920 he was elected to the Ukrainian Academy of Sciences (the same year Severtsov was elected to the Russian Academy of Sciences), where he created a laboratory of morphology and became involved in studies of embryology and growth. Elected to the Soviet Academy of Sciences in 1935 as Severtsov's heir apparent, Schmalhausen moved his group to Moscow in 1937 following Severtsov's death and directed the Severtsov Institute until 1948. Between 1938 and 1946, Schmalhausen published four classic books, only one of which has been translated into English, which establish him as one of the major evolutionary theorists of this century. The views of Severtsov, then, indicate a kind of morphology that had a great impact on one of the major evolutionary synthesizers.

A full explication of Severtsov's evolutionary viewpoint can be found in his biography (Severtsova, 1946) and most especially in his two major theoretical works available in Western languages (Severtsov, 1928, 1931).

Three questions about his views are important for the evolutionary synthesis. First, what did Severtsov understand to be the task of morphology and the status of its laws in relation to evolution, especially in light of the critiques of morphology from experimental biologists? Second, what did Severtsov understand to be the causes or mechanisms underlying the evolutionary process, and in particular, what stance did he take throughout his career on orthogenetic or neo-Lamarckian mechanisms? Third, what role did he see genetics playing in evolutionary theory, and how did his legacy become transformed in the works of Schmalhausen?

Laws, Causes, and the Tasks of Morphology

Severtsov selected morphology as a scientific career in the late 1880s. He entered the field after Haeckel and Gegenbaur had published their most influential work, and his view of morphology was largely shaped by them. Like them, he believed that "the primary obligation imposed by Darwin was less the ever-wider extension and validation of the hypothesis of natural selection than a demonstration and detailed elaboration of the historical record of life" (Coleman, 1976, pp. 149-150). As Severtsov said, "It is obvious that the first task which lies before the investigator, once he has become convinced that evolution really occurs in living nature, consists of attempting to establish this history of changes of living beings in as much fullness and detail as possible, i.e. in other words, to use the terminology of Haeckel, to solve the phylogenetic problem" (1914, p. 4). For him, this was one of two principal tasks facing the evolutionist: "(1) the study of the course of evolution and its laws, and (2) the study of the causes of phylogenetic changes in organisms" (1939, p. 80). The first was the task of morphology, using the three available methods set out by Haeckel—comparative anatomy, comparative embryology, and paleontology; the second was the task of physiology, the study of heredity, and the experimental sciences. In Severtsov's view, the two tasks were separate and distinct, but the primary task of the evolutionist was the first, and it was the one that he addressed throughout his career.

The Primacy of Morphology

Throughout his theoretical writings Severtsov reiterated his view that morphology was the fundamental evolutionary discipline, and as such was independent of any theory as to the cause of evolution. He first stated the matter plainly in the preface to his 1912 book:

> As much as possible I will try not to connect my conclusions to the acceptance of any theoretical explanation of the causes of evolution, i.e. Lamarckism or Darwinism, or any of the other theories of evolution common at the current time. I am not doing so because it seems to me that the theoretical investigation of evolutionary morphology represents an independent field of investigation, and the acceptance or non-acceptance of its theoretical results must not be made dependent on the acceptance of any theory on the causes of evolution . . . For a perfectly analogous reason, I do not consider it possible to make the results of morphological research of the type described depend on any particular theory of heredity (1921, pp. ix, x).

In expounding this conception of the status of morphological science, Severtsov used three lines of reasoning.

First, any theory as to the causes of evolution must be secondary in that the phenomena for which it is attempting to assign the causes must first be established by the morphological sciences: an account of the facts of evolution must precede an account of the causes of such facts.

Second, it followed that whereas morphology could be independent of any theory as to the causes of evolutionary change (because its task was to establish the historical evolutionary record), any theory as to the causes of evolution could *not* be independent of morphology, because it had to account for the actual changes that have taken place in evolutionary history. Morphology was not the only test of the adequacy of a theory about the causes of evolution, nor were the facts of morphology always helpful in choosing between such theories. But where it was germane, morphology provided the most pertinent test. Severtsov was careful to point out that in asserting the *independence* of morphology from causal theories, he was not suggesting its *irrelevance:* "I am in no way rejecting the possibility that the results are in agreement with the conclusions of Darwinism, Lamarckism, etc." (1921, p. x). But he held that the theories giving a causal account of the origins of the animal world, with which "many biologists are busily engaged . . . can function merely as abstract conceptions in the absence of a phylogenetic basis" (1939, p. 79).

Severtsov's third justification was that the morphological sciences were far ahead of the experimental ones in dealing with the evolutionary problem. He attributed this lead to the unpopularity of phylogenetic studies among a new generation of experimental biologists who, following Bateson and Johannsen, rejected such approaches as speculative and had failed to keep up with the latest results (1928, p. 60).

In his earliest theoretical books (1912, 1914) Severtsov was clearly aware of the challenge to morphology posed by the new experimentalist mood and seemed anxious to defend his approach. After discussing briefly various hereditary theories of European experimental biologists, he explained, "It is not possible for me to pause on these questions, despite their profound theoretical importance, since their consideration would divert us from our direct task: here I am not concerned with the question of *why* evolution proceeds, because for me the important question is *how* it proceeds" (1921, p. 97). This distinction between the "why" and "how" of evolution recurs in every subsequent theoretical work by Severtsov.

By 1914 he seemed even more sensitive to the challenge: by then, experimental biology was receiving considerable public play in Russia, thanks to the efforts of Filipchenko and Kol'tsov, and Western works in the field were being reviewed and translated in substantial numbers. In his 1914 book Severtsov felt compelled to explain that he was "consciously putting aside questions of the causes of evolution, of the Mutation theory, Mendelism, etc." because "they have become rather commonplace, since we have excellent surveys and popular accounts in this field, not only in the foreign literature but also in Russian; I am deliberately choosing to deal not with them, but with questions of a morphological character which are equally important and interesting, but much less well known" (1914, p. 3).

In defending morphology against the rising tide of experimentalism, Severtsov even tried to show that its practical implications for the human future were as grandiose as the claims made for the "new" biology:

> Let us imagine that by investigating the evolution of some multicellular organism that interests us, we succeed in determining which particulars of a given complex organ or organ system are primary variations and which are correlated with them. It is obvious that with such knowledge, to a considerable degree we control the process of evolution, and shall, under the circumstances, submit it to our will and our molding much more than now . . . It is completely evident that with such knowledge the power of man over organic nature, and perhaps over the evolution of man himself, would increase very significantly (1914, pp. 154-155).

In his postrevolutionary works, Severtsov never again held out the tantalizing prospect that the development of morphology might lead to the control of human evolution. But throughout his career he continued to believe that the laws of evolution were discernible by the science of morphology and that they were as valid as any other scientific law.

Experimentalist Critiques of Morphology: Filipchenko

In a historical essay written in 1931 Severtsov sought to account for the antipathy shown by experimental biologists to traditional morphology:

> Among experimental zoologists the opinion became widespread that valuable and, most important, exact results in the biological sciences could only be achieved by using the experimental method, and that the results arrived at by the descriptive sciences—comparative anatomy and embryology, paleontology, and zoogeography—were not precise, not scientific, many even said fantastical. This evaluation touched chiefly on phylogenetic theory. Even very great representatives of the experimental tradition, whose opinion a serious scientist had to reckon with, embraced these views. Hence the most outstanding geneticist in England—Bateson—expressed the thought that evolutionary theory is a subject whose truth is scientifically undemonstrable, and an analogous opinion was also expressed by that well known investigator of pure lines, Johannsen (1939, p. 74).

Although he did not mention them, Severtsov was perfectly aware that challenges to the validity of morphology had been posed much closer to home. Just as he was turning to evolutionary theory in 1910, his fellow morphologist Kol'tsov was rejecting morphology as a dead subject and mounting a campaign for the new experimentalist approach in zoology.

Later Kol'tsov wrote that "purely comparative and descriptive methods had lost their possibilities and their problematics" by that period, asserting that "only by uniting with the experimental methods of the new biological disciplines—especially developmental physiology and genetics—can the old comparative anatomy and embryology be reborn as creative sciences" (1936, p. 14).

Severtsov's characterization of experimentalist attitudes toward evolutionary theory fits even more perfectly Filipchenko, Russia's leading geneticist between 1910 and 1930. Filipchenko had been concerned with evolutionary theory since his first book in 1915, in which he had rejected all "experimentalist" approaches then current as explanations of macroevolution, and then, using analogies to the development of the solar system and a chick egg, asserted his preference for autogenesis (1915), a preference he maintained right up until his death. Nor was he ignorant of morphology: indeed, in 1923 he published a 288-page book on the history of evolutionary theory from Lamarck to his day, and in its second edition in 1926, he criticized Severtsov's 1912 book as being "a return to Geoffroy St.-Hilaire" and insufficiently based on experimental results.

For Filipchenko, reasoning by analogy was quite acceptable in macroevolutionary matters, because he felt that "evolutionary theory always was and always will be only hypothetical, since the transformation of species is not one of those phenomena which can be directly observed . . . Evolutionary theory to this day has a purely speculative character" (1929, pp. 249, 256). A year before, in a survey of Soviet genetics during its postrevolutionary decade, he had attributed the progress of genetics to its "excluding from its field of vision all those disputes over the evolution of organisms, their genealogical relationships, etc. which until this very day have literally torn apart both systematics and morphology" (1927, p. 3).

In 1929 Filipchenko made clear that he regarded the microevolutionary question as largely settled, thanks to the input of genetics—"in the current state of affairs there can not be much space for disagreement"—however, regarding macroevolution "by contrast, so far we know nothing at all precise, and this opens the field for the broadest speculation" (1929, p. 257). Then Filipchenko asked, "Do we need all of this evolutionary speculation?" He answered in the affirmative and presented a three-stage model of the development of science reminiscent of both Comte and Whewell: the investigation of each new problem begins with a speculative phase, moves on to a descriptive stage in which the problem "is illuminated by observational data alone," and finally in the third stage "achieves the status of precise experimental investigation . . . It is entirely probable that the problem of macroevolution will proceed through these three stages, just as the problem of microevolution did. And it is undoubtedly useful in this case for the given problem to go through a speculative stage, and one resting on observational data alone, since only after this can we hope to apply an experimental method to it. It seems to us that this constitutes the significance of current evolutionary propositions, and it is not really necessary to criticize them very sharply, even though such propositions are by now completely alien to genetics" (1929, p. 258). Given the intrinsically speculative character of the subject, then, Filipchenko felt justified in accepting an autogenetic viewpoint even though it lacked scientific proof.

For Severtsov, such a position could not have been more wrongheaded. For him, fact and argument, not the "experimental method," lay at the basis of science. Autogenetic views such as those to which Filipchenko seemed intuitively attracted *were* subject to critical scientific evaluation. Indeed, on numerous occasions Severtsov argued against such views, occasionally citing Filipchenko by name, to show their logical implausibility and their incompatibility with the facts of morphology and the laws of phylogeny.

It would be tempting to see Filipchenko's distinction between micro- and macroevolution as parallel to Severtsov's distinction between the causes of evolution and its laws, but to do so would miss a crucial difference: for Severtsov, determining the laws of evolution is the evolutionist's first and primary task and such laws had to be the basis for any view about the causes of evolution. These laws were anything but speculative: they could be precisely determined by comparing the wealth of data available to the studious morphologist. By contrast, it was theories of the causes of evolution that tended to be speculative, because the evolutionary record—the only firm evidential basis for evaluating competing theories—could not necessarily clearly and directly demonstrate which theory was true.

Two Conceptions of Scientific "Law"

Underlying the different perspectives of Filipchenko and Severtsov are differing concepts of the nature of science and of scientific law. Born in 1882 and educated in St. Petersburg, Filipchenko was Severtsov's junior by sixteen years and his views were those of a young experimentalist trained in Europe at the Naples Station and with Richard Hertwig at Munich in 1911-1913, fifteen years after Severtsov's first European pilgrimage. For Filipchenko, the contrast is not between law and cause, but between law and hypothesis or speculation. For him, a law can only be the product of an experimental science, as is clear from the following passage in which he contrasted evolutionary studies and genetics: "The first, at least at present, can never be free of the struggle between different views and contradictions, since the 'genesis of organisms' is not something strictly concrete which can be observed at will. By contrast, variation and heredity, on which the genesis of organisms is based, are sides of the general living activity of organisms which can be studied very precisely, and here there cannot and must not be any conflicting personal views and so forth; instead there must be sharply defined laws, permitting the prediction of phenomena and even their control" (Filipchenko, 1928, p. 4).

For Filipchenko, laws can emerge only from carefully controlled experiments designed to investigate carefully delimited problems. They have to be like Mendel's laws or those of genetics: precise, limited, predictive, subject to causal analysis by experiments on individual organisms, and thereby permitting control of the process they describe. For him, Severtsov's evolutionary laws were really simply speculative hypotheses, because they were not subject to experimental confirmation, could not lead to accurate evolutionary predictions, and could not lead to control of the evolutionary process. Filipchenko had the same notion of law used by the new generation of experimental biologists.

Severtsov's concept of law was very different: for him, the operating distinction was not between law and speculation, but rather between law and cause. For him, a law was a generalized pattern that had to be empirically determined in the best Baconian fashion. Once established, such a law set what had to be explained causally. Hence, the laws of evolution could *only* be established by phylogenetic studies: only the *causes* of these laws were subject to experimental investigation. This view of law and the distinction between law and cause have gone out of fashion today, but they were common in the nineteenth century. Darwin used the words in roughly the same way, as did Ernst Haeckel and Herbert Spencer. We can see the distinction clearly, for example, in Spencer's essay "Progress: Its Law and Cause" (1863). The law of progress is the historical tendency for things to become more heterogeneous and complex, as derived from the enumeration of countless examples from the sciences and history; the cause of this law, as given by Spencer, is that every cause produces multiple effects, and he attempted to show that this cause is sufficient to account for the law in all its various instances.

The terms Severtsov used for "law" were the Russian word *zakonomernost'* and the German *Gesetzmässigkeit*. Aside from the nineteenth-century English meaning of the word "law," these terms have no current exact English equivalent, but might be variously rendered as "lawfulness," "pattern," or "regularity." Throughout his writings Severtsov's concern is to elaborate the morphological or morphobiological laws of evolution in this sense—hence his 1931 title *Morphologische Gesetzmäs-sigkeiten der Evolution*, or the 1939 *Morfologicheskie zakonomernosti evoliutsii*. He discussed briefly what he meant by law in the 1931 preface:

> It is possible to ask whether one can speak here of some sort of laws. I emphasize that I use this term in exactly the same sense that my predecessors used it. Von Baer's law, A. Dohrn's principle of *Funktionswechsel*, etc. introduce definite order into our notions of how the process of phylogenetic changes occurs in many organs; it frequently is assumed that this process is the result of a few regular phenomena. In my opinion, it is not essential which term we use for these or similar generalizations: whether we call them rules, principles, or something else. *The significance of such generalizations consists in the fact that they permit us to classify better and understand better the course of phylogenetic change of animal organs than could have been done earlier* (1931, p. ix).

A brief overview of the laws of evolution that Severtsov presented during his career illuminates the nature of Severtsov's conception of morphology and his approach to the macroevolutionary problem.

Severtsov's Laws

Severtsov's theoretical writing deals with three subjects. The first is a theory of correlative variability, based on a modification and extension of the work of Haeckel and other subsequent researchers on the biogenetic law and the relationship between ontogeny and phylogeny. The second is an attempt to specify the laws or principles manifested in the changing structure and functions of organs and organ systems as they evolve over time. The third is an attempt to classify the overall patterns manifested in phylogeny. Severtsov formulated laws relating to the second and third of these topics.

Severtsov divided the progressive changes in organs and their functions into two basic types: those changes in which the organ maintains the same primary function or functions, but in which these functions change quantitatively; and those changes which are qualitative, in which the primary function of the organ changes. In his final version of his theory (1939) the first group consisted of six principles or laws, of which three were his own creation: (1) the intensification of function (Plate); (2) the fixation of phases; (3) the loss of intermediate functions; (4) the substitution of organs (Kleinenberg); (5) physiological substitution (Fedorov); and (6) a reduction in the number of functions. The second group, in which the primary function of the organ changes, included seven principles, of which Severtsov claimed to have originated five: (1) the broadening of functions (Plate); (2) the replacement of functions

(Dohrn); (3) the substitution of functions; (4) the activation of functions; (5) the immobilization of functions; (6) the simulation of functions; and finally, (7) the dissociation of organs and functions. These general principles were exemplified in his discussions of the evolution of major groups and the morphological changes they had undergone in evolution.

Severtsov's more general laws of phylogenesis are an attempt to classify and understand the modes of biological progress. For him, biological progress meant *"the survival of the species and the favorable existence of their descendants,"* which was characterized more specifically by three features: "(1) *by the increase of numbers of individuals of the given systematic groups*; (2) *by progressive migrations, i.e., by the hold of new geographical areas*; (3) *by the breaking up into new subordinate systematic groups"* (1928, p. 27). Such biological progress could take place in any of four general ways, according to Severtsov.

The first is *aromorphosis*, which he also called "morphophysiological progress." In this mode descendants are characterized by an increase in general (versus specific) adaptation, a rise in the intensity, variety, and complexity of organ functions, and a general rise of vital energy and activity, all of which taken together raise the organism to a "higher level of organization." One example Severtsov gave of aromorphosis was the development of external gills, permitting more complete oxidation of the blood. A second mode is *idioadaptation*, in which organisms adapt to specific and definite environmental conditions within which they live, but in such a way that there is no increase in general activity or vitality and no increase in the variety or complexity of organ functions not associated with the fixed environment. Specialization is, for Severtsov, one form of idioadaptation. A third mode is *degeneration*, or "morphophysiological regress," the loss of all that is gained in aromorphosis—a decrease in the intensity, variety, and complexity of organ functions. Two examples Severtsov gave are the development of parasitism and the development of sessile characteristics in formerly free-moving forms—both of which can constitute biological "progress," in his sense of the word.

Severtsov frequently illustrated these principles by the use of two charts reproduced here (figures 1 and 2). The first showed a general phylogenetic branching tree with the various animal groups indicated, in a manner popularized by Haeckel. The second added a third dimension to the diagram, with movements upward to a higher plane representing aromorphosis, downward representing degeneration, and branchings within a given plane representing idioadaptations, on which sharply and obliquely jutting lines are specializations. By the late 1920s, Severtsov had added a fourth mode: *coenogenesis*, or embryonal adaptation, in which changes have occurred in embryological development without any

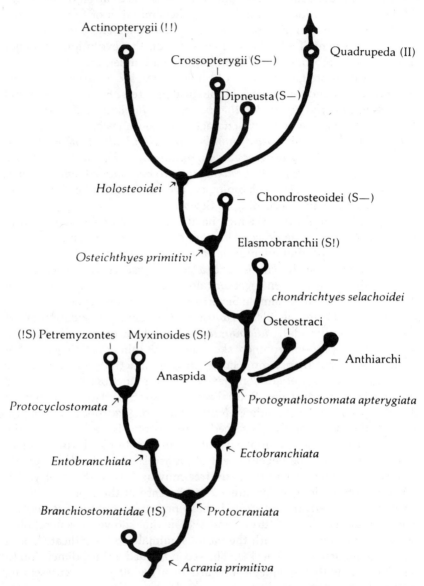

Figure 1   Two-dimensional phylogenetic tree. From "Die Factoren der progressiven Entwickelung der niederen Virbeltiere," *Russkii zoologicheskii zhurnal* 4, nos. 1 and 2 (1924):55.

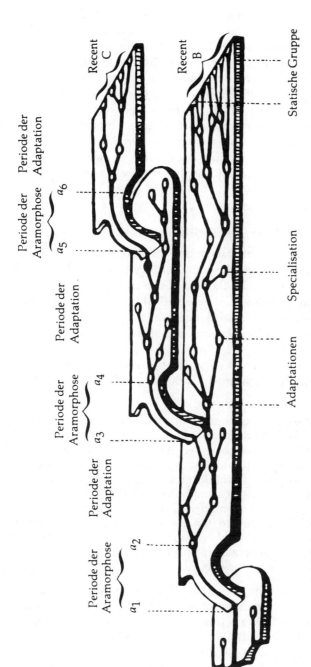

Figure 2 The earliest published version of Severtsov's three-dimensional phylogenetic tree, illustrating his concepts of aromorphosis, adaptation, and specialization. From "Die Factoren der progressiven Entwickelung der niederen Wirbeltiere," *Russkii zoologicheskii zhurnal* 4, nos. 1 and 2 (1924):56.

changes having occurred in the organization of adult descendants. However, he never tried to include this mode in any diagram.

Severtsov used this language in describing the overall course of evolution of the animal kingdom, and especially the emergence of the vertebrates. By examining successive changes in modes of biological progress which various groups and their descendants are undergoing in successive periods, he noted a general law: no group ever went from specialization to aromorphosis, or from degeneration to aromorphosis, whereas all other directions were possible and had occurred.

These, then, were the sorts of laws to which Severtsov devoted his career, laws whose explication was, in his view, the primary task of the evolutionist.

Severtsov on the Causes of Evolution

Although Severtsov devoted most of his theoretical writing to deriving, elaborating, and discussing the actual history of animal life and its laws, he did discuss his views on the various causal theories of evolution in brief passages in all of his major theoretical books, usually at the beginning. These discussions included a general classification of such theories and critical evaluations of them in terms of their compatibility with morphology.

In 1912 Severtsov presented a classification of theories about the causes of evolution that framed his thinking on the subject throughout his career:

> Relative to the causes of the evolution of adaptations in adult animals, it is possible to make two assumptions, both of which have been set forth and have their defenders in the scientific literature: first, it is possible to suppose that evolution proceeds on the basis of a few *immanent* principles of development, located within the developing organisms themselves and independent of changes in the surrounding environment (von Baer, Cope, etc.); second, it is possible to suppose that the key to the evolutionary process is given by *changes in the external environment*, in the broadest sense of these words—that is the general principle of the theories of Lamarck and Darwin and all the newer variants of these theories (1912, p. 96).

Later in the same passage, Severtsov distinguished between "the indirect influence of the surrounding environment on animal organisms" (Darwinian factors), the "direct action of the environment" (Buffon), and "the use and disuse of organs" (Lamarck). By 1925 he employed the term *autogenesis* for theories postulating immanent internal principles of development, and *ectogenesis* for environmental theories, dividing the latter into three types: *neo-Lamarckism* (use and disuse), *ectogenetic orthogenesis*

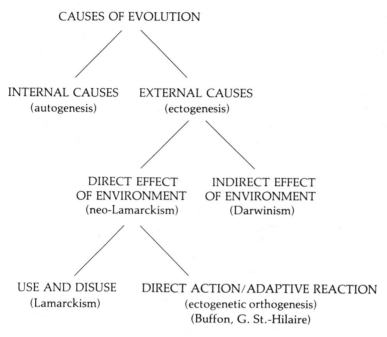

Figure 3 Severtsov's classification of theories on causes of evolution.

(direct effect of the environment), and *Darwinism* (indirect effect of the environment by natural selection) (1925, pp. 68, 69). Severtsov's classification is schematized in figure 3. Beginning in 1912, and throughout his career, Severtsov dismissed autogenesis and supported ectogenesis, regarding Darwinian natural selection as by far the most powerful causal theory in accounting for evolution: "I consider that the indirect influence of the surrounding environment on animal organisms (Darwinian factors of evolution) has a far greater significance for phylogenetic evolution than the results of the direct action of the environment and the use and disuse of organs (the factors of Buffon and Lamarck)" (1912, p. 97).

Other morphologists of Severtsov's generation tended to support non-Darwinian theories, which had widespread popularity at the time. Severtsov's critique of them is intriguing because of the way in which he assembled the facts of comparative morphology and phylogenetic history against them.

Autogenesis
In a lengthy footnote in his 1912 book Severtsov argued against autogenesis on logical grounds. He regarded his argument as fatal to the theory:

I consider the internal (immanent) principle of evolution improbable (though not *a priori* impossible) on the basis of the incontestable general law of the *adaptation of organisms to their surrounding environment*. In accepting this immanent principle of evolution, we must also accept along with it the notion that in nature there exists some sort of *pre-established harmony* (in Leibniz's sense) *between the evolution of organisms and changes in the environment*, i.e., it would be necessary to suppose that every change in the environment (and they are innumerable) corresponds to a previously established, highly distinctive, and appropriate change of organization. By hypothesis, these changes must be independent of one another, i.e., using a well-known example, we would have something analogous to two clock mechanisms which are not connected with one another but are so regulated from the beginning by a prime watchmaker that they always move in agreement, each neither moving ahead of, nor falling behind, the other. But in living nature we must suppose an improbably large number of the most diverse sorts of clock mechanisms (living organisms), of which each is regulated by a whole series of different mechanisms of different constructions (changing external conditions)—and all of this such that all these clock mechanisms not only run perfectly, but also run fast or slow in harmony with one another. Such an account of the evolutionary process is at the very least improbable, and, if we ponder it, we will see that as an "explanation" it gives us nothing, since in essence we find that it only expresses the fact of organic adaptation, not an explanation of it (pp. 96-97).

This same text appeared verbatim in Severtsov's subsequent books, each time followed or preceded by the statement "To this day I have found no grounds, either in the literature or in my own research, for changing my point of view" (see, for example, 1925, p. 68).

In 1925, however, Severtsov added a second argument against autogenesis, this time derived from phylogenetic laws. Pointing out that autogenesis cannot be used "in accounting for changes in the direction of the evolutionary process, i.e., the fact that in a given period of its evolution, a group undergoes progressive evolution, whereas in a subsequent period it regresses," Severtsov asked: "Must we then assume that the very same principle acts in one period of time in a given direction, and in the next period in the opposite direction?" In a highly suggestive footnote to this passage, he made clear that he regarded mutationist or saltationist theories such as those set forth by Korzhinsky, de Vries, and Filipchenko to be autogenetic in character and subject to the same refutation. After criticizing autogeneticists for the vagueness of their formulations, he commented: "In the best construction, one could come to the position that there exist definite directions of hereditary variability (in current terminology, mutations); but even in this case, it would be possible to conclude that in two subsequent epochs the direction of change of a given organ or organ system should be one and the same" (1925, p. 69). In 1925 he added: "The question as to whether hereditary changes are truly autogenic, or whether this process is directly dependent on changes in the external environment so far still remains open"; in the 1934 edition, presumably aware of Muller's work on the mutagenic effect of X rays, he dropped this sentence (1934, p. 122).

Finally in 1927, in "The Historical Trend in Zoology," Severtsov added his final argument against autogenesis, also derived from his knowledge of phylogenetic history: "This view can hardly be reconciled with the often observed fact that one line of descendents of some group which has achieved an acknowledged level of high organization continues to develop progressively, at the same time that some other branch, descending from the very same ancestors, begins to regress and degenerate. We know many such examples . . . It is hardly possible to suppose that the general ancestors of these divergent evolved groups were invested simultaneously with both a principle of progressive evolution and a principle of degeneration" (1927, p. 179).

At the beginning of the second chapter of his 1939 book Severtsov dismissed autogenesis as "having no current significance at the present time" (1939, pp. 81, 82). When Severtsov wrote this line, Filipchenko had been dead for six years, and had been castigated for his autogenic views—both before and after his death—by Marxist philosophers and political ideologues who found them bourgeois, vitalistic, and idealist. By contrast, many Marxists had found neo-Lamarckian views philosophically appealing (Gaisinovich, 1968), and Lysenko and his followers were mounting an offensive against established genetics and in favor of their own "Michurinist" version of neo-Lamarckism.

Ectogenesis: Use and Disuse
In his books of 1914 and 1939, Severtsov discussed at some length the importance of use (*uprazhnenie*) and disuse (*neuprazhnenie*) of organs in evolution. His views did not significantly change between the publication of these books. In both places he used two basic arguments.

First, Severtsov argued that there existed no evidence whatever for the inherited effects of use and disuse: "The first and chief [difficulty for Lamarckism] consists of the fact that, in spite of the most careful investigation, we know of no verified cases of inheritance of the results of the use and disuse of organs . . . All the facts which could lend themselves to such proof can without exception be accounted for without assuming that such effects are inherited. To the contrary, in those cases where we might expect such inheritance, we do not find it" (1914, pp. 101-102). He elaborated on this final point using three examples. First, if the results of use were inherited, domestic animals, such as work horses, would be expected to develop stronger muscles independently of artificial selection, but "we do not find this to be the case and for developing new breeds, distinguished by their power and ability to carry loads, we must rely, not on exercising the stock, but on artificial selection, often omitting exercise entirely" (1914, p. 102). Second, "Mechanical injury, even in those cases where it is performed over the course of many generations, is not inherited" (p. 102). Finally, the degeneration of eyes of animals that have lived in the dark for many generations has not become hereditary: "As it happens, the Proteids raised not in darkness but in light develop normal eyes, i.e., degeneration occurring over long periods of time in the course of an incredibly large number of generations living in darkness has not become hereditary, and the apparently degenerated organ reappears as soon as does the stimulus for its development (light)" (1914, p. 104).

Severtsov's second basic argument is the most interesting in helping us to see his use of the facts of phylogenetic history and his expertise in morphology to test Lamarckian views. For he maintained that, even if the inherited effects of use and disuse were to be demonstrated, they could not play a major role in the evolutionary process. Let us consider, as he did, the various organ systems that might be involved.

PASSIVE ORGANS: "Is it really possible to imagine that the carapace of many fishes, the scales of Crossopterygii and Lepidosteoidei, the shells of tortoises, the exoskeletons of arthropods and insects, the shells of molluscs, etc., developed as a result of use and disuse? There are a great number of such passive organs and they have an enormous biological significance, and undoubtedly they have changed greatly in the course of the evolutionary process. We simply cannot imagine that such organs can be exercised" (1914, p. 100).

RARELY USED ORGANS: Severtsov argues that use and disuse cannot explain "the development of organs which are rarely employed." "It is well known that the males of many insects perish after the sex act, i.e., this act is done only once in a lifetime, but they have a large number of secondary sexual characteristics which undoubtedly develop progressively; we can hardly consider seriously that actions completed once in a lifetime by animals that die immediately thereafter can be considered 'use' sufficient for progressive evolution" (1914, p. 101).

THE NERVOUS SYSTEM: In his 1939 book Severtsov added to his arguments that use cannot play a role in the evolution of nervous systems, because "nerve cells don't multiply in adults" regardless of how much they are used.

SEGMENTARY ORGANS: He also pointed out that segmentary organs, such as muscle and skeleton, often develop evolutionarily by metamerism— that is, by an increase (or decrease) in the number of segments, a development that can hardly be explained by the exercise of the segments already present.

Severtsov titled a section of a chapter of his 1939 book "The Inadequacy of the Principle of Use as an Explanation for Progressive Evolution." He concluded by remarking that, in the most positive construction, use could account for only "minor aspects of the skin, bones, muscles, and the digestive system," all of which could be better explained in other ways. In the next, briefer section, titled "The Inadequacy of the Principle of Disuse as an Explanation of Regressive Evolution" (1939, pp. 92-94), he made four basic points. First, he reiterated his argument concerning the atrophy of the eyes of animals living many generations in the dark, this time citing experiments with *Drosophila* and other organisms to demonstrate that when their progeny are returned to light, the eyes develop normally. Second, he pointed out that, where reduction occurs in an organ over time, "it frequently occurs at the most used section." Third, he pointed out that because some organs function continuously— for example, the heart and circulatory system—reduction in these organs cannot be explained by disuse because the organs are used and exercised throughout the life of the organism. Finally, he argued that the atrophy of organs in evolutionary history always occurs with a "maintenance of function"—that is, other structures take up the function formerly performed by the atrophying or reducing organ.

All these considerations led Severtsov consistently to reject the notion that inherited effects of the use and disuse of organs were a plausible explanation of evolutionary change: "These and analogous facts argue against accepting that the changes resulting from the use and disuse of organs are inherited . . . It is clearly evident that we must . . . entirely reject his position, which seemed so plausible to earlier investigators."

Ectogenesis: Direct Effect of the Environment

In his 1914 discussion, Severtsov clearly appears to accept the view that the direct action of the environment can lead to heritable changes: "A large number of changes in the environment act directly on organisms and produce changes in their structure. Frequently such changes are not inherited, but in cases where they are profound and reach the sex cells, changing them correspondingly, there also appear *hereditary changes resulting from the direct action of external conditions.* At the present time we have a large number of examples of such changes, brought about experimentally, and in many cases these changes are made hereditary. I recall the experiments of Standfuss and Fischer on the action of temperature on the coloration of butterflies, the experiments of Kammerer on changes in the coloration of salamanders under the action of light, and very many other cases" (1914, p. 134).

However, as he made clear in an earlier footnote in the same book, Severtsov doubted that such factors have any general significance for explaining the evolutionary process, and particularly adaptive evolution:

> Apparently, recent investigations show that several results of changes in conditions are undoubtedly hereditary and play an important role in the evolutionary process. But this position is incredibly difficult to accept as a principle to explain the evolutionary process as a whole in its entirety, precisely in view of the fact that the influence of changes in the environment cannot lead to an understanding of changes of an adaptive character, in particular complex adaptations. It is difficult, for instance, to imagine how and as a result of what changes in the environment the limb of a lizard, or the paw of a mammal, changed into the wing of a bird or of a bat, leaving aside of course the hypothesis of the use and disuse of organs . . . which does not survive criticism for other reasons. Further, the explanation of the interrelationships between different organ systems in different beings is a great difficulty for this hypothesis, for instance, changes in the mouth parts of insects and corresponding changes in the structure of flowers whose pollen they carry, and so forth (1914, pp. 105-106).

Thus, although accepting the inheritance of acquired characteristics as something that can occur, Severtsov doubted its general applicability in explaining evolution.

Not until his final work, written shortly before his death, did Severtsov take up the question again in any detail. There, in two sections of a new chapter "Neo-Lamarckism," he dealt with what he now termed "the principle of adaptive reaction" of organisms to changes in the external environment. "The essence of this hypothesis," as Severtsov understood it, "consists of the proposition that animals and plants have a fundamental ability to react adaptively or purposefully to changes in external conditions, i.e., to spontaneously elaborate the adaptations to the changed conditions of existence necessitated by each new circumstance" (1939, p. 86). After elaborating the case for this view in some detail, concentrating on the views of Weidenreich, Severtsov argued that although changes in the external environment do result in changes in animal and plant structure, these changes are not necessarily adaptive. After again accepting the results of Standfuss and Kammerer, Severtsov posed three questions: "Can the changes called forth by the direct action of the environment be inherited?" He answered that "at the present time this question cannot be regarded as settled, and even so Neo-Lamarckists have failed in their attempt to convince their opponents that the inheritance of these so-called acquired characteristics is the rule rather than the exception" (1939, p. 98). "To what degree are changes produced by changed environmental conditions adaptive [*prisposobitel'nye*]?" He responded that this thesis, "so important for Neo-Lamarckism, can scarcely be regarded as proved." "Do all environmental changes, both inorganic and biological, directly influence organisms by changing their form?" Here too, his conclusion was negative.

He elaborated the difficulties in applying the principle of adaptive reaction to complex changes in organisms, asserting that "for more complex changes, this explanation—if one does not give it some mystical significance—is very difficult indeed to apply," and discussed the difficulty in the case of the membrane covering birds' eyes and the musculature that controls it. His chapter concluded: "We see that this principle, like the principles of organ use and disuse, contributes very little to an explanation of evolutionary phenomena" (1939, p. 102).

Severtsov summarized his view on the status of neo-Lamarckism with regard to evolutionary theory. After criticizing neo-Lamarckists for "their insufficient attention to experiments and observations," he concluded that neo-Lamarckism "cannot presume to be a single, general theory accounting for animal evolution," adding: "In the best case, we could explain on the basis of neo-Lamarckism only a few very simple examples of the evolutionary changes in organs, examples which can be entirely accounted for on the basis of other principles, for example on the basis of the theory of natural selection" (1939, p. 102).

Severtsov's developing treatments of various theories about the causes of evolution exemplify his concept of morphology as the primary evolutionary science. Phylogenetic history and the laws governing that history, which he sought to elaborate throughout his career, could be established scientifically, notwithstanding the doubts of experimentalists like Filipchenko. Besides, these laws were and had to be the ultimate arbiter of theories explaining that history. In arguing against autogenesis, Severtsov first combined reasoning with general observations on adaptations, but later would marshal the facts and laws of morphology against it. Throughout his career, Severtsov would do the same for neo-Lamarckism. At times Severtsov seemed to grant that an inheritance of acquired characteristics could take place and seemed to accept Kammerer's conclusions at face value. But as a morphologist, Severtsov was simply not interested in neo-Lamarckism as a theory of heredity or the inheritance of acquired characters as a kind of hereditary phenomenon: he was interested in them only as theories of the causes of evolution. And for Severtsov, morphology spoke against all such theories as general accounts of the evolutionary process. For him, there was only one general theory of the causes of evolution that could account for all evolutionary phenomena: Darwinism and natural selection.

Severtsov's views of these theories help us to understand why his brand of morphology was preadapted to the coming synthesis. If one characteristic of the synthesis was that Darwinian natural selection, as seen to operate in population genetics, was sufficient to account for evolutionary phenomena—that no alternative theories were called for—then the synthesis met in Severtsov's works a form of morphology rendered in exclusively Darwinian terms. All that was missing was the population genetics.

Genetics, Morphology, and the Synthesis

Severtsov was sixty-one years old when Chetverikov's classic paper of 1926 was published. Then only four years from retirement at Moscow University, Severtsov would still publish three major theoretical books and numerous articles before his death in 1936, in addition to preparing a fourth book that was published posthumously in 1939. These works show an increasing recognition of the importance of genetics for evolutionary problems and express the hope that genetics would make important contributions to evolutionary theory.

In his 1932 essay "Evolutionary Studies after Darwin (1859-1930)" (republished in slightly altered form as the first chapter of the 1939 Russian edition of *Morphological Laws of Evolution*), Severtsov gave his view of the history of evolutionary theory and its changing interrelations with genetics:

> The gradual deepening of researches of the experimentalist trend, chiefly of geneticists, led investigators to the same problems which faced zoologists and botanists of the 1850s, i.e., the problem of evolution and its causal explanation . . . It was precisely the growing research of geneticists and the enormous successes which they achieved which led them to pose the evolutionary question . . . It

was possible either to consciously renounce these problems, or to try to apply the laws of genetics to the evolutionary process. The study of pure lines (Johannsen) at first led geneticists to the view that hereditary factors do not change over time, i.e., to the hypothesis of the constancy of the species. However, the study of spontaneously arising mutations and the discovery of the dependence of hereditary mutations on changes in the external environment (temperature, humidity, X rays, etc.) led once again to the possibility of approaching the evolutionary problem (1939, pp. 75-76).

Severtsov's main critique of the experimentalists had always centered on their attitude toward evolution, and his interest in their work increased with their involvement in evolutionary questions.

Furthermore, as Severtsov made clear, he had come to regard genetics as an unexpected ally of Darwinism: "At the present time we must not forget the enormous positive significance of the researches of geneticists for the explanation of the evolutionary process, i.e., for Darwin's theory of natural selection . . . Currently intensified work has begun on dominant and recessive mutations, and on the frequency and causes of their occurrence. Evidently, these studies are the beginnings of a revival of the Darwinian approach, to which paleontologists, attached to Neo-Lamarckist views, were very negative . . . We have every reason to think that the broad ranging and valuable material which geneticists have already obtained, and the results of further researches relating to mutations, will prove useful to the supporters of natural selection" (1932 [1939], pp. 75-76).

Two years later, in the 1934 edition of his 1925 book *Principal Directions of the Evolutionary Process*, Severtsov once again called on the results of genetics against Lamarckian views and in support of Darwin: "At the present time, under the influence of more detailed and penetrating studies of the laws of heredity, it has become clear that the chief postulate of neo-Lamarckism—the heritability of the effects of the use and disuse of organs—is not confirmed by the results of the experimental study of heredity, and consequently first place has once again been assumed by the theory of natural selection," and in 1934 added the phrase, "greatly strengthened in recent years by all the well-known work of geneticists" (1925, p. 70; 1934, pp. 133-134).

In one of his last publications, "Modes of Phyloembryogenesis" (1935), Severtsov gave his final view of what a general theory of evolution would look like and the roles that morphology and genetics might be expected to play in its formulation: "Up until this time we have followed the approach of morphology . . . Of course, one must recognize that this theory is only a *first approximation* of a full theory of the means of evolution. At the present time, we do not have this full theory, and we can only guess how it will be created. It seems to us that in the near future, ecologists, geneticists, and developmental biologists must move toward the creation of such a theory, using their own investigations, based on ours. The theory created by them will probably be a *second approximation* to a complete theory of evolution. We still do not know when this will come about, since works of this type are only now beginning and have yet to come up with results" (1945, p. 523).

Despite the results already arrived at by population genetics and the role that Severtsov expected genetics to play in evolutionary theory, he did not see genetics as having accomplished much. He reiterated this theme in his last major work: "Despite the brilliant successes in hereditary theory, the results of genetic research have brought little to the solution of evolutionary questions. Experimental embryologists have done even less in this direction" (1939, p. 80).

Severtsov's attitude toward genetics was less a product of his own reorientation toward experimentalism than a response to the changing orientation of geneticists toward evolutionary theory. He had bemoaned the antagonism of such geneticists as Bateson, Johannsen, and Filipchenko toward Darwinism, but came to applaud genetics when it began to disprove neo-Lamarckian views, confirm natural selection, and favor evolutionary theory as he knew it. However, genetics had still not really provided a theory of the causes of evolution in Severtsov's terms because genetics, and even population genetics, had still not made any progress toward explaining the major macroevolutionary phenomena: the general paths of phylogenetic history and the changes in their modes; the origin of higher groups; the nature, causes, and sequential appearance of progressive or regressive evolution in the histories of various animal groups; or the relationship between ontogeny and phylogeny. These were the large, traditional evolutionary problems of the nineteenth century, and they remained the problems of central concern to a morphologist like Severtsov.

At no place in his writings did Severtsov make any attempt whatever to integrate genetic explanation into his morphological theories: morphology as he practiced it had not as yet become part of an evolutionary synthesis. Yet it is indicative of the degree to which he prepared the way for the synthesis in his own country that Schmalhausen, his star pupil, attempted just such a synthetic integration in four major theoretical works published within a decade after his teacher's death.

Schmalhausen and Lysenko: The Fate of the Synthetic Tradition
Throughout his early career until the mid-1930s, Schmalhausen's research closely mirrored his teacher's. Although Severtsov's studies in

comparative morphology were concerned primarily with the origins of fishes, Schmalhausen concentrated also on the origins of amphibia and reptiles. Like his teacher, he saw his primary task to be the clarification of phylogeny, concentrating on the evolutionary origins of major animal groups using comparative anatomy, embryology, and paleontology. However, during the Kiev period (1920-1937), Schmalhausen had established a laboratory of phyloembryogenesis, and had become increasingly interested both in experimental embryology and in the implications of genetic research for evolutionary theory. With a few exceptions (published in Ukrainian), this interest was not manifest in any of his publications prior to Severtsov's death, however, when Schmalhausen was very much his teacher's protégé and being groomed as his heir apparent. His publications during this earlier period rarely stray from the problematics established by Severtsov and from his approaches, and all give major play to him as a great founder of evolutionary morphology.

However, as the new director of the Severtsov Institute following Severtsov's death, Schmalhausen immediately launched himself on an attempt to recast his teacher's work in new synthetic terms. Schmalhausen became a full academician of the Soviet Academy of Sciences in 1935 and director of the Severtsov Institute in 1937. Within the next ten years, in addition to preparing new editions of his classic textbook, *The Fundamentals of the Comparative Anatomy of Vertebrates* (1947), he published four wholly new theoretical books. *Problems of Darwinism* (1946a) was a university textbook used throughout the Soviet system until 1948, and probably represented the most ambitious and successful attempt made anywhere at that time—East or West—to integrate all areas of evolutionary studies into a single synthesis with natural selection as its core. The other three books represented a deliberate attempt to recast the Severtsov tradition in modern terms: each took up in turn the subject of a comparable theoretical book that Severtsov had published.

Severtsov's first major theoretical work had been subtitled "Individual Development and Evolution" (1912) and had taken as its central problem the relationship between ontogeny and phylogeny and the elaboration of his theory on correlation. Schmalhausen's first major theoretical book, *The Organism as a Whole in Individual and Historical Development* (1938), took up precisely the same problem. Indeed, the major differences between the two derive from Schmalhausen's own work in embryology and are reflected in the literature that each cited in its arguments. Schmalhausen's book cites papers in population genetics by E. I. Balkashina, N. P. Dubinin, R. Goldschmidt (four), C. Gordon, V. S. Kirpichnikov (two), H. J. Muller, F. B. Sumner, N. W. Timoféeff-Ressovsky (two), and S. Wright (three), as well as seven other papers on *Drosophila* genetics.

Severtsov's second major theoretical work, *Principal Directions of the Evolutionary Process: Progress, Regress, and Adaptation* (1925), was a highly developed version of his 1914 book, dealing with the overall patterns of phylogenetic change. Schmalhausen's second major theoretical book was *Trends and Laws of the Evolutionary Process* (1939) which dealt with precisely the same problem set. Indeed, in its preface Schmalhausen makes the parallelism explicit: remarking that his first book had been written "under the influence" of Severtsov's earliest work, Schmalhausen stated that this second book was written "under the influence" of Severtsov's 1925 book. Schmalhausen's second book is even more infused with the results of genetics than his first one and expressly mentions Chetverikov's theoretical and experimental papers. It also includes the first appearance of Schmalhausen's theory of "stabilizing selection."

Severtsov's final book (1931, 1939) had been his most general presentation of his evolutionary theories as a whole. Like this work, Schmalhausen's *Factors of Evolution (The Theory of Stabilizing Selection)* (1946b) is also Schmalhausen's broadest theoretical statement and shows the greatest divergence from his teacher: over half its references are to works in genetics or population genetics, with an average of six citations apiece to Alpatov, R. L. Berg, Gershenson, Dubinin, Kamshilov, Rapoport, and Timoféeff-Ressovsky. Dobzhansky is referred to for the first time. In the 1949 American edition of this book its translator and editor, Dobzhansky, characterized it as supplying "an important missing link in the modern view of evolution" (Schmalhausen, 1949, p. ix).

If we examine the state of evolutionary thinking, training, and research in the Soviet Union on the eve of World War Two, we cannot help being impressed with the degree to which ideas labeled in the West as the "evolutionary synthesis" had taken hold. In the first phase of the synthesis—population genetics and the microevolutionary problem—the Soviet Union had been a leader, both in early statements of a synthetic theory and in the pioneer work in experimental population genetics (Adams, 1968). In the second phase of the synthesis—extending it to the macroevolutionary problem—Darwinism was widely understood and accepted by most Russian researchers as a generally correct and sufficient account of the causes of the evolutionary process. Many traditional institutions within the academy paid little attention to evolutionary genetics but, under Schmalhausen's direction, the Severtsov Institute was busily engaged in extending the evolutionary synthesis into embryology and morphology. Indeed, it can be argued that he and his students were well launched on an attempt to integrate the results of population genetics with embryology and morphology in a more ambitious way than had been attempted in the West, or would be for decades to come. In 1939 it would have been difficult to find a country in the world where evolutionary theory was more modern or more widely diffused than it was in the Soviet Union.

The end of this success story is tragically ironic. Severtsov died on December 19, 1936, a date memorable in the history of Soviet biology for yet another reason: it was the date of the opening session of the first meeting of the Lenin Academy of Agricultural Sciences where Lysenko, Prezent, and their supporters first launched their attack on Soviet genetics. As their campaign against classical genetics gained momentum, its effects were felt increasingly wherever genetics had taken hold. And because the evolutionary synthesis had been developed with such vigor in the Soviet Union, Lysenko's campaign damaged institutions and disciplines that would surely have been regarded in the West as rather far afield from genetics and agriculture.

Their campaign culminated at the August 1948 meeting of the Agriculture Academy. Ironically, Severtsov was portrayed at the meeting by Lysenko's followers as a great hero of Soviet science in whose path they were following. According to Eichfeld: "Soviet agrobiologists are developing Severtsov's conclusions and are not ashamed of being consistent followers of A. N. Severtsov" (Lysenko, 1949, p. 70). Prezent even commented that Lysenko's work was an example of Severtsov's principle of "aromorphosis": "But we, modest Michurinists, have taken up and are studying this problem from the angle of the theory of biological progress elaborated by another eminent Russian scientist, A. N. Severtsov" (Lysenko, 1949, pp. 586-587).

In contrast to Severtsov, Schmalhausen had infused his work with genetics, and as a result he (together with Dubinin) was one of the major living objects of attack by Lysenkoists. He was attacked by the philosopher M. B. Mitin for accepting idealist genetics and by Prezent for being an autogenist and for betraying Severtsov's tradition: "If Severtsov did anoint Schmalhausen, then it must be admitted that the anointed has not justified the very honourable dignity conferred upon him. It may in truth be said that Academician Schmalhausen, under the guise of 'continuing' Severtsov's work, only multiplies and classifies words, pretending that he is developing Severtsov's theory" (Lysenko, 1949, p. 594).

Schmalhausen arose from a sickbed to defend himself on August 6. According to the version of his speech published in the "stenographic record" of the meeting, he read from his works, attempting to show that the criticism aimed at him had been unjustified and that he was a continuer of Severtsov's work, and dissociating himself from genetics: aside from his work on the phenogenetics of racial characters in domestic fowl, "none of my other work has, or ever has had, any connection with genetics, still less has my work any connection with formal genetics" (Lysenko, 1949, p. 489). However, it is difficult to know how to interpret such remarks, since they were published under Lysenko's editorship. According to Raisa Berg (personal communication, 1979), Schmalhausen's published "speech" was a total falsification; his actual remarks at the conference were a stirring defense of genetics. According to Berg, although Schmalhausen protested the falsification, it was published under his name.

When Lysenko announced at the final session that the Central Committee of the Communist party had considered his report "and approved it," the die was cast. Schmalhausen was removed as director of the Severtsov Institute, which was reorganized under G. K. Khrushchov, a Lysenkoist researcher in tissue culture. Schmalhausen himself took up a position at the Institute of Zoology under E. N. Pavlovsky. A traditional institute under the direction of a traditional parasitologist, the zoology institute had had few dealings with genetics or geneticists and was thus spared from Lysenkoist intrusion.

Meanwhile, even Severtsov's legacy would not remain untouched by the Lysenko affair. A five-volume edition of Severtsov's complete works was published between 1945 and 1950, originally under Schmalhausen's direction. By 1948 the second, third, and fourth volumes had appeared. But the fifth volume, which was to have been a republication of Severtsov's 1939 book, was issued under the editorship of Pavlovsky and omitted the significant first and second chapters in which Severtsov had presented his history of evolution theory and his critique of neo-Lamarkian theories (Bliakher, 1970).

In the 1950s Soviet evolutionary biologists tried to fight Lysenko's saltationist view of species formation and his denial of intraspecific competition by publishing articles in the *Bulletin of the Moscow Society of Naturalists* and *Botanical Journal*. Meanwhile, the intellectual climate encouraged work in evolutionary biology that had little relationship to genetics, and Soviet publications during the period took on an unusually traditional stripe. Schmalhausen was able to continue to do research and write, and began publishing once again in the mid-1950s. He had returned to his "safe" interest, charting the phylogeny of reptiles and amphibia using traditional approaches (1963). In the late 1950s he developed an interest in the application of cybernetics to evolutionary theory and began writing on the subject. Some of these works were published at the time, others only posthumously (1968).

During the 1925-1948 period, the evolutionary synthesis advanced more insistently and pervasively in the Soviet Union than in any other country. As a result of the events culminating in the infamous meeting of

1948, all those who had built and developed it most actively had, by 1965, died or been killed, lost their jobs, terminated their teaching and research, or had diverted their energies by fighting Lysenko and trying to preserve some protected enclaves of good biology. Thus, when Lysenko fell from influence in 1965, nearly two decades had been all but lost out of the Russian evolutionary tradition, and some of its most illustrious representatives had lost the most productive years of their careers. Considering the important works published in the West during those years by Huxley, Simpson, Dobzhansky, Mayr, Lerner, Wright, and many others, we cannot help but wonder what Soviet researchers might have done during the same period if they had not had to contend with the Lysenko problem.

Evolutionary biology is a traditional Russian strong point. Whether Soviet activity in this field will ever fully recover and resume a position of world leadership and how long this recovery might take remains to be seen. Since 1965 Schmalhausen's works have been widely republished (1968, 1969). It also remains to be seen whether Schmalhausen's students and successors will be able to complete his research program by extending the evolutionary synthesis to embryology and morphology in the way he had envisioned. In recent years there have been enormous successes in the West in developmental biology, while traditional morphological disciplines such as comparative anatomy and paleontology have faded in relative prominence. Despite these changes, many of the problems posed by Severtsov and Schmalhausen have yet to be settled.

References

Adams, M. B. 1968. The founding of population genetics: contributions of the Chetverikov school, 1924-1934. *Journal of the History of Biology* 1(1):23-39.
———— 1970. Towards a synthesis: population concepts in Russian evolutionary thought, 1925-1935. *Journal of the History of Biology* 3(1):107-129.
Bliakher, L. J. 1970. A. N. Severtsov i neolamarkizm [A. N. Severtsov and neo-Lamarckism]. *Iz istorii biologii* [From the history of biology]. Moscow: Nauka, vol. 2, pp. 112-122.
Coleman, W. 1976. Morphology between type concept and descent theory. *Journal of the History of Medicine* 31:149-175.
Filipchenko, Iu. A. 1915. *Izmenchivost' i evoliutsiia* [Variation and evolution]. Petrograd: Bios.
———— 1927. Uspekhi genetiki za poslednie 10 let (1918-1927) v SSSR [Achievements of genetics in the USSR over the past 10 years (1918-1927)]. *Trudy Leningradskogo obshchestva estestvoispytatelei* 62(1):3-11.
———— 1929. *Izmenchivost' i metody ee izucheniia* [Variation and methods for its study], 4th ed. Moscow/Leningrad: Gosizdat.

Gaisinovich, A. E. 1968. U istokov sovetskoi genetiki: I. Bor'ba s lamarkizmom (1922-1927) [The origins of Soviet genetics: I. The struggle with Lamarckism (1922-1927)]. *Genetika* 4(6):158-175.

Huxley, J. S. 1940. *The new systematics.* Oxford: Clarendon Press.

—— 1942. *Evolution: the modern synthesis.* London: Allen and Unwin.

Kol'tsov, N. K. 1936. *Organizatsiia kletki* [The organization of the cell]. Moscow/Leningrad: Biomedgiz.

Lysenko, T. D., and others. 1949. *The situation in biological science: proceedings of the Lenin Academy of Agricultural Sciences of the USSR session: July 31-August 7, 1948. Verbatim report.* Moscow: Foreign Languages Publishing House.

Mayr, E. 1942. *Systematics and the origin of species.* New York: Columbia University Press.

Menzbir, M. A. 1927. *Za Darvina* [For Darwin]. Moscow/Leningrad: Gosudarstvennoe izdatel'stvo.

Rensch, B. 1947. *Neuere Probleme der Abstammungslehre.* Stuttgart: Enke.

Schmalhausen, I. I. 1938. *Organizm kak tseloe v individual'nom i istoricheskom razvitii* [The organism as a whole in individual and historical development]. Moscow/Leningrad: Izdatel'stvo Akademii nauk SSSR.

—— 1939. *Puti i zakonomernosti evoliutsionnogo protsessa* [Trends and laws of the evolutionary process]. Moscow/Leningrad: Izdatel'stvo Akademii nauk SSSR.

—— 1942. *Organizm kak tseloe v individual'nom i istoricheskom razvitii* [The organism as a whole in individual and historical development], 2d ed. Moscow/Leningrad: Izdatel'stvo Akademii nauk SSSR.

—— 1946a. *Problemy darvinizma* [Problems of Darwinism]. Moscow: Sovetskaia nauka.

—— 1946b. *Faktory evoliutsii (teoriia stabiliziruiushchego otbora)* [Factors of evolution (the theory of stabilizing selection)]. Moscow/Leningrad: Izdatel'stvo Akademii nauk SSSR.

—— 1947. *Osnovy sravnitel'noi anatomii pozvonochnykh zhivotnykh* [The fundamentals of the comparative anatomy of vertebrates], 4th ed. Moscow: Sovetskaia nauka.

—— 1949. *Factors of evolution: the theory of stabilizing selection.* Philadelphia/Toronto: Blakiston.

—— 1964. *Proiskhozhdenie nazemnykh pozvonochnykh* [The origin of terrestrial vertebrates]. Moscow: Nauka.

—— 1968. *Kiberneticheskie voprosy biologii* [Cybernetic problems of biology]. Novosibirsk: Nauka.

—— 1969. *Problemy darvinizma* [Problems of Darwinism], 2d ed. Leningrad: Nauka.

Severtsov (Sewertzoff), A. N. 1912. *Etiudy po teorii evoliutsii: individual'noe razvitie i evoliutsiia* [Studies on the theory of evolution: individual development and evolution]. Kiev.

—— 1914. *Sovremennye zadachi evoliutsionnoi teorii* [Current tasks of evolutionary theory]. Moscow: Bios.

————— 1921. *Etiudy po teorii evoliutsii: individual'noe razvitie i evoliutsiia* [Studies on the theory of evolution: individual development and evolution], 2d ed. Berlin: Gosizdat.

————— 1924. Die Factoren der progressiven Entwickelung der niederen Wirbeltiere. *Russkii zoologicheskii zhurnal* 4 (1 and 2):12-60.

————— 1925. *Glavnye napravleniia evoliutsionnogo protsessa* [Principal directions of the evolutionary process]. Moscow: A. V. Dumnov.

————— 1927. Istoricheskoe napravlenie v zoologii [The historical trend in zoology]. *Nauka i tekhnika v SSSR: 1917-1927*, ed. A. F. Ioffe et al., vol. 2. Moscow: Rabotnik prosveshchenii, pp. 143-193.

————— 1928. Directions of evolution. *Acta Zoologica* 9:59-140.

————— 1931. *Morphologische Gesetzmässigkeiten der Evolution*. Vienna: Fischer.

————— 1934. *Glavnye napravleniia evoliutsionnogo protsessa* [Principal directions of the evolutionary process], 2d ed. Leningrad: Biomedgiz.

————— 1939. *Morfologicheskie zakonomernosti evoliutsii* [Morphological laws of evolution]. Moscow/Leningrad: Izdatel'stvo Akademii nauk SSSR.

————— 1945. *Sobranie sochinenii* [Collected works], vol. 3. Moscow/Leningrad: Izdatel'stvo Akademii nauk SSSR.

————— 1949. *Sobranie sochinenii* [Collected works], vol. 1. Moscow/Leningrad: Izdatel'stvo Akademii nauk SSSR.

————— 1950. *Sobranie sochinenii* [Collected works], vol. 5. Moscow/Leningrad: Izdatel'stvo Akademii nauk SSSR.

Severtsova, L. B. 1946. *Aleksei Nikolaevich Severtsov: biograficheskii ocherk* [Aleksei Nikolaevich Severtsov: a biographical essay]. Moscow/Leningrad: Izdatel'stvo Akademii nauk SSSR.

Simpson, G. G. 1944. *Tempo and mode in evolution*. New York: Columbia University Press.

————— 1953. *The major features of evolution*. New York: Columbia University Press.

Spencer, H. 1863. Progress: its law and cause. *Essays: Scientific, Political, and Speculative*, vol. 1. London: William and Norgate, pp. 1-60.

PART TWO

The Synthesis

in Different Countries

8 Soviet Union

The Birth of the Genetic Theory of Evolution in the Soviet Union in the 1920s

Theodosius Dobzhansky

When Darwin's *On the Origin of Species* was published in 1859, Russia was entering a period of political reforms and a ground swell of radicalism among its intelligentsia. The coincidence was fortuitous, but it left an impress on the intellectual tradition. Evolution was accepted not only as a scientific theory but also as a part of the liberal world view. Chernyshevsky and Pisarev, standard bearers of the radical youth, proclaimed that a valid personal philosophy must rest on a solid base of natural science, and evolution was a pivotal part of that. Professor Kutorga had lectured on Darwin's theory at Moscow University in 1860 and published an account of it in 1861. A Russian translation of Darwin's classic appeared in 1864. In 1865 Timiriazev, who was as effective an exponent of Darwinism in Russia as Huxley in England or Haeckel in Germany, published a collection of essays on evolution, which went through seven editions by his death in 1920.

Politically conservative circles took a dim view of Darwin's work because of its alleged political implications rather than evolution as a biological theory. As Danilevsky, one of Darwin's critics, indicates, the issue was taken extremely seriously: "It is clear how vitally important not only for zoologists and botanists, but for any moderately intelligent person is the issue whether Darwin is or is not right. It is so important that I am firmly convinced that no equally important problem exists either in any other field of knowledge or in any realm of practical life. Indeed, this is the problem of 'to be or not to be' in the strictest and broadest sense."

The Orthodox church was never very active in its opposition to evolutionism, and nothing like laws prohibiting the teaching of evolution or a Tennessee "monkey trial" occurred in Russia. The polemics of the debate were published neither in scientific nor in religious periodicals, but mostly in general literary and sociopolitical journals intended for broad circles of educated readers. This public did not usually have much understanding of biology as a science, but its bearings on human problems were considered important. When I was in my middle teens, a conservative lady of our acquaintance asked me: "Do you really believe in this horrid theory of Darwin?" I could have replied that those who do not credit this horrid theory seem to me rather dull-witted.

Critics and Defenders of Darwinism

Among those who accepted evolution as a part of the new gospel, some had reservations about certain parts of Darwin's theory. The struggle for existence seemed to have particularly undesirable connotations; already Chernyshevsky held the Lamarckian "transformism" superior to Darwin's natural selection. Kropotkin's attempt to give primacy as an evolutionary force to mutual aid, instead of competition and struggle, is sufficiently well known, because his essays were published in an English periodical (1902), while Kessler's "The Law of Mutual Aid" (1880), published in Russian, was virtually ignored. The checkered career of neo-Lamarckism in Russia has recently been well analyzed by Gaisinovich (1968) and Bliakher (1971). The polemics about Lamarckism versus Darwinism and genetics became a caricature of scientific discussion when the problem was taken over by Marxist philosophers in the 1920s and early 1930s. The Timiriazev Institute was working on "the study and propaganda of the scientific foundations of dialectical materialism." The Communist Academy had a Section of Natural and Exact Sciences. The Faculty of Medicine of Moscow University had a Society of Materialist Physicians. The criterion of validity of theories of evolution was their congruity with dialectical materialism as construed by different authorities. I remember the frustration I felt discussing some problem of genetics or evolution with Serebrovsky, an excellent geneticist and a convinced Marxist, in 1926 or 1927. His clinching argument to me was, "Your reasoning is undialectical." The debates among the high priests of dialectics were often impassioned but inconclusive. Both Lamarckians and Darwinians claimed to be faithful dialecticians. These polemics had however a wholly unintended effect: they prepared the ground for Lysenko's simplistic brand of dialectics, which for almost a generation swept away much of biology in Russia.

Most evolutionists in Russia were, of course, busy doing biological research rather than philosophizing. All approaches to evolutionary problems explored in other countries had their proponents in Russia. One tendency was particularly widespread and characteristic—a preoccupation with the organic diversity and the congruity of organic structures with the ways of living in different environments. Russia has a variety of climates from arctic to subtropical, vast plains and high mountains, forests and steppes, great rivers, lakes, and seas. Exclusively laboratory workers who neither possess nor wish to have any knowledge of living beings in nature were and still are a minority. Zoological and botanical systematics, comparative anatomy, phylogenetic studies on fossil and living forms, and later genetics were pursued in universities, as well as in museums and laboratories under the aegis of the Academy of Sciences. It was this tradition that yielded such evolutionists as Kovalevsky, Pavlov, and Mechnikov in the 1800s and Berg, Chetverikov, Karpechenko, Levitsky, Philipchenko, Schmalhausen, Severtzov, Vavilov, and many others during the first third of the 1900s. Some of the numerous amateur naturalists became very competent scientists. Genetics started tardily, but developed with great élan once it did start. The first course of genetics was given in the University of St. Petersburg by Philipchenko in 1913. I never had a course of genetics at the University of Kiev, although my teacher, Kushakevich, was an excellent cytologist and adherent of the chromosome theory of heredity. At that time I was an entomologist specializing in the taxonomy of *Coccinellidae* (lady beetles). Mendel's laws were occasionally mentioned, though not in evolutionary context.

The theory of natural selection reached the nadir of its repute among evolutionists in the early twentieth century. Russia was no exception, despite the aging Timiriazev thundering against what he believed reactionary forces bent on discrediting Darwinism. Perhaps the chief among the panoply of arguments advanced against natural selection was the absence of a clear correspondence between the organic diversity and the diversity of the environments. In particular, the characteristics that distinguish related species, and are utilized in taxonomic keys, appear to be neither useful nor harmful to their possessors. In other words, those characteristics are neutral traits, which natural selection can neither promote nor eradicate. Identical arguments, now in vogue among the so-called non-Darwinian evolutionists, who find no sense in the amino acid sequences in the proteins of different forms of life, are to me distinctly déjà vu. The eminent systematist and zoogeographer Semenov-Tian-Shansky compared nature to an artist who creates aesthetically appealing forms for no reason other than their beauty. If a viable organism can be built this way as well as that way, both kinds of organisms will exist. Natural selection can only interdict what is inviable. The concept of mutation was still rather unfamiliar to at least the older generation of biologists in the 1920s, even though the theories of de Vries and of his Russian anticipator Korzhinsky were not unknown. However, in 1919 and 1922 Philip-chenko published in *Priroda*, a journal analogous to *Science* or to *Nature*, excellent reviews of the works of the Morgan school on the genetics of *Drosophila*. To me these reviews were a revelation. To most senior biologists *Drosophila* mutants were a collection of monstrosities, of no significance for evolution.

Neo-Lamarckism

To criticize the theory of natural selection was easy; to put something better in its place very difficult. Some of the critics embraced neo-Lamarckism, the inheritance of acquired traits. Smirnov, a brilliant young zoologist at Moscow University, engineered in 1926 the invitation to Kammerer to come to Russia to head a laboratory in which they would have worked to establish the validity of Lamarckism, about which they had not the slightest doubt to begin with. The plan came to naught owing to Kammerer's suicide. The Lamarckists were a small minority among biologists (though not, for a time, among philosophers and politicians). (See Gaisinovich, 1968; Bliakher, 1971.)

The most widespread view was that evolutionary changes are induced by the environments in which the organisms live. But the evolutionary responses to the environment occur neither via natural selection nor via the inheritance of acquired traits. It is rather by means of direct influ-ences of the "geographical landscape" on the germ plasm of the inhabi-tants of a given territory. Berg gave a most articulate exposition of this view, which he accepted as a subsidiary to his nomogenesis theory. The geographical landscape subsumes "the external environment in the wid-est sense, physical as well as biotic factors united in one harmonious whole." The geographical landscape "affects organisms in an imperative manner, compelling all the individuals, so far as the organization of the species permits, to vary in a determined manner. There is no place here for chance: consequences follow with the same fatal constancy as chemi-cal reactions or physical phenomena." Just how the "harmonious whole" changes the genotype was left quite vague. Those who attempted to ex-plicate this puzzle (Berg was not one of them) postulated direct alteration of the genetic materials by environmental factors, rather than somatic or parallel inductions favored by the Lamarckians. Evidence that could be quoted in support was scanty in the extreme: the now discredited experi-ments of Tower on the beetle *Leptinotarsa*, and the also discredited in-duction of inheritable eye defects by antibodies in rabbits, alleged by Guyer and Smith. Muller's discovery of gene and chromosomal muta-tions in *Drosophila* induced by ionizing radiations was, of course, a firmly established demonstration of genetic changes by an environmental agency, but hardly relevant to the geographical landscape theory.

Nomogenesis

More interesting was Berg's theory of nomogenesis (1922). A man of splendid intellect and great personal charm, Berg developed his views in a book marshaling an abundance of evidence comparable to Darwin's *On the Origin of Species*. For a year or two after reading Berg's work I was close to becoming a partisan of nomogenesis. To Berg the phylogeny is basically similar to the ontogeny—the development follows paths predetermined by causes internal to the organism. The environment and natural selection permit or prohibit the survival of a new form but are not competent to bring it into existence. Evolutionary changes are subject to laws analogous to, and perhaps consubstantial with, those of embryonic development. With impeccable honesty, Berg admits that his putative "laws" leave unexplained the adaptedness of organisms to their environments. He postulates that adaptedness is an immanent property of living matter. Nomogenesis was attacked from all sides. Dialectical Marxists rightly saw that nomogenesis borders on vitalism. Lamarckians were displeased that Berg not so much denied as ignored acquired traits and somatic induction. His book has no reference to Kammerer's works being hotly debated at that time. Genetics also receives inadequate attention, which is hardly surprising because Berg wrote his book mostly before biological literature published abroad since 1914 had reached the blockaded Russia.

The Rise of Genetics

Scientists in Russia had an extraordinary experience of being almost wholly isolated from world science for about seven years, and then obtaining access at once to what was accomplished during this period by their colleagues abroad. To those interested in genetics and evolution this amounted to a sudden revelation. The review of the work of the Morgan school on *Drosophila* published by Philipchenko in 1922 was an eye-opener to many. Muller visited Russia in 1922 and left a fair number of cultures of *Drosophila* mutants at Koltsov's Institute of Experimental Biology in Moscow—the seed from which grew the Russian school of *Drosophila* genetics. Moscow geneticists were generous with the gift; I was enabled to start working with *Drosophila* mutants in Kiev early in 1923, as were other converts to *Drosophila* studies elsewhere. I myself left Russia on the morning of December 4, 1927, having received a post-graduate fellowship to work one or two years in Morgan's laboratory. I did not suspect at all, when I left, that I might stay on after the expiration of the fellowship.

There were three nuclei of genetics research in the 1920s. By far the largest were the Institute of Applied Botany headed by Vavilov, with headquarters in Petrograd (later Leningrad), Koltsov's Institute of Experimental Biology in Moscow, and Philipchenko's department of genetics at the University of Leningrad. Vavilov, soon joined by Karpechenko, Levitsky, and numerous other botanical and agricultural geneticists, maintained his familiarity with current ideas and research in genetics and evolution, despite what to almost anybody else would have been a crushing load of administrative responsibilities. Koltsov's personal interests were in cell biology, in what later became molecular genetics, and finally in eugenics. Two of the several sections in his Institute of Experimental Biology were headed by Chetverikov and Serebrovsky. They attracted a number of talented young researchers, some of whom became eventually leading figures in evolutionary biology in Russia and abroad. Among them were Timoféeff-Ressovsky, Astaurov, Romashov, and Dubinin.

Sergei Chetverikov

Chetverikov's classic paper (1926) showed that evolution is brought about by natural selection acting on a store of genetic variability generated by mutation. Between 1930 and 1932, the same theory was expounded with greater mathematical refinements by Fisher, Wright, and Haldane. The four "founding fathers" thus provided the cornerstone of what later became the biological, or synthetic, theory of evolution. Chetverikov's work would have received much greater renown had it not been published in Russian in a journal with limited distribution abroad. As it was, Fisher, Wright, and Haldane were simply unaware of Chetverikov's contribution. It did, however, stimulate a great deal of research activity in Russia and eventually abroad. As a background against which to see the crosscurrents in evolutionary biology in Russia in the 1920s and 1930s, brief sketches of some of the personalities involved in the apparently sudden blossoming of evolutionary studies may be useful.

Chetverikov started as a naturalist working on systematics and ecology of Lepidoptera. He had a contemplative turn of mind and, although he was a steady worker, he published little. His interest in the expansions and contractions of the populations of various species in time led to the publication in 1905 of "Waves of Life," a rather short but allusive article. His great 1926 essay is only 52 pages long, although it contains enough material to make an at least medium-sized book. Chetverikov was at his best in face-to-face discussions with colleagues and students. His sharp and rigorous analysis of evolutionary problems was accompanied by constructive ideas and suggestions for research. He shared his ideas with his interlocutors with complete selflessness. Trying to establish his priority seemed to him unworthy of a scientist.

I had the privilege of several visits with him in his laboratory in Moscow. Though at that time I was only a bit more than half as old as he was, Chetverikov was unstinting with his time and inspiring with his advice. He disliked writing letters, however, which proved to be a blessing in disguise for me. Otherwise I might have shared the fate of Timoféeff-Ressovsky in Germany, instead of coming to work with Morgan in the United States. Chetverikov was an experimentalist as well as a theoretician. Even before the publication of his 1926 theoretical paper, he initiated a study of what later became known as genetic loads, in natural populations of *Drosophila melanogaster*. The results were reported briefly in a paper read in 1927 at the International Genetics Congress in Berlin, and in a very short paper published in Russian in 1928. In 1929 Chetverikov was exiled, because, it was rumored, he was denounced to the political police by one of his students. This incident occurred before Stalin's terror reached its peaks of brutality. Although Chetverikov was never able to develop the work begun so auspiciously with his many students in Moscow, he remained alive until 1959. Recognition, in the form of the Darwin medal award by the German Academy Leopoldina, reached him shortly before he died at the age of seventy-nine.

Alexander Serebrovsky

Serebrovsky accepted Chetverikov's theory of evolution. His own work developed quite independently and in somewhat different directions, however. He was an adherent of the now almost forgotten presence-absence theory, originated chiefly by Bateson in England, which for reasons beyond my comprehension seemed to Serebrovsky particularly congenial with Marxist dialectics. This theory postulated that most mutations are minute deficiencies, little holes in the chromosomes. Serebrovsky attempted to measure the size of the assumed "hole," by estimation of the map distance in centimorgans between two marker genes on both sides of a mutant in *Drosophila melanogaster*. Together with his student Dubinin, he interpreted the so-called step allelism at the gene locus of scute in *Drosophila* in terms of overlapping deficiencies in a series of linearly arranged subgenes.

Of more lasting interest and significance are Serebrovsky's studies on gene geography, first published in Russian in 1927. Chicken populations in the Caucasus (Dagestan) are highly polymorphic, and the gene frequencies differ considerably in nearby mountain valleys, as well as elsewhere in the Soviet Union. This led Serebrovsky to formulate the concept of the gene pool, which he called the gene fund ("genofond"). His interest in the human gene pool landed him in trouble. Independently from Muller, Serebrovsky advanced a plan of improvement of human populations by means of artificial insemination of women with the semen of eugenically desirable donors. This plan was too much for the then poet laureate Demian Bedny (eventually himself destroyed in Stalin's terror). Serebrovsky was blasted and ridiculed in *Pravda* and elsewhere by totally ignorant hacks. When Lysenko became the ruler of biology, Serebrovsky turned to applied genetics and put forward the idea of pest control by means of release of masses of individuals that would produce sterility or lethality in the recipient populations. This idea was surely ahead of his time, but he had no opportunity to make even small-scale tests of its applicability. When genetics, particularly *Drosophila* genetics, was banned, Serebrovsky, the ardent believer in communism and dialectical materialism, died a forlorn and embittered man, just when Lysenko reached the peak of his glory.

Juri Philipchenko

Philipchenko built a school perhaps smaller than those of Chetverikov and Serebrovsky, but his impact on the evolutionary and genetical thought in Russia was, if anything, greater. He wrote six textbooks and numerous review articles on various aspects of genetics and evolution that were used in institutions of higher learning until Lysenko's pogrom. The first course of genetics in Russia was given by Philipchenko from 1913 until he was forced to resign his professorship at the University of Leningrad a year before his death in 1930. His books and lectures were didactically flawless: the depth and breadth of his scholarship were as admired by his students and colleagues as they were feared by his opponents. And yet this man of superb intellect fell short of fulfillment of his capabilities, in part because of his untimely death, and in part because of the skeptical and overcritical turn of his mind. He was always able to see a host of alternative explanations for any phenomenon and to poke holes in any theory. Yet while a deficiency of critical ability is fatal to a scientist, its excess may be frustrating; life is too short to test every alternative, and one has to accept the most probable one, subject, of course, to possible falsification.

Philipchenko welcomed Chetverikov's mutation-natural selection theory, but stressed that it might be only a part, and a rather minor part, of the story. The genes and the Mendelian inheritance may be pertinent only to superficial traits, distinguishing individuals within a species or species within a genus; fundamental features, those of orders, classes, and phyla, may be inherited and changed by quite different and as yet unknown mechanisms. This view was a minority position but scientifically quite respectable in the 1920s. Similar opinions were held, for example, by the Danish geneticist Johannsen. Though a zoologist by training, in the last decade of his life Philipchenko worked on the genetics and ontogeny of certain characteristics in species of wheat and other grasses. The results were neither in accord with his expectations nor flatly contradictory. Had he lived longer he might well have changed his evolutionary credo. As it was, he still played a most useful role not only as a teacher but also as a stimulant of critical thought and research in evolutionary biology.

Between 1924 and 1927 I was the counterpart of an assistant professor in Philipchenko's department in Leningrad. Moscow is just a night railway journey away, so I visited there rather frequently. At that time at least, the second-class tickets were reasonably cheap. I did not, however, participate in the actual work done at the Koltsov Institute. When I visited there, I looked at what they were doing but I cannot claim the honor of having been one of Chetverikov's students. The subject of my own research in Philipchenko's laboratory in the years 1924-1927 was the effect of pleiotropic genes in *Drosophila*. However, I never lost my interest in *Coccinellidae*.

As an entomologist I had read *On the Origin of Species*, I believe in 1915, when I was all of fifteen years old. I also read German entomological literature because my main foreign language then was German. One of my beloved books was that of Standfuss, describing experiments on butterflies and moths. The transition from that to *Drosophila* was very easy. Berg's nomogenesis was a temporary infatuation: it sounded very remarkable and it took me some time to recover my independent judgment. Berg was a very impressive and certainly brilliant man.

The Great Institutes

Vavilov headed the Institute of Applied Botany (later the Institute of Plant Industry, still later the Lenin Academy of Agricultural Sciences). Koltsov was the director of the Institute of Experimental Biology. In its heyday Vavilov's institute had local sections and experiment stations in all parts of the Soviet Union and included among its research workers, if only on a part-time basis, nearly every geneticist and cytologist working with plants as materials. Koltsov's institute was not quite so large, but it had a powerful group of both young and old research workers. Research and thinking of the then avant-garde evolutionary biology were active in both institutions.

Vavilov and Koltsov were eminent scientists in their own right. As administrators, they necessarily sacrificed much of their personal research potentials for the sake of furthering the work of others. In a sense, their greatest discoveries were the outstanding creative scientists whom they chose as their colleagues and leaders of various research projects. In 1922 Vavilov established the law of homologous series in genetic variabilities. Notably in related species of cultivated plants, but also in wild species, the mutation process generates alleles of homologous genes that produce phenotypically similar variants. Artificial and natural selections then choose from this assortment the genotypes suitable for survival and exploitation. Some overenthusiastic reporters went so far as to compare this law with the periodic system of chemical elements. Vavilov's other major contribution, based on his extensive familiarity with cultivated plants of the world, was locating the geographic "centers of origin" of the domesticated forms. He was inclined to equate these centers with regions of maximum genetic variability in the respective species. For a few years preceding his arrest and death in prison, Vavilov was stripped of administrative work and devoted his time to studies of the evolution of cultivated plants; his publications of that period seem to lack much of his former ardor.

Levitsky and Karpechenko were two outstanding evolutionary cytogeneticists among Vavilov's cohorts. Levitsky, originally professor of botany at the Faculty of Agriculture in Kiev, published *The Material Basis of Heredity* (1924) and *Cytological Basis of Evolution* (1939). Karpechenko obtained fertile allopolyploid hybrids of radish and cabbage, a classic example of emergence of a full-fledged new species, by doubling the chromosomal complements in otherwise sterile interspecific hybrids. Both Levitsky and Karpechenko were articulate speakers and writers and stimulated a great deal of genetic and evolutionary research among botanists, not only in Leningrad but elsewhere in the far-flung system of Vavilov's experiment stations. Inevitably they ran afoul of Lysenkoism, were arrested in 1941, and died in prisons or concentration camps.

Koltsov was a man of multifarious interests and knowledge, of imposing presence, and with the eloquence of a spellbinding orator. His public lectures were events memorable to his peers and to beginners alike. His main field of research was cell biology; some Russian writers claimed that Koltsov discovered the main features of the genetic code, although he believed (as did biologists in general in his time) that the genetic information is stored in proteins rather than in the nucleic acids. He realized clearly the basic importance of Mendelian genetics for the elucidation of evolutionary processes and gave full support to Chetverikov, Serebrovsky, and other geneticists working in his institute and elsewhere. One of the Koltsovians, Timoféeff-Ressovsky, had a remarkable life; for about two decades after leaving Koltsov's institute he worked with great success in Germany, then spent some years in Soviet prisons, and finally emerged the most eminent evolutionary geneticist in the Soviet Union.

Eugenics seemed to be an important application of genetics to Koltsov. He was the organizer and the moving spirit of the Russian Eugenics Society and of its periodical, a perilous undertaking in the Soviet Union, especially when eugenics was among the slogans of the waxing Nazism in Germany. No matter how much effort Koltsov and Philipchenko made to separate their eugenics from the racist teachings of the Nazis, they were easy marks for the accusations of the Lysenkoists that all eugenics, and in fact all genetics, are pernicious inventions of the capitalists. The last years of Koltsov's life, though he avoided arrest and imprisonment, were poisoned by unceasing harassments. He was lucky to die of a sudden heart attack; his wife committed suicide within a few hours after his death.

The Flourishing State of Russian Genetics

One of the most impressive aspects of Russian genetics in the 1920s was the size of the genetics establishment. In England, few were interested in evolutionary genetics besides Bateson, Fisher, Haldane, Huxley, and Ford. In Germany, there was some interest in physiological genetics, but aside from Baur there was no population geneticist until Timoféeff-Ressovsky arrived from Russia.

In contrast, three schools flourished in Russia, each with numerous investigators. Numerically the largest school, of course, was that of Vavilov, although much of its research was applied. Chetverikov's group consisted in 1927 of twelve people doing *Drosophila* research, as well as five visitors to the group (Zhivago, a cytologist, Frolova, Serebrovsky, Sacharov, and Koltsov). Serebrovsky had his own group that included Dubinin among others. Frolova has the distinction of having discovered the difference between European *Drosophila obscura* and U.S. "*obscura.*" When she was unable to cross the two stocks, she examined the chromosomes and found them to be quite different. She and Astaurov described the American sibling species as *D. pseudoobscura*.

It is sometimes remarked that Russian geneticists seemed to be much more of the naturalist type than pure experimentalists, perhaps because of the great size and environmental diversity of the country. New and unusual animals and plants were steadily collected and brought for study in university laboratories, zoological and botanical museums, and marine institutes. Most biologists had experience working in the field and observing living beings in their habitats. Most young biologists traditionally spent a year or more abroad in "postdoctoral" work (although there were no doctorates in our sense), usually including visits to Naples' marine zoological station or similar institutions. It was taken for granted that a biologist must know animals or plants, or both.

Severtsev and Schmalhausen

Severtsev and Schmalhausen were evolutionary morphologists, working on comparative anatomy and embryology of vertebrates. Severtsev's interests centered on the relationships between phylogeny and ontogeny, and led him to conclusions in many ways resembling those of Rensch and Simpson published two decades later. Severtsev viewed the mechanisms that bring about evolution in the light of classical Darwinism; he regarded natural selection the leading agent and considered the inheritance of acquired traits neither proved nor completely ruled out. He was certainly not opposed to the genetical theories of evolution being developed by Chetverikov and others, but rather felt that geneticists should do their business while he and his collaborators were doing theirs. To enthusiastic young geneticists in the mid-1920s this attitude seemed narrow-minded if not outright retrograde. Philipchenko was among those who pointed out that the narrow-mindedness was more likely the foible of the young enthusiasts because evolutionary biology is multidimensional, not only a specialized branch of genetics.

Schmalhausen was one of Severtsev's pupils, but by the late twenties while a professor of zoology in Kiev, he became interested in the genetic factors bringing about evolutionary changes. As Severtsev's successor as director of the Institute of Evolutionary Morphology in Moscow, Schmalhausen published a series of books on evolution that belong to the period of the emergence of the synthetic theory rather than to its incipience. Schmalhausen's analysis of the varieties of natural selection is fully modern and has in no way lost its significance today. As could have been expected, Schmalhausen was attacked by Lysenkoists and expelled from the directorship of his institute, but fortunately avoided the arrest and imprisonment that then commonly befell biologists who refused to compromise their scientific and personal integrity. In his last years Schmalhausen became interested in cybernetics and its applications to evolutionary biology.

The Emergence of a Paradigm

In his essay on the Russian school of evolutionary genetics, Adams (1968) remarks: "The Russian school is important both because of what it ended and what it began. Many authors have alluded to the estrangement between two traditions in biology which characterized its history in the early decades of this century: the 'experimentalist' and the 'naturalist' traditions. It is significant, then, that the Russian school is one of the earliest to draw from both traditions in order to clarify the evolutionary process." Russian biologists habitually viewed the evolutionism in two contexts: that of its philosophical implications, and in the light of life as it exists in pristine nature. To be acceptable, a paradigm had to pass scrutiny from these points of view, in addition to being valid on purely factual grounds.

Why had such a paradigm emerged when it did and not earlier or later? Interest in evolutionary problems was traditional among Russian biologists since the nineteenth century. Except for the years during World War One and the subsequent revolutionary upheaval, Russia was not a scientific backwater. Current scientific literature in major European languages was not only available but it was read rather more systematically than is the foreign-language literature now in the United States. Acquaintance with the experimental work of the Morgan school, and with the findings of other geneticists in Europe and in the United States, became possible only in about 1921. However, this exposure would have been insufficient without Chetverikov's insight and without the experiments on the genetics of natural populations that he promptly undertook to test the validity of his theoretical construct.

It now seems almost unbelievable that quite authoritative biologists in the 1920s and 1930s in Russia as well as in Europe and the United States contended that the phenomenon of mutation has no bearing whatever on evolution. *Drosophila* mutants described by the Morgan school were monstrosities, and they were all found in laboratory bottles, not in natural populations, products of some kind of disruption of hereditary materials in highly artificial environments. The claim of de Vries and his followers that new species arise by single mutational leaps seemed incompatible with everything that systematists and paleontologists had learned about species in living and fossil forms. It was therefore quite logical and necessary that the experimental work of Chetverikov and his collaborators and students was directed toward testing for the presence of mutants in natural populations of *Drosophila*. They explicitly considered their work to be the decisive test of the validity of the theory, and the results of the test turned out to be positive. After Chetverikov ran afoul of the political police and was banished from Moscow, the work went ahead under the direction of Timoféeff-Ressovsky in Germany, and by Dubinin, Romashov, and others in the Soviet Union.

It is tempting to represent a scientific discovery of a given person or of a group of persons at a given time too neatly as an automatic consequence of a certain scientific tradition. After all, the basic features of the mutation-natural selection theory of evolution were discovered almost simultaneously and certainly quite independently from Chetverikov by Fisher, Wright, and Haldane. The personal backgrounds and the scientific traditions of this Anglo-American trinity were quite different from Chetverikov's, as well as from each other. It is well known that Mendelism and genetics in general were, by a singular miscomprehension, regarded as contradictory to Darwin's theory of natural selection in the early decades of the twentieth century. The short but important theoretical paper by Hardy in 1908 should have dispelled this miscomprehension, and in fact it served as a point of departure for Chetverikov as well as for the other founding fathers (the parallel work of Weinberg was not then considered). By the 1920s the mutation-natural selection theory was definitely in the air, and four persons independently seized it. Analogously, in the thirties and forties, what we now call the synthetic theory was in the air, a logical complement to the Chetverikov-Fisher-Wright-Haldane theoretical base, and at least eight biologists of different backgrounds and traditions have brought it down to earth. Some of them, fortunately, are still alive in the 1970s and can tell their stories.

References

Because of Dobzhansky's death before his selection had been fully prepared for publication, there is no list of references such as appears after the other selections. Many of the books and articles cited by Dobzhansky appear, however, in other chapters of this book, and complete information is given there.   —Eds.

Sergei Chetverikov, the Kol'tsov Institute, and the Evolutionary Synthesis
Mark B. Adams

Between 1920 and 1940, researchers associated with the Institute of Experimental Biology in Moscow made major contributions to the emerging evolutionary synthesis. Those contributions included one of the first theoretical papers synthesizing biometric, naturalist, and genetic approaches to evolution in a Darwinian framework (Chetverikov, 1926); the first systematic genetic analyses of wild populations of *Drosophila melanogaster* and other *Drosophila* species (Timoféeff-Ressovsky and Timoféeff-Ressovsky, 1927; Chetverikov, 1928a and b; Gershenson, 1934; Balkashina and Romashov, 1935); the first formulation of the gene pool concept ("genofond") (Serebrovsky, 1926, 1928, 1930); the concept of the genotypic milieu and its significance for evolution (Chetverikov, 1926); an independent formulation of the concept of genetic drift (Romashov, 1931; Dubinin, 1931; Dubinin and Romashov, 1932) and its study in wild *Drosophila* populations (Dubinin and others, 1934, 1936, 1937); and the first analysis of the frequency of lethals in natural populations (Dubinin and others, 1934; see also Adams, 1968, 1970, 1980).

The importance of these Russian contributions to the evolutionary synthesis may come as a surprise to those who associate the synthesis with the Anglo-American trio of Fisher, Haldane, and Wright, whose classic mathematical papers of the 1920s were so important in reconciling the biometric approach to evolution with particulate, Mendelian genetics (Provine, 1971). However, if Haldane, Fisher, and Wright established a mathematical framework of analysis for the new synthesis by creating theoretical population genetics, it was the Russian school—and especially Chetverikov—who made the link with the concerns of naturalists and field biologists and largely created "experimental" population genetics, based on the genetic analysis of actual natural populations, notably *Drosophila*. Thanks largely to the efforts of Dobzhansky, Lerner, Dunn, and Mayr, the trio has become a quartet, and Chetverikov's name is now cited with increasing frequency as a founder of population genetics and a key figure in the evolutionary synthesis (see, for example, Allen, 1975).

Yet the more we recognize the important role played by Chetverikov and his colleagues, the more puzzling their contributions become when we consider when they worked, where they worked, and who they were. First, Chetverikov and his group were able to develop a substantially modern view of the evolutionary process and to create experimental population genetics between 1922 and 1926. The timing is remarkable in two respects: it predated comparable work on the evolutionary synthesis in other countries, and it occurred in Russia immediately following a devastating world war, two revolutions, and a bloody civil war.

Second, if we had looked at Soviet research centers in 1920 and tried to predict where population genetics might emerge in conjunction with evolutionary theory, we would hardly have expected major contributions from the Kol'tsov Institute. Kol'tsov had only a peripheral interest in evolutionary theory, and genetics occupied a very minor role in his plans for his institute, which had been designed to emphasize physicochemical biology, embryology, and cellular morphology. By contrast, both Filipchenko and Vavilov were excellent and experienced geneticists with strong interests in evolutionary problems, and both headed major research centers. Yet by 1929 no Russian institution had contributed more to the evolutionary synthesis than had the Kol'tsov Institute, where experimental population genetics had been born. Why did more not come from the Filipchenko or Vavilov groups? Why did so much come from the Kol'tsov Institute?

Third, all of Chetverikov's publications in the years 1900-1925 dealt with butterfly collecting, systematics, and taxonomy. His only teaching (1906-1919) had been a course in entomology and prior to 1920 he had only a passing knowledge of biometrics and almost no knowledge of genetics. Why did Kol'tsov, in 1921, invite a butterfly taxonomist into his institute of laboratory biology? Why did Chetverikov take up genetics? Why was he able to succeed?

Until quite recently there was no basis for answering these questions. During the last decade, however, new information removed much of the obscurity surrounding Chetverikov and his research enterprise. This information comes principally from recent Soviet publications about Chetverikov, Kol'tsov, and their fellow researchers; from materials that I have examined in the archives of Chetverikov, Kol'tsov, and Filipchenko; and from interviews with members of the Chetverikov group, most notably Timoféeff-Ressovsky. On the basis of these materials, it has become possible to reconstruct the conditions and events that led to the origins of experimental population genetics in Moscow between 1920 and 1926 and to explain why Russian workers at the Kol'tsov Institute were so well advanced toward the evolutionary synthesis at such an early date.

Postrevolutionary Conditions Favoring the Synthesis

In the decade following the October revolution in 1917, four factors proved conducive to the emerging evolutionary synthesis. Two were ideological in character; two institutional. First, the new government was strongly committed to science, both in word and in deed. This commitment was reflected in its political philosophy, which saw Marxism as the science of society and sought to build "scientific" (as opposed to "utopian") socialism in Russia. Seeking to replace religion with the authority of science, the new government saw its mandate as spreading literacy and popularizing scientific results. In addition, Lenin emphasized that the new regime had to cooperate with "bourgeois specialists," despite their class backgrounds and ideological propensities, to build the country technologically and economically.

As a result, scientists and scientific institutions frequently received special treatment. For example, on August 16, 1920, Kol'tsov himself was called in for questioning by the "Special" Division of the Executive Revolutionary Tribunal. On August 19 he was arrested and was held for 38 hours without food. Kol'tsov was one of many being investigated for conspiring to organize a new Moscow city government when it appeared that the White forces under Denikin would succeed in capturing the city from the Reds. The investigation led to the execution by firing squad of 24 persons, but Kol'tsov was released. The event is recorded in a laconic footnote to a paper by Kol'tsov on the effect of malnutrition on body weight, in which he gave details of his arrest, together with his caloric intake, amount of sleep, and weight during the troubled days (Kol'tsov, 1921a, pp. 28-29).

The new government's pro-science orientation helps to explain some of the remarkable achievements of Russian science during the 1920s. Even during the worst of times, scientists and scientific institutions fared better than most. When more normal times returned under the New Economic Policy (NEP), the government was relatively lavish in its support of science, and sought to make its own specialists happy and to lure back to Russia the scientific talent that had left during the civil war years.

A second factor favoring the evolutionary synthesis was the new status enjoyed by Darwinism. Even in tsarist days, Darwinism had been a major component of the world view of the Russian intelligentsia, though it had also been criticized for its atheistic, materialistic implications (Vucinich, 1970). But following the revolution, with the advent of an avowedly atheistic, materialist ideology, religious opposition was largely silenced and Darwinism enjoyed an unparalleled heyday. Such early champions of Darwinism as Timiriazev and Menzbier became popular national figures. Although Timiriazev died shortly after the revolution, Menzbier lived until 1935, and in the 1920s he edited a new edition of the complete works of Darwin with introductory essays evaluating the current status of Darwinism and defending it from the attacks against it by mutationists, geneticists, and others (Menzbier, 1926).

Menzbier had taught at Moscow University from the late nineteenth century until 1911 (when he left in protest against the direct takeover of the university by Kasso, the tsarist minister of education) and, as head of the Department of Comparative Anatomy, he had made it the major Russian center of training in evolutionary biology. Among his students were Kol'tsov, Chetverikov, the evolutionary morphologist A. N. Severtsov, and the biogeographer Sushkin—all strong Darwinians, and all thereby enjoying a certain prestige following the revolution despite their varying political inclinations (for example, Severtsov was relatively conservative).

If Darwinism tended to be selectively favored as a "materialist" theory of the living world, non-Darwinian or anti-Darwinian tracts were for the first time strongly selected against. In 1922, when the ichthyologist and geographer Leo S. Berg sought to publish his *Nomogenesis*—seen in the West as one of the best anti-Darwinian books ever written—it apparently had difficulty passing the censor precisely because of its anti-Darwinian character (Muller, 1923). After it was published, Berg's book was subjected to widespread attack on both biological and ideological grounds (see, for example, Kozo-Poliansky, 1923, and Taliev, 1926). Although the official sanctioning of Darwinism did not create an evolutionary synthesis, it did tend to favor the work of Darwinians, and it did encourage attempts to integrate or reconcile Darwinism with scientific results that, in other contexts, might have been seen as antagonistic.

A third factor was the influence and popularity in the 1920s of dialectical materialist language and with it, the model that history moves by the dialectical interaction of opposites ("thesis" and "antithesis") which produces "synthesis." Discussions over biological issues held in the middle and late 1920s in the Communist Academy involved attempts by a number of Communist biologists to formulate a theoretical synthesis of various apparently conflicting positions (Gaisinovich, 1968). For example, one Marxist presented a paper arguing for a "synthesis" of Lamarckism and Darwinism. In defending genetics from the accusation that it was not "dialectical," Serebrovsky argued in 1926 that genetics represented a "synthesis" between the Weismann "thesis" that heredity was immutable, and the Lamarckian "antithesis" that it could easily be altered—a synthesis that maintained that heredity could indeed be altered through mutations, but that these mutations were infrequent and unpredictable (Serebrovsky, 1926). Thus, the currency of dialectical language was general and did not exclusively favor the kind of evolutionary synthesis that later emerged, nor did it create all of the synthetic efforts of the 1920s from nothing. In particular, in designing his Institute of Experimental Biology, Kol'tsov had emphasized the "synthetic" character of experimental biology from prerevolutionary days. But the emphasis in the 1920s on dialectics did tend to encourage synthetic enterprises such as Kol'tsov's, and also to encourage the formulation of scientific results in "synthetic" language.

After the revolution there was also a new institutional emphasis on practical results. Before the revolution a surprising amount of Russia's research effort was privately funded. For example, in Moscow the Scientific Research Institute Society had succeeded in obtaining large endowments for the establishment of independent research institutes, including Kol'tsov's. These endowments had been obtained in part from the Moscow city government, but the largest amounts had come from the bequests of wealthy entrepreneurs, such as the railway magnate G. M. Mark. Following the revolution and the seizure of private capital, research enterprises such as Kol'tsov's had to find new sources of funding, primarily the Academy of Sciences Commission for the Study of Natural Productive Forces of the USSR (KEPS), the Commissariat of Agriculture (Narkomzem), and the Commissariat of Public Health (Narkomzdrav) (Adams, 1980).

One result of the new importance of official government patrons in the postrevolutionary decade was the growth of genetics research. Unlike other experimental biological sciences, genetics did not require complex or expensive equipment that could only be obtained from abroad. More important, animal and plant breeding were directly tied to agricultural productivity and, in particular, to avoiding famines such as those which Russia had suffered in the late years of the civil war. It is natural that the new government, with its great faith in science, should turn to science—and in particular genetics—to put its agriculture on a sound footing. Hence genetics was easy to legitimate from a practical standpoint and easier to fund through KEPS and Narkomzem than more esoteric or theoretical fields. This necessity of cultivating new patrons, and the practical concerns of those patrons, helps to explain the explosive growth in genetics research following the revolution.

Of course, these conditions did not affect the Kol'tsov Institute uniquely. Indeed, the primary beneficiary was Vavilov and his Institute of Applied Botany. Taking over the government's Bureau of Applied Botany with the death of its director Robert Regel' in 1922, Vavilov was able to build it into a giant enterprise, sending expeditions all over the world to collect plant specimens and directing the work of breeding stations all over the Soviet Union. This institute became the core of the Lenin All-Union Academy of Agricultural Sciences, which was created largely by Vavilov himself in the late 1920s and early 1930s. Of course, all this work was done with strong government backing and funding, and it was done principally because of its agricultural promise. Another major beneficiary was Filipchenko's group in Leningrad. With funding from KEPS, he and his students mounted summer expeditions to survey the animal wealth of the Soviet Union's southern and eastern regions, and his genetics research enterprise was heavily supported by KEPS and Narkomzem. The Kol'tsov Institute also benefited, although its profile was predominantly theoretical. With funding from KEPS and the Commissariat of Agriculture, Kol'tsov created the Central Genetic Station for Animal Breeding at Anikovo. Kol'tsov headed the Moscow section of the genetics division of KEPS, and put two of his students—Lebedev and Serebrovsky—in charge of the work at the station, although officially he remained its director.

Thus, in the postrevolutionary decade, certain ideological and institutional factors proved conducive to the synthesis. Ideologically, Darwinism was favored as a materialist theory of the living world, and dialectical language involving conceptualizing "syntheses" became a favored mode of expression. Institutionally, the new government strongly supported science, became the sole patron of scientific activity, and favored research that promised practical benefit—resulting in an efflorescence of work in animal and plant breeding. But surely these conditions applied equally to the enterprises of Vavilov, Filipchenko, and Kol'tsov. Why, then, did the Kol'tsov Institute make such disproportionate contributions to the synthesis? Part of the answer lies in the approaches of Vavilov and Filipchenko to evolutionary theory and genetics.

Filipchenko and Vavilov

In 1921 Iurii Filipchenko (1882-1930) and Nikolai Vavilov (1887-1943) were Russia's leading geneticists. Both had begun their genetic studies before World War One—Filipchenko in Germany, Vavilov with Bateson, Punnett, and Biffen in England (Adams, 1978). Both had published genetics research in foreign journals well before the revolution, and in the subsequent decade Filipchenko wrote extensively on evolution and genetics. Yet, with almost a decade's head start, neither would contribute to the evolutionary synthesis nearly to the degree of Kol'tsov's group. The reasons lay principally in their conceptions of the nature of genetics as a scientific discipline.

After his return from Germany in 1913, Filipchenko began teaching genetics as a docent at Petrograd University. The lectures for his course were published in two books (Filipchenko, 1915, 1917). The earlier of these, *Variation and Evolution*, presented a view of the relationship of genetics to evolutionary theory that shaped his thinking throughout the remainder of his career. In Filipchenko's opinion, the view of the English biometricians that evolution occurs through the selection of continuous heritable variation had been successfully refuted by Johannsen's work on pure lines, demonstrating the limits of such selection. After discussing de Vries's mutation theory and the subsequent work on the genetics of *Oenothera*, Filipchenko concluded that, although mutations probably play some role in evolution, they do not explain progressive evolution or macroevolutionary phenomena. In discussing Lotsy's theory of hybridization, Filipchenko concluded that recombination undoubtedly plays only a minor role in evolution because it cannot account for the appearance of evolutionary novelties. In discussing the inheritance of acquired characteristics as a possible explanation of evolution, Filipchenko remained skeptical of the validity of the work of Kammerer and others. In his view, such researchers were looking in the wrong place for the cause of evolution.

For Filipchenko the evolution of animals and plants was a developmental process analogous to the development of a chick embryo or the solar system: all evolve through internal factors. Citing as a "general law of development that each whole develops primarily under the influence of its own internal causes and impulses which may be affected only secondarily by external ones," he concluded that "in general *organic evolution originates primarily under the influence of causes lying within organisms which the action of their surrounding environment can only affect in a purely secondary way*" (Filipchenko, 1915, pp. 78-80).

This autogenetic approach to evolution, formed at a time when genetic results seemed antagonistic or irrelevant to Darwinism, continued to dominate Filipchenko's work until his death in 1930, although it gained sophistication in later presentations. In 1923 he argued that intraspecific variation was different in kind from traits characterizing genera and higher taxa, which exhibit "less variation" and "appear significantly earlier during individual development." Unlike "traits characterizing the species," for which "it is known exactly that their carriers are genes localized in the chromosomes of the sex cells . . . for the traits of a generic character, not only has no one proved this, but an entirely different proposition is considerably more likely, namely that they are present in entirely special carriers located, not in the nucleus, but in the plasm of the sex cells" (1923, p. 213). In the 1929 edition of the same work, he distinguished between *microevolution* [*mikroevoliutsiia*] (the evolution of biotypes, Jordanons, and Linnaean species), which could be elucidated by genetics, and *macroevolution* [*makroevoliutsiia*], which lay outside the scope of genetics.

Filipchenko clearly embodied the "experimentalist" view toward evolutionary theory discussed by Allen (1975). For him, "evolutionary theory always was and always will be only hypothetical, since species transformation is not one of the phenomena that can be directly observed" (Filipchenko, 1929, pp. 249-250). In contrast, "genetics . . . has proceeded along the path of perfectly exact scientific research." In his view, "what facilitated the achievements of genetics more than anything else" was that it "has excluded from its field of vision all those disputes over the evolution of organisms, their familial relationships, etc., which until this very time have literally torn apart both systematics and morphology." For Filipchenko, "evolutionary theory is one thing, and genetics altogether another" (1927, p. 3).

Thus, for Filipchenko, the central evolutionary problem was accounting for macroevolutionary change and, by their very nature, explanations of macroevolution were hypothetical and speculative; his own predilection was for autogenesis. By contrast, genetics was a precise experimental science—but its results could only illuminate microevolutionary phenomena. This viewpoint, expressed repeatedly in his historical books and textbooks, dominated the research orientation of the genetics group he headed at Leningrad University, which explains why the evolutionary synthesis found such little development or even sympathy there.

If Filipchenko was an archetypal university academic geneticist, the career of Nikolai Vavilov was dominated by grandiose practical work. Like Filipchenko, he had learned his genetics before the revolution, studying first at the Moscow Agricultural Academy (which led to a position at the government's Bureau of Applied Botany) and then in England with Bateson, Punnett, and Biffen. In 1920 he began a meteoric rise because of his paper, "The Law of Homologous Series in Variation," at a Saratov conference he had organized. Afterward, he focused his considerable energies on organizing and administering a rapidly expanding system of plant breeding stations and collecting expeditions. Evolutionary theory was less important to Vavilov than the "science of selection," which he defined as "evolution directed by the will of man" (Adams, 1978).

Like most of Russia's intelligentsia in the 1920s, Vavilov would have called himself a Darwinian. Yet his genetical works of the 1920s seem curiously noncommittal on specifics. For example, his law of homologous series in variation sought to establish the principle that closely related species of agricultural plants have the same series of intraspecific genetic variations (1920). From the practical point of view, the rule would open the possibility of seeking and finding specific desirable variations in a given domesticated variety of plant that had not been observed in that variety before, but only in closely related species—thus opening the way for "sculpting" plants to have characteristics desirable to the breeder. From a theoretical point of view, however, Filipchenko saw this "law" as evidence that intraspecific variation was qualitatively different from interspecific, because clearly different species and even genera could manifest the same intraspecific variation (1925). Leo Berg also found evidence for his own theory of nomogenesis in Vavilov's concept (1922). Although Vavilov published a criticism of Berg's theory in 1929, there is indirect evidence that he sympathized with it when it first appeared.

Similarly, Vavilov's brilliant work on the centers of origin of cultivated plants led him to the "rule" that the centers of origin have a preponderance of dominant characters, whereas the periphery has more recessive characters (1926, 1927). Vavilov discussed geographic isolation as a cause of this phenomenon only in the later versions of his theory. Again, his theory was closely related to practical needs because it would have provided an indication of where to seek the greatest amount of genetic variation to be used for breeding, but its early form seems out of keeping with a Darwinian evolutionary approach.

Vavilov may have been prejudiced against evolution theory by Bateson, whom he greatly admired. In any case, with his incredibly busy schedule, Vavilov had little time for evolutionary theorizing or even for keeping up with the genetic literature not strictly relating to his agricultural breeding work, for which Kol'tsov publicly chastised him (Targul'ian, 1937, p. 243). In retrospect, Vavilov and those working around him seem to have been largely unconcerned with evolutionary theory as such.

Headed by Filipchenko and Vavilov, who had been well schooled in the pre-Morgan classics, Leningrad researchers had long been aware that the giants of genetics—de Vries, Johannsen, and Bateson—had undermined the simplistic evolutionary viewpoints of speculative Darwinism. True to their training, Filipchenko and Vavilov avoided what they regarded as the inevitable pitfalls of the evolutionary morass in favor of the demonstrably successful strategy of pursuing their new, experimental, truly scientific discipline of genetics. Kol'tsov's Moscow group did not know their genetics nearly so well, and hence suffered under no such handicap.

Kol'tsov and His Institute

Nikolai Kol'tsov (1872-1940) has been called "brilliant . . . probably the best Russian zoologist of the last generation, an amiable, unbelievably cultured and clear thinking scholar, admired by everybody who knew him" (Goldschmidt, 1956, p. 106). Although he was trained as a morphologist in Moscow University's department of comparative anatomy under the great Darwinian ornithologist and biogeographer Menzbier, (1866-1935), Kol'tsov's research trips to European marine biological stations led him to break with morphology in about 1903 to become Russia's leading exponent of the new "experimental biology." His researches on the fine structure of cells (Kol'tsov, 1905, 1908, 1911) led to his election in 1916 as a corresponding member of the Russian Academy of Sciences. Today he is remembered primarily for two highly suggestive monographs proposing a chemical structure for the gene (Kol'tsov, 1928, 1935).

Excluding his early morphological publications (such as Kol'tsov,

1901), his prolific scientific output included only two articles on evolutionary theory: a critical summary of Lotsy's theory of evolution (Kol'-tsov, 1915a) and a discussion of the causes of progressive evolution from a biochemical point of view (Kol'tsov, 1933). His scientific papers on genetics dealt chiefly with eugenics and the chemistry and physiology of the gene; none dealt with population genetics. Thus Kol'tsov's works, unlike those of Filipchenko and Vavilov, show little interest in evolutionary theory or the genetics of populations. Yet population genetics was developed within his institute by his students. In order to understand how this came about, we must examine Kol'tsov's conception of his institute and the field it was intended to embody.

Kol'tsov dated his conversion to the new biology to his experiences at European marine biological stations (1897-1903). While at the Russian station at Villefranche (1900), he became close to Richard Goldschmidt and Max Hartmann, and the three young dreamers laid plans to establish a permanent institute for the experimental study of the living cell to operate in conjunction with the station (Kol'tsov, 1921b, pp. 2-5). When the project fell through, each resolved to establish such an institute in his home country. After his return to Russia in 1903, Kol'tsov devoted most of his efforts to establishing an institute of experimental biology in Moscow.

These efforts were largely shaped by his conception of the new field he hoped to introduce in Russia. In his view, the traditional nineteenth-century disciplines of comparative anatomy, morphology, and systematics had provided a solid groundwork in the zoological "ABCs" and had to be known by modern biologists "like the multiplication tables," but they had ceased to be the cutting edge of research: they had largely "exhausted their research program and their possibilities. Only by uniting with the experimental methods of new biological disciplines . . . can the old comparative anatomy and embryology be reborn as active, creative sciences" (Kol'tsov, 1936a, p. 14). It was time for biology to move from observation and description to experimentation.

Kol'tsov saw the new "experimental biology" as fundamentally different from physiology. Although the two shared the experimental approach, for Kol'tsov physiology was primarily a medical school specialty whose analytic approach posed ever narrower and narrower problems, "studying only one specific function of the organism and striving to break it into simple and—if possible—physico-chemical phenomena . . . By contrast, experimental biology strives to keep before it the total problem of life as a whole and to be not only an analytic, but also a synthetic science" (Kol'tsov, 1927, pp. 40-41). Elsewhere he maintained that "in every science, the best results are to be obtained when the same theme is treated by two quite different methods belonging to two different scientific branches" (Kol'tsov, 1924, p. 497).

Kol'tsov used the great physiologist Pavlov as an example of the necessity of such a synthetic approach. Pavlov's laboratory had published an experimental "proof" of the inheritance of acquired characteristics. After conversations with Kol'tsov, Pavlov acknowledged that the results did not support such a conclusion. Kol'tsov attributed the physiologist's error to the fact that his analytic approach had led him to conduct experiments principally on one organism using purely physiological methods: his ignorance of other fields and other organisms was responsible for his mistake, a mistake precluded by the broadly synthetic, interdisciplinary approach Kol'tsov envisioned for experimental biology (Kol'tsov, 1927, 1936b).

True to his synthetic concept, Kol'tsov argued that the experimental biologist must be trained in biochemistry, biophysics, physical chemistry, anatomy, histology, morphology, systematics, genetics, and biometry; must know how to conduct ecological observations and experiments under natural conditions; and in addition might find general philosophical preparation useful, especially in epistemology and psychology (1927, pp. 40-41). After their educational groundwork was laid, Kol'tsov urged the brightest students to specialize in some particular discipline, ultimately teaching supplementary courses in connection with his *Grosse Praktikum*. Ultimately, the best students who had gone through this training would take up research in this field, and each field would become a division of the Institute of Experimental Biology that Kol'tsov hoped to create. Although each division would develop its own subject matter and methods, they would maintain interaction through seminars and workshops and by jointly working on common research problems, each using their own disciplinary approach.

From 1905 through 1917 Kol'tsov was active as a scientific entrepreneur, seeking to train students and establish experimental biology while working against severe constraints. Originally he sought to use his home base at Moscow University, but these efforts were largely abortive. As a liberal, he was drawn into the political struggles in 1905, and his book documenting tsarist atrocities against university students—*In Memory of the Fallen* (1906)—could hardly have endeared him to the government. A few years later, he argued for broad-scale reform in the Russian university system in *On the University Question* (1909), which earned him other enemies. Kol'tsov effectively ceased teaching at Moscow University in 1909 and, following the direct takeover of universities in 1911 by Kasso, the tsarist minister of education, he officially left the university along with sixty other teachers, including his mentor Menzbier.

Fortunately, by 1909 Kol'tsov had found two alternative institutional bases for his work. The Shaniavsky University (officially the Moscow City People's University) was a private higher educational institution that had opened in 1906 with endowments from the will of P. I. Shaniavsky, a Polish count and successful industrialist. Kol'tsov began lecturing there in 1906, and by 1911 had established an outstanding program in zoology supplemented by a research laboratory. Kol'tsov's second research base was the Beztuzhev Advanced Courses for Women—also known as the Moscow Women's University—where he began teaching in 1908 and succeeded in establishing a research laboratory.

Meanwhile Kol'tsov was actively promoting his proposal for an Institute of Experimental Biology. Both private universities were heavily funded by entrepreneurs, and it was to these entrepreneurs that Kol'tsov turned for financial support. In 1914 the Moscow Scientific Research Institute Society was formed with civic and business support by a group of outstanding young experimental biologists including Kol'tsov. Its purpose was to sponsor and help raise private funding to establish a series of autonomous institutes. The new Russian science journal *Priroda*, of which Kol'tsov was an editor, announced in 1916 that the society had received a number of contributions, notably 1.2 million rubles (equivalent to roughly half as many contemporary dollars) from the will of the Russian railroad entrepreneur G. M. Mark. As a result, a new Institute of Experimental Biology was to be founded, to be headed by Kol'tsov and located on land contributed by the Moscow City Duma (Adams, 1980).

By the time his new Institute of Experimental Biology opened in March 1917, Kol'tsov had trained students and established laboratories at Shaniavsky and Beztuzhev and had already published a volume of their original researches (Kol'tsov, 1915b). These students became the staff for the new institute, and their research specialities give a good indication of the character of the institute Kol'tsov sought to establish. At Shaniavsky, he had trained S. N. Skadovsky (physicochemical biology, hydrobiology), G. O. Roskin (fine structure of cells), P. I. Zhivago (cytology), M. M. Zavadovsky and D. P. Filatov ("developmental mechanics" or embryology), I. G. Kogan (tissue transplantation), O. L. Kan' (tissue culture), V. N. Lebedev (pathology), G. V. Epstein (protistology), V. G. Savich, and V. V. Efimov. From the Beztuzhev Advanced Courses for Women came S. L. Frolova (karyology), V. Schröder and A. Tausend (physicochemical biology), and Kol'tsov's future wife, M. P. Sadovnikova (animal behavior). Only one student had specialized in genetics— A. S. Serebrovsky—and Kol'tsov commissioned him to write a paper for *Priroda* on the current status of the mutation theory (1915). By the time the article appeared, however, its author had already been mobilized into the Russian army.

Genetics and the Kol'tsov Institute

As of 1917, Kol'tsov's plan for the institute had made no provision for evolutionary theory and had assigned genetics a very minor role. Within six years, however, genetics had become one of the central activities of Kol'tsov's enterprise. The new prominence of genetics was the result of three converging factors in postwar Russia. First, after the October 1917 revolution, Kol'tsov's institute—which he had finally managed to endow —lacked funding and Kol'tsov had to find new patrons for his organization. At that time he oversaw four separate but interrelated research centers: the laboratories at Shaniavsky and Beztuzhev, his nascent institute, and the Hydrobiological Station at Zvenigorod, which had been donated to Shaniavsky University's zoology department in 1910. The day was temporarily saved by the Russian Academy of Sciences.

In late 1917 Kol'tsov was appointed the head of the Moscow division of the genetics section of KEPS. In the spring of 1918, with KEPS funding, Kol'tsov created a genetic agricultural research station at Anikovo, some 40 kilometers outside Moscow (Kol'tsov, 1921b, 1922). The creation of this station for the study of chicken and sheep breeding legitimated the financial support of the institute by KEPS and provided additional salaries and ration cards during hard times. Kol'tsov selected Lebedev to run the station. Trained as a pathologist under Kol'tsov at Shaniavsky, Lebedev's only previous experience in genetics had come from assisting Kol'tsov with work on a genetic analysis of coat color in guinea pigs, work undertaken in late 1917 on the premises of the institute and reported to the Moscow Society of Naturalists early in 1919 (Kol'tsov, 1921d, pp. 87-97). As the head of the Moscow branch of KEPS's genetics section, Kol'tsov also funded an agricultural station located near the village of Slobodka, outside Tula, and appointed his student Serebrovsky as its director of research. The work on chicken genetics was transferred to the Tula station and put under Serebrovsky's direction. The location of the station was probably related to the fact that Tula was Serebrovsky's hometown.

In 1919 Kol'tsov sought larger and more secure funding for his multiple enterprises. The primary financial support for the institute was provided by the Commissariat of Public Health (Narkomzdrav) headed by Kol'tsov's friend Nikolai Semashko. In late 1919 or early 1920, together with the other institutes created by the Moscow Scientific Research Institute Society, Kol'tsov's institute became part of GINZ, a system of research institutes funded by Narkomzdrav. To support the Anikovo and Tula stations, Kol'tsov succeeded in obtaining funding from the Animal Husbandry Division of the Commissariat of Agriculture (Narkomzem) in late 1918 or early 1919. In 1920, as part of the plan, the two stations were

amalgamated at Anikovo, with Lebedev as director of the station and
Serebrovsky as the head of laboratories. The move from Tula was completed in 1921. Thus, the first reason for the new prominence of genetics
stemmed from the necessity of obtaining financial support from KEPS
and later Narkomzem.

In addition, the kind of research that Kol'tsov was most anxious to
develop—physicochemical biology and studies of fine cell structure—
required expensive laboratory equipment that was not available, and for
which there was no funding during and immediately after the civil war.
The only microscopes at hand were cheap American student models
(Timoféeff-Ressovsky, private communication, 1977). Kol'tsov was
almost apologetic in explaining his new emphasis of genetics research to
American readers in 1924:

> When in 1916 I began the organization of the new laboratory of the
> Institute of Experimental Biology, great difficulty was encountered
> in getting in time of war all the necessary instruments for providing
> a modern physico-chemical laboratory . . . In such circumstances we
> have been forced to concentrate our attention on such problems of
> experimental biology as may be studied without complex instru-
> ments and rare chemical preparations.
>
> In connection with these circumstances I was compelled to lead
> the most immediate investigatory work of the Institute in two new
> directions: firstly surgical experiments, especially in relation to en-
> docrine glands, and, secondly, genetic investigations (1924, p. 499).

These factors, especially the second, probably explain the earliest
work in genetics under Kol'tsov's direction. In addition to his own work
with Lebedev on coloration in guinea pigs in 1917, Kol'tsov had put his
young student D. D. Romashov to work on biometric studies. In 1919 he
assigned Romashov and another student the task of attempting to create
mutations in *Drosophila* (Romashov) and *Artemia salinia* (N. N. Gaev-
skaia), attempts that proved unsuccessful (Kol'tsov, 1930, p. 242). Aside
from Kol'tsov himself, his only associates who had done any genetics
work were Lebedev (trained in pathology), Serebrovsky (specializing in
poultry genetics), and Romashov (biometrics).

At this time, however, a third factor came into play. After the scientific isolation of Russian scientists during the latter years of World War
One, the revolutions, and the civil war, information began to reach Moscow of the Morgan school's work on *D. melanogaster*. The first public
information was provided by Filipchenko in two popular reviews of
Morgan's work, published in *Priroda* in 1919 and 1922 (Filipchenko,
1919, 1922). Around this time Kol'tsov was sent a copy of the 1915 classic by the "Four Musketeers," as the Russians called them—Morgan,

Sturtevant, Muller, and Bridges' *Mechanism of Mendelian Heredity*. As a result, Kol'tsov took an increased interest in *Drosophila* genetics as a possible research area for his institute. Filipchenko's 1919 article was almost certainly the stimulus for Romashov's attempts to induce mutations in *Drosophila* later that same year. Although fruit flies were not of agricultural interest, from Kol'tsov's point of view they were an ideal object of genetic study because they were widely available in nature, could be cultivated relatively easily and cheaply, and could be bred and studied without technical equipment that was expensive or difficult to obtain.

The urgency of developing *Drosophila* research was heightened by H. J. Muller's visit to the Kol'tsov Institute and its Anikovo genetics station in August 1922. He brought with him an invaluable gift: *Drosophila* cultures containing 32 of the famous mutants that had been the basis for the spectacular researches of the Morgan school. Muller's talk at the station, titled "Ten Years of *Drosophila* Research," was subsequently published in a journal edited by Kol'tsov. Muller's own account of his trip appeared in his article, "Observations of Biological Science in Russia" (1923). In that article he thanked Kol'tsov and noted that "Prof. Lebedeff" and Serebrovsky were in charge of genetics research at the Institute of Experimental Biology. The mutants brought by Muller were described in an article by Serebrovsky and Sakharov (1925).

In retrospect, it might seem surprising that Muller did not mention Chetverikov. But we do not even know if Chetverikov was at Anikovo during the afternoon of Muller's visit. He had been invited to join the institute staff late in 1921, only a year earlier. At that time, Chetverikov had no firsthand experience with genetics. It seems clear that Kol'tsov had invited Chetverikov to join the institute for one primary reason: of all the many students and colleagues in Kol'tsov's circle, he was the only one who knew a great deal about flies.

Sergei Chetverikov

Kol'tsov's appointment of Chetverikov to his institute was an important event in the history of population genetics and the evolutionary synthesis. Yet in some respects the appointment was problematical because Chetverikov knew almost no genetics when he was appointed. To understand why Kol'tsov chose to take the risk, and why in retrospect it proved a key event in the history of the synthesis, we must understand Chetverikov's background.

Sergei Chetverikov (1880-1959) was eight years younger than Kol'tsov. Like Kol'tsov, he was born into a family of successful entrepreneurs;

indeed, his father was a millionaire. At the turn of the century much of the Moscow merchantry was related by marriage, and they played a very active role in the economic, civic, and cultural life of Russia. Kol'tsov's great grandmother and Chetverikov's mother were both members of the Alekseev family, and both Chetverikov and Kol'tsov were second cousins of K. S. Alekseev, better known by his stage name "Stanislavsky," founder and director of the Moscow Art Theater (Astaurov, 1974; Astaurov and Rokitsky, 1975).

Chetverikov's father sent him to a *Realschule* to prepare him to become an engineer and take over his manufacturing enterprises. However, he learned Darwinism from his physical geography teacher V. P. Zykov (a docent at Moscow University) and became committed to a career in zoology. His father resisted his decision, sending him to a German *Technikum*, but finally relented. To enter Moscow University at that time, a student had to pass examinations based on gymnasium training. Chetverikov spent a year preparing for these examinations, during which time he absorbed a great deal of the German and English biological literature.

Chetverikov finally entered Moscow University at the age of twenty in 1900—the same year that Kol'tsov became a privat-docent in Menzbier's department of comparative anatomy. From that point on, their careers exhibit parallels. In 1905 Chetverikov took an active part in the student disturbances, and was arrested and jailed for two months (Astaurov, 1974); it was in sympathy with these disturbances that Kol'tsov had written his tract, *In Memory of the Fallen* (1906). When Chetverikov graduated in 1906, he became a graduate student in Menzbier's department, earning his "professor's" degree in 1909 for a painstaking study of the anatomy and development of *Asellus aquaticus* L., a widely distributed isopod crustacean. It is likely that he did this work under Kol'tsov's direction: Kol'tsov had opened a new course in invertebrate zoology at Moscow University in 1904.

In 1909 Kol'tsov had to give up teaching at Moscow University and so shifted his full attention to teaching zoology at Shaniavsky University (general zoology) and the Beztuzhev Advanced Courses (invertebrate zoology). That same year Chetverikov became an assistant (officially, *laborant*) to the zoology department at the Beztuzhev Courses, where he taught a course in entomology from 1909 to 1919. The coincidence of dates suggests the possibility that Kol'tsov had brought Chetverikov along with him and given him a post, although there is no direct documentary evidence for this. In 1916 Kol'tsov coedited *The Calendar of Russian Nature*, an anthology of observations of seasonal variation in the flora and fauna of the Moscow area: the article on butterflies (one of the longest) was written by Chetverikov (1916).

In 1918 the Beztuzhev Courses became the Second Moscow University, and Chetverikov was advanced from *laborant* to lecturer. The next year the two Moscow universities were united and Chetverikov automatically became a lecturer at Moscow University, teaching two courses: general entomology, and a new course, "Introduction to Theoretical Systematics," which dealt chiefly with biometrics. Chetverikov began to teach a genetics course there only in 1924, three years after he was appointed to the staff of the Kol'tsov Institute. He became a docent in 1926.

Kol'tsov and Chetverikov had had many opportunities to become acquainted: through family, social circles, the Menzbier department, and as fellow teachers of invertebrate zoology at Beztuzhev and later at Moscow University. Given their many connections, we may well wonder why Chetverikov had not been called on earlier to take a more active role in Kol'tsov's activities aimed at creating experimental biology as a discipline in Moscow. The answer lies in the fact that unlike Kol'tsov, Chetverikov was not an experimentalist, but a naturalist.

Even before studying at Moscow University in 1900, Chetverikov had become a passionate butterfly collector. Shortly after arriving there, he began to work for the Society of Natural History, Anthropology, and Ethnography and its study commission on the fauna of the Moscow province, for which he did the work on *Lepidoptera*. His delayed graduation from the university was the result not only of political events in 1905, but also of his participation in a collecting expedition to the Tarbagatai mountain range, under the direction of P. P. Sushkin, the ornithologist. Together with Sushkin's young wife, Anna Ivanovna, Chetverikov also completed expeditions to the Minusinsky region, the Western Sians, and Lake Zaisan where he avidly collected butterflies. Shortly thereafter Anna left Sushkin to become Chetverikov's wife.

The years between 1900 and 1919 have been termed Chetverikov's "Lepidoptera period" (Astaurov, 1974). Between 1902 and 1925, Chetverikov published nineteen scientific papers: his diploma work, three articles on how to collect insects, thirteen on butterfly systematics (including descriptions of new species), and two more general, theoretical works (Rokitsky, 1975, pp. 73-75). The first of these theoretical works, "Waves of Life" (Chetverikov, 1905), described seasonal and irregular radical fluctuations in the size of insect populations in nature and speculated about the cause and effect of such fluctuations; later Dubinin and Romashov cited this work in their papers on genetic drift (Adams, 1970). His second theoretical article, "The Fundamental Factor of Insect Evolution" (1915, 1918) noted that whereas the vertebrates tended to evolve toward larger forms, insects evolved toward smaller ones—a general trend he attributed both to selection pressures and, most importantly, to the morphological and mechanical consequences of the internal versus external skeletons of the two groups.

Chetverikov's experience and propensities as a naturalist help to explain why he had not become more involved in Kol'tsov's enterprise before 1921. It was precisely this natural history approach in systematics, comparative anatomy, and morphology that Kol'tsov regarded as outmoded and old-fashioned and that he hoped to replace, or reinvigorate, with the new experimental approach based largely on laboratory work. In 1919 Chetverikov already was showing his willingness to adopt new approaches by his new interest in biometrics.

Between 1917 and 1921 Kol'tsov became aware of the importance of *Drosophila* genetics and also of the necessity of obtaining funding for his workers through various academy and government commissions interested in surveying the natural wealth of the country. Chetverikov brought to the institute widespread naturalist experience, a newly adopted interest in biometrics, and an unparalleled expertise in insects. He was the one figure associated with Kol'tsov's network of students and colleagues with these characteristics.

Chetverikov's Group

If Chetverikov brought to the institute his expertise as a naturalist, he found there a group of students who had received some of the best zoological training then available anywhere. All had gone through Kol'tsov's *Grosse Praktikum*, an intensive two-year laboratory course that Timoféeff-Ressovsky has called the best such course he has ever seen, including those he saw in the following decades in Europe and America (personal communication, 1977). When Chetverikov joined the Kol'tsov Institute in 1921, he began to work with a group of young students who had been given research jobs. Unlike Serebrovsky's group, which was centered at Anikovo and was primarily concerned with agricultural breeding work, especially on sheep and poultry, Chetverikov's group consisted of students of a more theoretical inclination; centered at the hydrobiological station at Zvenigorod—7 or 8 kilometers from Anikovo—they were headed by Skadovsky, a specialist in hydrobiology.

The original group (1921-1922) consisted of Chetverikov; his wife, Anna Ivanovna; N. W. Timoféeff-Ressovsky; H. A. Fiedler (subsequently Timoféeff-Ressovsky's wife); D. D. Romashov (Kol'tsov's pupil in biometrics); S. R. Tsarapkin; and A. N. Promptov (an ornithologist). The group was joined in the next two years by bright undergraduates from Moscow University: B. L. Astaurov, E. I. Balkashina (later Romashov's wife), N. K. Beliaev, S. M. Gershenson, and P. F. Rokitsky. Following his synthetic strategy, Kol'tsov had brought together a group of well-trained general zoologists with a theoretical bent, urging each to take up genetic study on his or her own chosen group of organisms.

Timoféeff-Ressovsky has called this group a collection of "wet zoologists" [mokrye zoologi], specializing in the biota of the lakes that covered central Russia. The hydrobiological station had been commissioned by Glavryba (the Central Administration of Fisheries) to inventory the flora and fauna of Russian lakes in their region. The earliest genetic study of natural populations undertaken by Chetverikov's group, then, was not on Drosophila at all: it was on the fluctuations of albino mutations among lake and river populations of forget-me-nots. The group observed that at certain times there appeared to be no albino mutants; then shortly thereafter there would be very many. What was the genetic basis of this character, and in particular why did its frequency vary so greatly, both between different populations and within the same population over time? Early experiments were undertaken comparing the incidence of the mutant in wild populations with that in cultivated populations bred at the station (Timoféeff-Ressovsky, personal communication, 1977).

At this time none of the group could have been called a geneticist—and certainly not Chetverikov. According to Timoféeff-Ressovsky, Kol'tsov's copy of The Mechanism of Mendelian Heredity was divided up and studied chapter by chapter and line by line; although most students with an intelligentsia background could read German or French with ease, English had not been as popular, so the group had to learn their English by reading genetics works published in English.

Such was the state of the group and its work on the eve of Muller's visit. The gift of Drosophila from the Morgan laboratory presented the group with an opportunity that it was then not fully prepared to exploit. After consultations involving the group and the senior zoologists of the institute, including Kol'tsov, the decision was made that in addition to each student pursuing his or her own studies on their own groups of organisms, all would pitch in to exploit the Drosophila material. From 1922 to 1925 most of the work was directed at mastering the theory and practice of Morganist genetics. Informal discussions were held at the apartments of the members of the group (frequently at Romashov's apartment, which had more space than most), in which the members would report on genetics works they had read. During the winters, most of the time was spent in these informal seminars and working with Drosophila in the laboratory; during summer months, the studies were continued at Zvenigorod.

In 1924 Chetverikov ceased teaching his course in entomology at Moscow University and began to teach a course in genetics. This course was based on the reading and researches that he and his coworkers had undertaken during the preceding three years. That same year Chetverikov organized his seminar on a somewhat more formal basis, and it gained the name *Droz-So-or*, an acronym for *sovmestnoe oranie drozofil'shchikov* (literally, "the combined cacophony of drosophilists," or, more loosely, "the drosophilist screeching society"). Its program was the same as that of the earlier group meetings. Its membership was restricted to those within the Chetverikov group, although others attended when they could, notably the cytologists S. L. Frolova and P. I. Zhivago, Serebrovsky, Sakharov, and Kol'tsov himself. During 1924 and 1925 the group engaged in discussions that were the basis for both Chetverikov's theoretical paper of 1926 and the first studies of the genetic composition of natural populations of *Drosophila* (Chetverikov, 1974).

The work of this group has been fully discussed elsewhere (Adams, 1968, 1970; Astaurov, 1974; Rokitsky, 1974, 1975). Here we should merely note the outline of the reasoning that for the first time brought together genetic, biometric, and naturalist approaches. By virtue of his naturalist background, Chetverikov's primary concern was investigating the genetics of natural populations, and in particular using genetic analysis to gain a deeper understanding of actual evolutionary processes occurring in natural populations. In this respect especially, their work differed from that done on *Drosophila* by Serebrovsky, Sakharov, and others who were most concerned with physiological and structural genetics.

By virtue of their own work on the *Drosophila* Muller had brought, the Chetverikov group knew that mutations appear to occur randomly, in ways that are not the result simply of laboratory conditions. Indeed, in the two years since Muller had brought the flies, new mutations had appeared in their progeny (Serebrovsky and Sakharov, 1925). The group also knew from the genetics literature that most mutations in *Drosophila* are recessive. From their examination of the biometric literature, they knew (from Hardy's law) that, in the absence of selection, such mutations would be preserved in a population. Finally, they knew from the table prepared by H. T. J. Norton that selection would eliminate unfavorable recessives from the population far more slowly than dominants. Chetverikov had come across Norton's table in R. C. Punnett's book *Mimicry in Butterflies* (1915), a book he knew well precisely because it dealt with his specialty—butterflies. Taken together, this reasoning led the group to conclude that natural populations, to use Chetverikov's language, "soak up mutations like a sponge," and that such mutations constitute the hidden genetic variation on which natural selection can act to produce evolution.

This reasoning, set forth in Chetverikov's classic 1926 paper, was the basis for the first studies of natural populations of *Drosophila*. These studies began in the spring and summer of 1925 on four species of *Drosophila* found in the environs of Moscow: *D. phalerata* (Astaurov and Beliaev), *D. transversa* (Astaurov), *D. vibrissina* (Balkashina), and *D. obscura* (Gershenson). The initial purpose was to study the genetics of these species to compare with *D. melanogaster*, to study the structure of natural populations, and to note the extent to which Vavilov's "law of homologous series in variation" held true in *Drosophila*. However, these species live primarily on fungus and proved difficult to cultivate in the laboratory. For this reason, the work continued in 1926 on *D. melanogaster*. Gershenson and Rokitsky brought back approximately 300 females fertilized in the wild, captured from a natural population in Gelendzhik, on the Caucasian coast of the Black Sea—certainly far enough away so that there could be no question of their having been contaminated by laboratory strains. From these females, 239 inbred lines were established and studied, and 32 hidden mutations were revealed in their progeny.

The results of these investigations were not published immediately. Brief accounts were given by Chetverikov in his papers before the Fifth International Congress of Genetics (1928a) and the Third All-Russian Congress of Zoologists, Anatomists, and Histologists (1928b). Some of the investigations of *Drosophila* species other than *melanogaster* were published only a decade later (Gershenson, 1934; Balkashina and Romashov, 1935). To quote Astaurov, "By far the most abundant material was collected on *Drosophila melanogaster* in 1926, and to this very day it has not been published in any complete form" (1974, p. 64). Part of the delay in publication no doubt resulted from Chetverikov's thoroughness. According to the testimony of members of his group, he would not be hurried—an understandable caution considering his relatively recent entry into genetics research. For their part, the members of his group regarded the right to initial publication as his, despite the cooperative nature of the group's work, because an analysis of the significance of the results would have been impossible without using the ideas he developed in his 1926 paper.

It is not at all clear that Chetverikov's senior Russian colleagues understood the significance of his work. Filipchenko had heard Chetverikov's paper delivered in Leningrad in December 1927, yet in a letter to Dobzhansky later that month he indicated that he did not regard the paper as having any special significance, commenting only on the tiff that broke out between Chetverikov, with his "pushy" Moscow style, and E. S. Smirnov, the advocate of the inheritance of acquired characteristics. In all probability, Filipchenko regarded Chetverikov's work as elucidating only the microevolutionary, but not the macroevolutionary problem. Kol'tsov alluded to Chetverikov's work as "having in my opinion great theoretical interest, in as much as it connects experimental laboratory genetics with the problem of the evolution of organisms in nature" (1927, p. 61). He remarked in 1932 that "the fundamental synthetic problem in biology continues to be the problem of evolution" and went on to discuss Chetverikov's work; but when he again mentioned "synthesis" it was in connection with the roles played by genes and the environment in producing the phenotype (Kol'tsov, 1932, pp. 36, 38). Serebrovsky only began his studies of the genetics of populations of domesticated poultry in 1926—after having been party to the discussions in the *Droz-So-or* for three or four years—but in his many publications he alluded only rarely to Chetverikov's work, and then as only one of several investigations classifiable as part of his new science of "genogeography" (see, for example, Serebrovsky, 1930, p. 85).

It may well be that Chetverikov's Russian colleagues were waiting for Western reaction to his work. Unlike Kol'tsov, Filipchenko, Serebrovsky, and Vavilov—who published abroad frequently—Chetverikov had not published anything in Western genetics journals. The Chetverikov archives contain two handwritten manuscripts of an English text of Chetverikov's paper, apparently translated by him, with two major changes from the original 1926 Russian text: he alluded to and discussed Muller's discovery of the mutagenic effect of X rays; and he summarized his group's studies of natural populations and showed how they confirmed his theoretical reasoning (Chetverikov, 1929[?]). Apparently Kol'tsov had urged him to translate his article for Western publication, and evidence indicates that he had done so and was working over the version in 1929 to improve the English. However, this English version was never sent abroad. The primary reason was Chetverikov's untimely arrest.

Chetverikov's Arrest and Its Aftermath

In June 1929 Chetverikov was arrested by the secret police (OGPU) and spent several months in its Butyrka prison in Moscow. Thereafter he was exiled to Sverdlovsk for three years, then to Voronezh, during which time he taught high school mathematics and worked as a consultant to a project for creating a zoo. During the three years of his administrative exile he was forbidden to travel to any republic of the USSR other than Russia, to Moscow or Leningrad, or to the regions surrounding them.

Chetverikov's arrest came as a surprise both to him and to Kol'tsov, who was lecturing in Paris at the time. No formal charges were ever brought against him, and right up until his death Chetverikov indicated that he did not know the reason for his arrest. However, in his reminiscences dictated to V. Soifer shortly before his death, Chetverikov alluded to an incident some years earlier that may have played a role. In the mid-1920s the Communist Academy had been emphasizing the importance of research on the inheritance of acquired characteristics, with some members regarding it as the only truly Marxist approach to the problem of inheritance. In this connection it had invited Paul Kammerer to come to Moscow and head a laboratory. After Noble's publication in *Nature* suggested that Kammerer's results were fraudulent, Kammerer committed suicide. His obituary, published in *Izvestiia* (October 7, 1926) and the *Herald of the Communist Academy*, referred to a postcard signed by a Professor Chetverikov "congratulating the Academy on Kammerer's suicide," and stated that Chetverikov was "one of the reactionary obscurantists left behind in the USSR." On the next day (October 8), *Izvestiia* published a letter from Chetverikov, together with a supporting letter from Kol'tsov, stating that he had sent no such postcard. Despite this rebuttal the imprecation stuck and was subsequently cited as true. Chetverikov had been a strong opponent of the doctrine of the inheritance of acquired characteristics and had openly scoffed at Kammerer's work. He also had a very sharp temper when aroused, and had gotten into two shouting matches with Smirnov: one during a paper given before the Communist Academy by Smirnov on May 26, 1926, which Chetverikov characterized as a "political denunciation" and in other terms such that Smirnov refused to continue; and later at the Leningrad conference where Chetverikov had given his paper in 1927. Against the background of these events the accusation may have seemed plausible at the time. Verified or not, it may have gone into a secret police file to be brought to attention later. These occurrences have been briefly recounted by Soviet authors (Gaisinovich, 1968; Astaurov, 1974).

A second scenario, and one that may have actually triggered Chetverikov's arrest, has been alluded to by Kol'tsov. In a letter to Maxim Gorky (July 28, 1929), in which Kol'tsov sought his aid in obtaining Chetverikov's release, the following account appeared: "Early this year at the time of my stay in Paris, a stupid incident occurred among the graduate students at my Institute . . . Tempers flared, and some of the students took it all out on S. S. Chetverikov. At the University they drew up an indictment of him consisting of ten charges . . . The charges were published, and several days later Chetverikov was arrested and has been sitting in the Butyrka prison for the last two months" (Shvarts, 1975, p.

256). Shvarts's account, published in a New York émigré journal, includes this and other letters that have come into his possession, but they provide no further details concerning the nature of the "stupid incident."

Rumors at the time and later suggest that it may have had something to do with the *Droz-So-or*. Given the political atmosphere in mid-1929, it is easy to understand how Chetverikov's group must have aroused the suspicion of the authorities. It was a private group that met in private apartments and had a regular and closed membership of some fifteen people; furthermore, new candidates for membership were subject to blackballing, with one negative vote excluding their participation. In his own account of the way he set up and ran the *Droz-So-or*, Chetverikov insisted that these ground rules were necessary to keep the group small and facilitate open and freewheeling scientific discussions. Others have suggested that everything was discussed there, including politics. In any event, the membership rule was apparently strictly enforced. Various rumors suggested that Chetverikov was denounced by a student wishing to join the group who had been refused admission; by a member of the group whose wife had failed to be admitted; by the husband of a graduate student who had not been admitted; or by a colleague who was a careerist and envious of Chetverikov's popularity. In any case, no one at the time knew who was the guilty party, or whether it was the same person who had sent the postcard: until the secret police files are opened, we cannot know.

Whatever the cause of his arrest, its consequences are clear. Within a year, his entire group had been dispersed. Astaurov left for Tashkent to work at the Central Asian Silkworm Breeding Institute (1930-1936). N. K. Beliaev worked briefly at this institute (1929-1932), then at a similar one in Tbilisi (1932-1937). Gershenson took up posts at the Zoological Institute of Moscow University and at Serebrovsky's laboratory at the Timiriazev Institute (1930-1937). Promptov joined the staff of the Second Moscow Medical Institute. Rokitsky joined the Institute of Animal Breeding of the Agriculture Academy, directed first by Kol'tsov and then by Serebrovsky. N. W. and H. A. Timoféeff-Ressovsky had left to work in Berlin four years earlier, in 1925. Chetverikov's wife followed her husband. Of the original group, only Balkashina and Romashov remained at the Kol'tsov Institute.

In 1929 Chetverikov's group probably represented a third of the world's animal population geneticists (and over two thirds of its *Drosophila* population geneticists), and was in many respects a core group at the institute. As a result, their dispersal might have spelled the end of population genetics there. However, Kol'tsov devised a strategy to keep Chetverikov's research going despite his arrest and exile, a crippling po-

litical scandal, and the dispersal of his group. In 1932 Kol'tsov invited
Nikolai P. Dubinin to head up a new, greatly expanded genetics section.
His selection was hardly accidental. Dubinin had been an orphan, presumably of peasant origin, and a *vydvizhenets*—that is, someone pushed
ahead in education, despite inadequate qualifications, because of class
origins or political sympathies. As a student at Moscow University, he
had taken Chetverikov's courses, specialized in genetics, and worked
with Serebrovsky on experiments on the structure of the gene and on expeditions to establish the geographical distribution of genes in Central
Asian populations of domesticated fowl (Dubinin, 1973). In short, he
was a first-rate young geneticist with exactly the right class background
and political leanings.

Meanwhile, Romashov and Balkashina had been kept on, ensuring a
continuity between the work of the old and new groups in population
genetics. Romashov had published a paper in the institute's journal on
genetic drift (Adams, 1970; Balkashina and others, 1975). Dubinin was
invited by Kol'tsov to submit a paper on the same subject and published
a third, joint paper with Romashov. In 1932 Kol'tsov invited Romashov
to organize a seminar on evolution and genetics, much like that formerly
run by Chetverikov, and he largely ran it during much of the 1930s.
Meanwhile, Balkashina and Romashov were kept in relatively low profile, while Dubinin, together with his new group of some fifteen subordi-
nates, were given prominence for their outstanding investigations of the
genetics of natural populations of *Drosophila* (Dubinin and others, 1934,
1936, 1937; see Adams, 1970). Hence, the research program of Chetverikov's group was maintained despite an almost total turnover in person-
nel. By 1934, 35 people or more were working on genetics at the institute
(mostly on *Drosophila*), of whom at least 18 were involved in studies of
natural populations (Kol'tsov, 1934). By 1942 more than 40 Soviets had
written or coauthored papers in the field, including a new generation of
researchers such as R. L. Berg and Iu. M. Olenov.

Chetverikov's influence on the development of population genetics
and the evolutionary synthesis in the West came principally through
three major figures: Timoféeff-Ressovsky, Haldane, and Dobzhansky.
Timoféeff-Ressovsky was part of Chetverikov's group at the Kol'tsov Institute beginning in 1920. In 1925 Oscar Vogt invited him to Germany to
develop genetics, and especially population genetics, within that country. Between 1925 and 1945 Timoféeff-Ressovsky and his wife and co-
worker, Helene, worked at Buch just north of Berlin, becoming Germany's leading evolutionary theorists during the 1930s, and virtually its
only population geneticists. Their German, Italian, and English publications in these fields, which cite Chetverikov, Romashov, Dubinin, Sere-

brovsky, and other colleagues from the Kol'tsov Institute, clearly constitute a development of the work they had first undertaken there as part of the Chetverikov group (see, for example, Timoféeff-Ressovsky, 1939, 1940; Timoféeff-Ressovsky and Timoféeff-Ressovsky, 1927).

A second important line of influence runs through Haldane and his students. Haldane was both a leading mathematical evolutionist and an ardent Marxist, and in both capacities he took a great interest in the work of Russian researchers (1932). He seems to have met Chetverikov at the Fifth International Congress of Genetics in Berlin in 1927 where, according to one account, he embraced and kissed Chetverikov after the latter had given his paper (Fel'dman, 1976, p. 118). Apparently Haldane arranged that the Russian works of Chetverikov and his group be translated into English by a Mrs. A. Sproule, and these translations were available in Haldane's laboratory. The earliest studies in experimental population genetics to emerge from the Haldane group mentioned these translations and clearly stated their debt to the earlier Russian work (Gordon, 1936; Gordon, Spurway, and Street, 1939).

Dobzhansky also played a key role in spreading Chetverikov's influence. Dobzhansky had studied at Kiev with Kushakevich before Filipchenko invited him to come to Leningrad University in 1925. He worked there as a docent until 1927, when he left to work at the Morgan laboratory in the United States, adopting the country as his new homeland. Although he had often visited the Kol'tsov Institute while still in Russia, it is clear from his letters to Filipchenko between 1927 and 1930 (now in the Filipchenko archives, Leningrad) that Dobzhansky did not appreciate the work of the group around Chetverikov and tended to identify more with Filipchenko's approach; at the time he was clearly more interested in laboratory genetics than in population genetics (Adams, 1979). However, by the mid-1930s, when he was preparing his Jesup lectures, he drew heavily on the work of Chetverikov, Timoféeff-Ressovsky, and Dubinin. Their influence on his research grew as Dobzhansky increasingly turned to studies on the genetics of natural populations (Dobzhansky and Queal, 1938), undertaken with Sturtevant, Wright, and other collaborators (Provine, 1978) largely in response to the earlier Russian work.

Meanwhile, following the expiration of his term of exile, Chetverikov reestablished a biological career in the city of Gorky. On the initiative of Z. S. Nikoro, in 1935 Chetverikov was appointed to head the genetics department of Gorky University, and in 1940 became the dean of the biological faculty. Although he trained a few students there in population genetics (Nikoro, his colleague at Gorky, began publishing her works in the field shortly after his arrival), Chetverikov himself never again undertook population genetics research to any great extent. In 1937 he undertook research on silkworm breeding at the government's behest, and in 1944 he won a high state award (*Znak pocheta*) for his research because of its connection with the production of silk for parachutes. He was forced to resign from his posts in 1947-1948 because of heart attacks and generally ill health. The Lysenko meeting of August 1948 led to his final separation from the university and the loss of his pension. He died in 1959, shortly after hearing that he was to be awarded one of the Darwin plaques of the (East) German Academy of Sciences.

Chetverikov, the Kol'tsov Institute, and the Evolutionary Synthesis

The origins of experimental population genetics and the significant moves toward the evolutionary synthesis achieved at the Kol'tsov Institute in the 1920s by Chetverikov and his group were the result of an interaction of scientific traditions, personal networks, institutional structures, inspired individuals, and broader social trends. The primary interaction was the one between the institute that Kol'tsov had created and staffed and the traditions and approaches that Chetverikov brought to it.

Kol'tsov's conception of his institute had two primary components: the development of experimental research in biology; and the synthetic education of biologists in many disciplines, together with the creation of an institutional environment where research problems could be commonly addressed by workers in different disciplinary specialties. That institutional form was what Kol'tsov had striven to create before either genetics or Chetverikov had found a place within it. The students who had been trained in Kol'tsov's program in zoology had been prepared in advance for precisely the kind of work that they, together with Chetverikov, would undertake.

Timoféeff-Ressovsky has remarked at Muller's surprise that his early publications on microevolution included data on plants, birds, fish, and other organisms as well as *Drosophila*. When Muller asked where he had found out about all those organisms, he answered, according to his reminiscences, that he had taken it in "with mother's milk" in Kol'tsov's *Grosse Praktikum*, where he and the other members of Chetverikov's group had learned their biology (Timoféeff-Ressovsky, personal communication, 1977). Kol'tsov had been trained with Menzbier, the great Darwinian. Although he had parted with his teacher over the importance of the traditional nineteenth-century disciplines of comparative anatomy, morphology, biogeography, and systematics, for Kol'tsov "experimental biology" did not entail a rejection of those disciplines as it did for some other experimental biologists, both in Russia and the West. Rather, the well-trained experimental biologist had to know those areas "like the multiplication tables." Thus, even as an experimental biologist, Kol'tsov had remained a pupil of Menzbier and a Darwinian. So, indeed, had Chetverikov, when he took up biometrics (1919) and still later genetics.

When Kol'tsov created a Moscow center for genetics research, he drew on the students who had been trained in this synthetic way. Those like Vavilov and Filipchenko, who had known the genetics of Johannsen, Bateson, and de Vries and had come to share their views of evolution, would have had to unlearn those views to arrive at a synthetic understanding of the evolutionary process. By the mid- and late 1930s, Vavilov had managed this; Filipchenko, who died in 1930, did not live long enough. Even Serebrovsky, the member of Kol'tsov's group with the longest experience in genetics, worked primarily on domesticated animals and was influenced by Bateson's presence-absence theory and had some unlearning to do. Chetverikov did not. Indeed, without the Kol'tsov Institute and its stimulus to take up genetics, its group of excellent students, and its supportive institutional setting, Chetverikov might never have taken up genetics. By the same token, without Chetverikov, it seems unlikely that the Kol'tsov Institute would ever have addressed the evolutionary problem.

Unlike geneticists or experimental biologists, who had to find their way back to Darwinism and natural populations, Chetverikov and his group were Darwinian naturalists who had to learn their genetics. When they did, one of their primary interests was in using it to explicate microevolutionary processes in nature. This fact helps to explain why Chetverikov was in a position to write his 1926 paper. His central concern was precisely expressed in its title: "Several Aspects of the Evolutionary Process from the Standpoint of Modern Genetics." He drew on genetics to elucidate processes of central concern to the evolutionary naturalist—natural variability, speciation, selection. With his background in natural history and his broad field experience as a lepidopterist, he could take up biometrics and genetics to bring together a view of the evolutionary process that drew on the three isolated traditions whose union would constitute the synthesis: naturalist, biometric, and experimentalist.

As part of the Kol'tsov Institute, Chetverikov and his group were attempting to do precisely what Kol'tsov had urged: to reinvigorate traditional zoology with experimental approaches. When Chetverikov presented the results of the first studies of his group on natural *Drosophila* populations in Leningrad in 1927, he titled his paper "An Experimental Solution of One Evolutionary Problem" (1928b). The *Droz-So-or* was exactly the kind of interdisciplinary seminar that Kol'tsov had envisioned. Even if he himself did not seem to realize that an evolutionary synthesis was in the making right within his institute, that synthesis was nonetheless made possible because of his organizational foresight in creating an institutional structure that encouraged synthetic work. Once Chetverikov and his group worked within that context, they had at their disposal the collective wisdom of some of Russia's best young biologists in a wide array of fields. The presence of not only Kol'tsov, whose mastery of the biological literature was encyclopedic, but also Serebrovsky with his experience in chicken genetics, Zhivago and Frolova with their cytological and karyological expertise—not to mention the other specialists who could provide answers to virtually any question to which Chetverikov could not—provided an ideal environment for the development of their work.

It is interesting to compare the Chetverikov group with another group involved in a different kind of synthesis—the Morgan school. Both achieved their major breakthroughs over a five-year period (1910-1915 versus 1922-1927). Both were involved in synthetic enterprises (breeding and cytology, versus Darwinism, biometrics, and genetics). Both took place in a metropolitan laboratory under the general institutional aegis of an older cytologist in his early fifties (in 1910 Wilson was fifty-four; in 1922 Kol'tsov was fifty). At the beginning of the active period of their researches Morgan was forty-four, Chetverikov forty-two. Comparing the works of 1915 and 1926, both are theoretical general statements of the collective work of the groups involved. Both groups were formed with a small core of graduate students and bright undergraduates. The number of active participants in both laboratories ranged over the periods from roughly five to twelve. Both groups had to make do with relatively primitive apparatus.

The coincidences are striking and suggest that at least two famous biological schools in the twentieth century were similar in structure. Of course, the major differences between the groups are equally impressive. For one thing, a third of the participants in Chetverikov's group were women—thanks to the Beztuzhev Courses and the coeducational post-revolutionary composition of Moscow University. For another, Morgan had first undertaken his work on *Drosophila* in search of de Vriesian mutations, and as an experimentalist convert was largely trapped in certain anti-Darwinian attitudes about evolution. By contrast, Chetverikov took up experimentalism to elucidate the evolutionary process. True, he was cautious in evaluating his own achievement, and admitted that "it is still too early to speak of a synthetic formulation of the evolutionary process" (1961, p. 193). Serebrovsky, always more enthusiastic in his formulations, stated the case for him in a different context: "We must emphasize how amazingly close modern genetics has come to the Darwinian understanding of evolution. To the brilliant thesis of Darwin arose the brilliant antithesis of studies of mutation, and to some degree Mendelism. At one time, it appeared that this point of view utterly excluded Darwinism . . . And now we have a synthesis so harmonious that it is necessary to declare that ever since Darwin we have only been deepening Darwinism" (1925, p. 75). Serebrovsky later developed views of the evolutionary process that are fully in accord with the evolutionary synthetic views he had absorbed and helped to develop in his interactions with the Chetverikov group.

With the benefit of hindsight, we can see what Kol'tsov, Serebrovsky, and Chetverikov appreciated only later. As the synthetic approach Chetverikov pioneered and the population genetics he helped to initiate took root at home and abroad, it gradually became clear that Chetverikov will be remembered for these two closely interrelated contributions to the history of biology. A third—and related—contribution for which he will be remembered, especially in the Soviet Union (Astaurov, 1974), is his collection of over 300,000 butterflies which he donated to the Academy of Sciences' Zoological Museum in Leningrad just before his death.

References

Adams, M. B. 1968. The founding of population genetics: contributions of the Chetverikov school, 1924-1934. *Journal of the History of Biology* 1(1):23-39.
———— 1970. Towards a synthesis: population concepts in Russian evolutionary thought, 1925-1935. *Journal of the History of Biology* 3(1):107-129.
———— 1978. Nikolay Ivanovich Vavilov. *Dictionary of scientific biography*, vol. 16, pp. 505-513.
———— 1979. From "gene fund" to "gene pool": on the evolution of evolutionary language. *Studies in the history of biology*, vol. 3. Baltimore: Johns Hopkins University Press, pp. 241-285.
———— 1980. Science, ideology, and structure: the Kol'tsov Institute 1900-1970. *The social context of Soviet science*, ed. L. Lubrano and S. Solomon. Boulder: Westview Press, pp. 173-204.
Allen, G. E. 1975. *Life science in the twentieth century*. New York: Wiley.
Astaurov, B. L. 1974. Zhizn' S. S. Chetverikova [The life of S. S. Chetverikov]. *Priroda* (2):57-67.
———— and P. F. Rokitsky. 1975. *Nikolai Konstantinovich Kol'tsov*. Moscow: Nauka.
Balkashina, E. I., and D. D. Romashov. 1935. Geneticheskoe stroenie populiatsii Drosophila. I. Geneticheskii analiz zvenigorodskikh (Moskovskoi oblasti) populiatsii *Drosophila phalerata* Meig., *transversa* Fall. i *vibrissina* Duda [The genetic structure of *Drosophila* populations. I. Genetical analysis of populations of *Drosophila phalerata* Meig., *transversa* Fall. and *vibrissina* Duda of Zvenigorod (Moscow Province)]. *Biologicheskii zhurnal* 4:81-107.

———— V. N. Beliaeva, K. A. Golovinskaia, and N. B. Cherfas. 1975. D. D. Romashov i ego rol' v razvitii genetiki [D. D. Romashov and his role in the development of genetics]. *Iz istorii biologii* [From the history of biology], vol. 5. Moscow: Nauka, pp. 76-91.

Berg, L. S. 1922. *Nomogenez ili evoliutsiia na osnove zakonomernostei* [Nomogenesis or evolution determined by laws]. St. Petersburg: Gosudarstvennoe izdatel'stvo.

Chetverikov, S. S. 1905. Volny zhizni. Iz lepidopterologicheskikh nabliudenii za leto 1903 g. [Waves of life. Observations of Lepidoptera in the summer of 1903]. *Izvestiia Imperatorskogo obshchestva liubitelei estestvoznaniia, antropologii i etnografii* 98, *Trudy zoologicheskogo otdeleniia obshchestva* 13, *Dnevnik zoologicheskogo otdeleniia* 3 (6):1-5.

———— 1915. Osnovnoi faktor evoliutsii nasekomykh [The fundamental factor of insect evolution]. *Izvestiia Moskovskogo entomologicheskogo obshchestva* 1:14-24.

———— 1916. Babochki [Butterflies]. *Kalendar' russkoi prirody na 1916 g.* [Calendar of Russian nature, 1916]. Moscow: Priroda, pp. 166-210.

———— 1918. The fundamental factor of insect evolution. *The Smithsonian Report* 2566, pp. 441-449.

———— 1926. O nekotorykh momentakh evoliutsionnogo protsessa s tochki zreniia sovremennoi genetiki [On several aspects of the evolutionary process from the viewpoint of modern genetics]. *Zhurnal eksperimental'noi biologii,* ser. A, 2 (1):3-54.

———— 1928a. Über die genetische Beschaffenheit wilder Populationen. *Zeitschrift für induktive Abstammungs- und Vererbungslehre,* 46(2):1499-1500.

———— 1928b. Eksperimental'noe reshenie odnoi evoliutsionnoi problemy [An experimental solution of one evolutionary problem]. *Trudy III Vserossiiskogo S"ezda Zoologov, Anatomov i Gistologov v Leningrade 14-20 dekabria 1927 g.* [Proceedings of the Third All-Russian Congress of Zoologists, Anatomists, and Histologists, Leningrad, December 14-20, 1927]. Leningrad: Izdanie glavnogo upravleniia nauchnykh uchrezhdenii, pp. 52-54.

———— [1929]. Some aspects of the process of evolution from the standpoint of modern genetics [handwritten manuscript]. Archives of the USSR Academy of Sciences, *fond* 1650, *opis'* 1, *delo* 6, *listy* 1-182.

———— 1961. On certain aspects of the evolutionary process from the standpoint of modern genetics (trans. Malina Barker, ed. I. Michael Lerner). *Proceedings of the American Philosophical Society* 105 (2):167-195.

———— 1974. Iz vospominanii [From the memoirs]. *Priroda* (2):68-69.

Dobzhansky, Th. G. 1937. *Genetics and the origin of species.* New York: Columbia University Press.

———— and M. L. Queal. 1938. Genetics of natural populations. I. Chromosome variation in populations of *Drosophila pseudoobscura* inhabiting isolated mountain ranges. *Genetics* 23:239-251.

Dubinin, N. P. 1931. Genetiko-avtomaticheskie protsessy i ikh znachenie dlia mekhanizma organicheskoi evoliutsii [Automatic genetic processes and their significance for the mechanism of organic evolution]. *Zhurnal eksperimental'noi biologii*, ser. A, 7 (5/6):52-95.

———— 1973. *Vechnoe dvizhenie* [Perpetual motion]. Moscow: Izdatel'stvo politicheskoi literatury.

———— and D. D. Romashov. 1932. Geneticheskoe stroenie vida i ego evoliutsiia: 1. Genetiko-avtomaticheskie protsessy i problema ekogenotipov [The genetic structure of the species and its evolution: 1. Automatic genetic processes and the problem of ecogenotypes]. *Biologicheskii zhurnal* 1 (5-6): 52-95.

———— and 14 coauthors. 1934. Eksperimental'nyi analiz ekogenotipov *Drosophila melanogaster* [The experimental analysis of ecogenotypes of *Drosophila melanogaster*], parts 1 and 2. *Biologicheskii zhurnal* 3 (1):166-216.

————, M. A. Geptner, Z. A. Demidova, and L. I. D'iachkova. 1936. Geneticheskaia struktura populiatsii i ee dinamika v dikikh naseleniiakh *Drosophila melanogaster* [The genetic structure of populations and its dynamics in wild *Drosophila melanogaster*]. *Biologicheskii zhurnal* 5 (6):939-976.

————, D. D. Romashov, M. A. Geptner, and Z. A. Demidova. 1937. Aberrativnyi polimorfizm u *Drosophila fasciata* Meig. (Syn.—*melanogaster* Meig.) [Aberrant polymorphism in *Drosophila fasciata* Meig. (Syn.—*melanogaster* Meig.)]. *Biologicheskii zhurnal* 6 (2):311-354.

Dunn, L. C. 1965. *A short history of genetics*. New York: McGraw-Hill.

Fel'dman, G. E. 1976. *Dzhon Berdon Sanderson Kholdein 1892-1964* [John Burdon Sanderson Haldane 1892-1964]. Moscow: Nauka.

Filipchenko, Iu. A. [Philiptschenko]. 1915. *Izmenchivost' i evoliutsiia* [Variation and evolution]. Petrograd: Biblioteka naturalista.

———— 1917. *Nasledstvennost'* [Heredity]. Moscow: Priroda.

———— 1919. Khromozomy i nasledstvennost' [Chromosomes and heredity]. *Priroda* (7-9):327-350.

———— 1922. Zakon Mendelia i zakon Morgana [Mendel's law and Morgan's law]. *Priroda* (10-12):51-66.

———— 1923. *Izmenchivost' i metody ee izucheniia* [Variation and methods for its study]. St. Petersburg: Gosizdat.

———— 1925. O parallelizme v zhivoi prirode [On parallelism in living nature]. *Uspekhi eksperimental'noi biologii* 3 (3/4):242-258.

———— 1927. Uspekhi genetiki za poslednie 10 let (1918-1927) v SSSR [Achievements of genetics in the USSR over the past ten years (1918-1927)]. *Trudy Leningradskogo obshchestva estestvoispytatelei* 62 (1):3-11.

———— 1929. *Izmenchivost' i metody ee izucheniia* [Variation and methods for its study], 4th ed. Leningrad: Gosizdat.

Gaisinovich, A. E. [Gaissinovitch]. 1968. U istokov sovetskoi genetiki: 1. Bor'ba s lamarkizmom (1922-1927) [The origins of Soviet genetics: 1. The struggle with Lamarckism (1922-1927)]. *Genetika* 4 (6):158-175.

Gershenson, S. M. 1934. Mutant genes in a wild population of *Drosophila obscura* Fall. *American Naturalist* 68 (719):569-571.

Goldschmidt, R. B. 1956. *Portraits from memory: recollections of a zoologist*. Seattle: University of Washington Press.

Gordon, C. 1936. The frequency of heterozygosis in free-living populations of *Drosophila melanogaster* and *Drosophila subobscura*. *Journal of Genetics* 33:25-60.

————, H. Spurway, and F. A. R. Street. 1939. An analysis of three wild populations of *Drosophila subobscura*. *Journal of Genetics* 38:37-90.

Haldane, J. B. S. 1932. *The causes of evolution*. London: Longmans, Green.

Kol'tsov, N. K. 1901. *Razvitie golovy minogi. K ucheniiu o metamerii golovy pozvonochnykh* [The development of the head of *Petromyzon planeri*. On the study of metamerism in vertebrate skulls]. Moscow: Universitetskaia tipografiia.

———— 1905. *Issledovaniia o spermiiakh desiatinogikh rakov v sviazi s obshchimi soobrazheniiami otnositel'no organizatsii kletki* [Investigations on the sperms of decapods in connection with general considerations relating to cell organization]. Moscow: Universitetskaia tipografiia.

———— 1906. *Pamiati pavshikh. Zhertvy iz sredy moskovskogo studenchestva v oktiabr'skie i dekabr'skie dni* [In memory of the fallen. Losses from among the Moscow student body in the days of October and December]. Moscow: Burche.

———— 1908. Studien über die Gestalt der Zelle. 2. Untersuchungen über das Kopfskelett des tierischen Spermiums. *Archiv für Zellforschungen* 2:1-65.

———— 1909. *K universitetskomu voprosu* [On the university question]. Moscow: Tipografiia Russkogo tovara.

———— 1911. Studien über die Gestalt der Zelle. 3. Untersuchungen über die Kontraktilität des Stammes von Zoothamnium alternans. *Archiv für Zellforschungen* 7:344-423.

———— 1915a. Vzgliady Lotsi na evoliutsiiu organizmov [Lotsy's views on the evolution of organisms]. *Priroda* (10):1253-1264.

———— 1915b. ed., *Uchenye zapiski Moskovskogo gorodskogo narodnogo universiteta imeni Shaniavskogo, Otdelenie est.-ist., Trudy Biologicheskoi laboratorii* [Scientific communications of the People's University of Moscow named for Shaniavskii, Natural History Division, Works of the Biological Laboratory]. 1 (1 and 2).

———— 1921a. Ob izmenenii vesa cheloveka pri neustoichivom ravnovesii [On changes in human weight under conditions of disturbed equilibrium]. *Izvestiia Instituta eksperimental'noi biologii* 1 (1):25-30.

———— 1921b. Institut Eksperimental'noi Biologii, Moskva [The Institute of Experimental Biology, Moscow] [handwritten manuscript]. Archives of the USSR Academy of Sciences, *fond* 450, *opis'* 4, *delo* 1, *listy* 1-10.

———— 1921c. Predislovie [Preface]. *Izvestiia Instituta eksperimental'noi biologii* 1 (1):3-6.

———— 1921d. Geneticheskii analiz okraski u morskikh svinok [The genetic analysis of coloration in guinea pigs]. *Izvestiia Instituta eksperimental'noi biologii* 1 (1):87-97.

———— 1922. O rabotakh geneticheskogo otdela Instituta eksperimental'noi biologii i ego Anikovskoi geneticheskoi stantsii [On the works of the genetic section of the Institute of Experimental Biology and its Anikovo genetic station]. *Uspekhi eksperimental'noi biologii* 1 (3/4):404-411.

———— 1924. Experimental biology and the work of the Moscow institute. *Science* 59(1536):497-502.

———— 1927. Eksperimental'naia biologiia v SSSR [Experimental biology in the USSR]. *Nauka i tekhnika SSSR: 1917-1927*, ed. A. F. Ioffe et al., vol. 2. Moscow: Rabotnik prosveshchenii, pp. 37-64.

———— 1928. Fiziko-khimicheskie osnovy morfologii [The physicochemical foundations of morphology]. *Uspekhi eksperimental'noi biologii*, ser. B, 7 (1):3-31.

———— 1930. Ob eksperimental'nom poluchenii mutatsii. Rech' na torzhestvennom zasedanii pri otkrytii Vsesoiuznogo s"ezda zoologov, anatomov i gistologov v Kieve 13 maia 1930 g. [On the experimental induction of mutations. A speech at the Gala session opening the All-Union Congress of Zoologists, Anatomists, and Histologists in Kiev, May 13, 1930]. *Zhurnal eksperimental'noi biologii* 6 (4):237-249.

———— 1932. Problemy biologii [Problems of biology]. *Sorena* [Sotsialisticheskaia rekonstruktsiia i nauka] (9/10):23-45.

———— 1933. Problema progressivnoi evoliutsii [The problem of progressive evolution]. *Biologicheskii zhurnal* 2 (4/5):475-500.

———— 1934. Raboty Instituta eksperimental'noi biologii Narkomzdrava. K XVII s"ezdu VKP(b) [The works of the Institute of Experimental Biology of Narkomzdrav. For the XVIIth Congress of the All-Union Communist Party (Bolshevik)]. *Biologicheskii zhurnal* 3 (1):217-232.

———— 1935. Nasledstvennye molekuly [Hereditary molecules]. *Nauka i zhizn'* (5):4-13; (6):6-15.

———— 1936a. *Organizatsiia kletki. Sbornik eksperimental'nykh issledovanii, statei i rechei 1903-1935 gg.* [The organization of the cell. A collection of experimental researches, articles, and speeches 1903-1935]. Moscow/Leningrad: Biomedgiz.

———— 1936b. Trud zhizni velikogo biologa [The life labor of a great biologist]. (I. P. Pavlov. 1849-1936). *Biologicheskii zhurnal* 5 (3):387-402.

Kozo-Poliansky, B. M. 1923. *Poslednee slovo antidarvinizma* [The last word in anti-Darwinism]. Krasnodar: Burevestnik.

Lerner, I. M. 1961. Introductory note [to S. S. Chetverikov, On certain aspects of the evolutionary process from the standpoint of modern genetics]. *Proceedings of the American Philosophical Society* 105 (2):167-169.

Menzbir, M. A. [Menzbier]. 1926. Pervye 65 let v istorii teorii podbora [The first 65 years in the history of the selection theory]. *Polnoe sobranie sochinenii Charl'za Darvina*, vol. 1, book 2. Moscow/Leningrad: Gosudarstvennoe izdatel'stvo, pp. 1-53.

Morgan, T. H., A. H. Sturtevant, H. J. Muller, and C. B. Bridges. 1915. *The mechanism of Mendelian heredity*. New York: Holt.

Muller, H. J. 1923. Observations of biological science in Russia. *Scientific Monthly* 16 (35):539-552.

Provine, W. B. 1971. *The origins of theoretical population genetics*. Chicago: University of Chicago Press.

———— 1978. The role of mathematical population geneticists in the evolutionary synthesis of the 1930s and 1940s. *Studies in the History of Biology*, vol. 2. Baltimore: Johns Hopkins University Press, pp. 167-192.

Punnett, R. C. 1915. *Mimicry in butterflies*. Cambridge: Cambridge University Press.

Rokitsky, P. F. 1974. S. S. Chetverikov i evoliutsionnaia genetika [S. S. Chetverikov and evolutionary genetics]. *Priroda* (2):70-74.

———— 1975. S. S. Chetverikov i razvitie evoliutsionnoi genetiki [S. S. Chetverikov and the development of evolutionary genetics]. *Iz istorii biologii* [From the history of biology], vol. 5. Moscow: Nauka, pp. 76-91.

———— and E. T. Vasina-Popova. 1978. Razvitie genetiki sel'skokhoziaistvennykh zhivotnykh v SSSR [The development of the genetics of agricultural animals in the USSR]. *Istoriko-biologischeskie issledovaniia*, vol. 6. Moscow: Nauka, pp. 5-27.

Romashov, D. D. 1931. Ob usloviiakh "ravnovesiia" v populiatsii [On the conditions for "equilibrium" in populations]. *Zhurnal eksperimental'noi biologii* ser. A, 7 (4):442-454.

Serebrovsky, A. S. 1915. Sovremennoe sostoianie teorii mutatsii [The current status of the mutation theory]. *Priroda*:1239-54.

———— 1925. Khromozomy i mekhanizm evoliutsii [Chromosomes and the mechanism of evolution]. *Zhurnal eksperimental'noi biologii*, ser. B, 2 (1): 49-75.

———— 1926. Teoriia nasledstvennosti Morgana i Mendeliia i marksisty [Marxists and the theory of heredity of Morgan and Mendel]. *Pod znamenem marksizma* 3:98-117.

———— 1928. Genogeografiia i genofond sel'skokhoziaistvennykh zhivotnykh SSSR [Genogeography and the gene fund of agricultural animals in the USSR]. *Nauchnoe slovo* (9):3-22.

———— 1930. Problemy i metod genogeografii [The method and problems of genogeography]. *Trudy s"ezda po genetike, selektsii, i plemennomu zhivotnovodstvu* [Proceedings of the Congress on Genetics, Selection, and Animal Breeding], pt. 2, pp. 71-86.

———— and V. V. Sakharov. 1925. Novye mutatsii y *Drosophila melanogaster* [New mutations in *Drosophila melanogaster*]. *Zhurnal eksperimental'noi biologii*, ser. A, 1:75-91.

Shvarts, A. 1975. Dve sud'by [Two fates]. *Novyi zhurnal* (December):248-269.

Taliev, V. I. 1926. *Organizm, sreda, i prisposoblenie* [The organism, the environment, and adaptation]. Moscow/Leningrad: Gosudarstvennoe izdatel'stvo.

Targul'ian, O. M., ed. 1937. *Spornye voprosy genetiki i selektsii* [Controversial questions of genetics and selection]. Moscow/Leningrad: Izdatel'stvo VASKhNILa.

Timoféeff-Ressovsky, N. W. 1939. Genetik und Evolution [Genetics and evolution]. *Zeitschrift für induktive Abstammungs- und Vererbungslehre* 76 (1/2): 158-218.

———— 1940. Mutations and geographical variation. *The new systematics*, ed. Julian Huxley. Oxford: Clarendon Press, pp. 73-136.

———— and H. A. Timoféeff-Ressovsky. 1927. Genetische Analyse einer freilebenden *Drosophila melanogaster* population. *Roux Archiv Entwicklungsmech.* 109 (1):70-109.

Vavilov, N. I. 1920. *Zakon gomologicheskikh riadov v nasledstvennoi izmenchivosti* [The law of homologous series in variation]. Saratov: Gubpoligrafotdel, 3e otdelenie.

———— 1926. *Tsentry proiskhozhdeniia kulturnykh rastenii* [Centers of origin of cultivated plants]. Leningrad.

———— 1927. Geograficheskie zakonomernosti v raspredelenii genov kulturnykh rastenii [Geographical regularities in the distribution of genes in cultivated plants]. *Trudy po prikladnoi botanike, genetike i selektsii* 17 (3):411-428.

Vucinich, A. 1970. *Science in Russian culture 1861-1917.* Stanford: Stanford University Press.

9 Germany

Darwin acknowledged a few years after 1859 that *On the Origin of Species* had received a better reception in Germany than in any other country, including England. The theory of evolution was soon adopted in Germany by nearly every qualified biologist and met little resistance until Haeckel used evolution as a weapon against the Christian dogma. Even then, there were few knowledgeable opponents. The last two were probably the Erlangen zoologist Fleischmann (1901) and the Bamberg Thomist O. Kuhn (1947).

Darwin's mechanism of evolution was not adopted as readily, however. Almost every other conceivable theory was promoted by one or another German biologist (Montgomery, 1974) in preference to Darwinian selectionism. The situation deteriorated even further when Weismann (1883, 1885) not only proposed his germ-track theory but also enumerated abundant evidence in conflict with the thesis of an occurrence of any inheritance of acquired characters. Weismann's uncompromising neo-Darwinism found few adherents. To be sure, there were many German biologists, perhaps most, who acknowledged the importance of selection; but nearly all of them added that there were "of course" also many phenomena that could not yet be explained by selection but required subsidiary mechanisms. The important split in Germany was not between those who acknowledged the existence of natural selection and those (like O. Hertwig) who did not, but between the consistent neo-Darwinians who admitted no other direction-giving evolutionary factor than selection, and their opponents who postulated additional factors, particularly in their explanations of macroevolutionary phenomena. If this difference is overlooked, the picture of the state of evolutionary biology in Germany will be wrong. The publication of de Vries's mutation theory of speciation led to an open split among German biologists, a split still unbridged when the German paleontological and genetic societies met in Tübingen on September 8, 1929 (Weidenreich, 1929; Federley, 1930).

279

Rensch (see his selection in this chapter) and Stern (see the biographical essays) give further details on the situation in the 1920s and 1930s.

Between 1873 and 1900 there was an enormous interest in genetic questions (Naegeli, Strasburger, Koelliker, Hertwig, His, Roux, Boveri, Weismann, and Ziegler, among others) but the emphasis was almost totally on developmental genetics. This focus continued after 1900, as is evident in the writings of Haecker, the Kühn school (which included Henke and Caspari), and much of Goldschmidt's work. No school of transmission genetics developed in Germany, such as those of Morgan and Castle in the United States and of Bateson in England. There is no good analysis yet of the reasons why there was not more interest in evolutionary genetics in Germany. Perhaps geneticists specialized too early in unorthodox genetic mechanisms, like cytoplasmic inheritance (Correns, Michaelis) and the chromosome rings of *Oenothera* (Renner). Although there was no deficiency of good genetics texts (Baur, Haecker, Goldschmidt), the average German biologists were unquestionably backward in their understanding of genetics.

Kühn's 1939 edition of his textbook of genetics unequivocally rejects an inheritance of acquired characters, yet he emphasizes that much about the causation of evolution is still a mystery. In particular, he feels that it is not at all certain that macroevolution is effected by the same forces as microevolution. Population genetics had a brilliant beginning in plant science with Baur's work on the Spanish populations of *Antirrhinum* (1924, 1932), tragically soon snuffed out by Baur's heavy administrative burden and untimely death. Although Goldschmidt made a good start, he did not get very far because of difficulties in maintaining *Lymantria* in the laboratory, and also because of his well-known predilection for unorthodox theories. Genetics as a whole was rather eclipsed in Germany in the early decades of the century by the success of some other branches of experimental biology (such as *Entwicklungsmechanik* and physiology), and there simply were not enough chairs at the German universities to take care of all branches of biology. As a result, to the best of my knowledge, not a single chair at any of the twenty-four German universities was occupied in the years 1900-1934 by an *Ordinarius* who specialized either in transmission genetics, evolutionary genetics, population genetics, or ecological genetics.

Some people in Germany were apparently aware of this deficiency in German biology. For example, O. Vogt, director of the Kaiser Wilhelm Institute for Brain Research, advertised in Russia for a young geneticist. Dobzhansky felt he had had a good chance for this position, except that Chetverikov was slow in writing the necessary letter of recommendation. As a result, Timoféeff-Ressovsky got the appointment. Vogt studied individual and geographic variation in lady beetles and bumblebees to discover regularities of pattern variation and pattern spreading, because he thought he could apply such rules ("eunomies") to the spreading of brain diseases on the cortex. He wrote a book on these problems. Apparently he found no one in Germany to assist him in the study of color variation in insects, and because he had close relations with some Russian neurologists he asked them for suggestions. As a result, Timoféeff and Zarapkin came as geneticists to the Institute for Brain Research. (In exchange, it is said, Vogt helped in the study of Lenin's brain.)

The lack of an active group of evolutionary geneticists in Germany greatly impeded an exchange of ideas between naturalists and experimentalists. The situation was aggravated by the fact that the two authors who wrote the most important German language treatises in the field of evolutionary genetics, de Vries (Holland) and Johannsen (Denmark), developed evolutionary theories that were quite incompatible with the experiences of the naturalists. The work done in Britain (for example, by Bateson, Ford, and Fisher) and in the United States (by Castle, Morgan, and Sewall Wright), even though some of it got into the genetics texts, was largely ignored (see Buchner, 1938; Remane's statements as cited by Rensch). Rensch shows very clearly the developments that finally overcame this communication gap.

One other reason may account for the slow spread of the new genetic knowledge in Germany. At most institutes there was only a single full professor, aided by two or at most three assistants. There was no staff to give more than very few special courses in the frontier areas of biology. Each professor had to cover all of biology in his major courses. Professors had only small, fixed salaries and their main income consisted of tuition fees. The senior professors, therefore, had the privilege of teaching the large comprehensive courses that attracted the greatest number of students and provided the largest income. The young assistants and *Privat-Dozenten*, who were perhaps best acquainted with the most recent advances, either assisted in the practical courses or taught special fields.

In the United States, generally one of the younger professors was asked to give the introductory course, which permitted senior faculty members to give advanced courses, more or less in their own field of specialization. I believe that the American system, despite certain evident shortcomings, speeds up the dissemination of new findings and ideas far more successfully than the traditional system maintained in Germany and some other European countries. In some American colleges and universities it was, of course, also the senior professor who taught the large introductory classes.

Selectionist thinking finally began to spread in Germany in the 1930s.

Of the contributors to Heberer's *Die Evolution der Organismen* (1943) not a single one defended Lamarckian ideas. They all accepted a more or less selectionist interpretation.

I should like to take this opportunity to present some evidence on the German situation of evolutionary biology in the 1920s on the basis of my own personal experiences (for more detail see the biographical essays). I consider this eyewitness account rather important, because a number of historians have accepted the most enlightened statements of some outstanding investigators (for instance, Erwin Baur) as representative of the prevailing intellectual climate of the period. Actually, one must also consult the leading textbooks in order to get a balanced picture.

In addition to reading Haeckel's *Welträtsel* and various popularizations of biology (such as Boelsche) at home, I had a rigorous course in biology at the gymnasium in Dresden. The most popular school biology text of the period was written by Kraepelin. The fourth edition (1919) had 331 pages (without index) and featured a remarkably good treatment of ecology occupying a third of the volume (pp. 163-270). In contrast, Kraepelin spent only 17 pages on evolution, most of them devoted to proofs of evolution. Lamarckism was rather favorably discussed and 2 pages were devoted to five major objections to the theory of natural selection.

After graduation I studied for four semesters at the University of Greifswald, where I enrolled in a course in genetics under a botanist. However, when I discovered after the second lecture that he strictly followed the recommended text—I presume it must have been Baur's—I no longer attended the lectures but simply studied the text. My professors of zoology at Greifswald were first Alverdes, who was anything but a Darwinian, and later Buchner. The zoology text used by everybody at that time was the famous Claus-Grobben (9th ed., 1917), which discusses Darwin but concludes that "we must confess that Darwin's selection theory . . . is far from explaining the ultimate causes of numerous adaptations." Nevertheless, he agrees that selection is a factor of considerable importance. He discusses various evolutionary theories that he seems to consider either as alternatives or supplementary to the selection theory, including Wagner's migration theory, orthogenesis, Naegeli's mechanicophysiological theory, Weismann's neo-Darwinism, Roux's theory, and saltationist theories (pp. 68-84). Grobben does not seem to have fully understood selection, variation, or inheritance.

Buchner, a student of Richard Hertwig, was a first-rate biologist who made major contributions to cytology and the study of intracellular symbiosis, on which he was the world's foremost authority. He was a fine teacher with great competence in the classical areas of zoology. In 1938 he published a well-received textbook of zoology that contained a chapter titled "Lamarckism and Darwinism," in which he treats the two theories as equally legitimate. In the next chapter he justifies his defense of Lamarckism. He rejects the evolutionary significance of gradual variation with arguments that reflect many nongeneticists' interpretation of Johannsen's work: "Systematic selection experiments, particularly by de Vries and Johannsen, have demonstrated convincingly that selection is unable to affect such variation. One can select extreme individuals as long as one wants and cross them with similar ones and yet their descendants will again vary around the parental mean value. The only condition is that the individuals used in these experiments are members of pure lines." That selection in nature deals with populations and not with pure lines apparently never occurred to Buchner. Still, he admits that the geneticists have now become the staunchest supporters of evolution by natural selection. On pp. 352-356 he lists what he considers to be evidence for an inheritance of acquired characters. He finally concludes the chapter by predicting that the final solution would side with Darwin and allow both for selection and for an internal drive that would lead to improvement and adaptation. Buchner was either unaware of or unimpressed by the major advances in genetics made between 1910 and 1938 —which certainly illustrates the lag phenomenon in exemplary fashion.

At that time I was a medical student and took a number of courses in human anatomy, histology, and embryology. My professor, Karl Peter, was a superb scientist. His primary research objective was to discover and demonstrate what we would now call teleonomic processes during ontogeny (*Die Zweckmässigkeit in der Entwicklungsgeschichte*, 1920). This outstanding work, of high evolutionary significance (and very easily interpreted in terms of selection), is unfortunately far too little known and appreciated, presumably because Peter chose the wrong interpretation (see also Peter, 1957). He was frankly unable to conceive how the extraordinarily harmonious and *zielstrebige* development of the vertebrate embryo could be the result of "accidental mutations."

As I have documented (see chapter 4), I did not encounter a single straight Darwinian in my student days (1923-1926). Although selectionism was featured in some of the genetics texts, both my teachers and fellow students believed that these laboratory experiments dealt with phenomena that had little to do with such processes in nature as adaptive geographic variation and geographic speciation.

These impressions should be compared with those of V. Hamburger (in this chapter and chapter 3) and C. Stern (biographical essays), who conducted their research in institutions with very different traditions and intellectual milieus. The major presentation in this chapter on the synthesis in Germany is by Bernhard Rensch, who prepared the manuscript below but could not present it in person because of illness.   E.M.

References

Baur, E. 1924. Untersuchungen über das Wesen, die Entstehung und die Vererbung von Rassenunterschieden bei Antirrhinum majus. *Bibliographia Genetica* 4:1-170.

―――― 1932. Artumgrenzung und Artbildung in der Gattung Antirrhinum. *Zeitschrift für Induktive Abstammungs- und Vererbungslehre* 63:256-302.

Buchner, P. 1938. *Allgemeine Zoologie*. Leipzig: Quelle und Meyer.

Claus, C., and K. Grobben. 1917. *Lehrbuch der Zoologie*, 9th ed. Marburg: Elwert.

Federley, H. 1930. Weshalb lehnt die Genetik die Annahme einer Vererbung erworbener Eigenschaften ab? *Palaeontologische Zeitschrift* 11:287-317.

Fleischmann, A. 1901. *Die Deszendenztheorie*. Leipzig: Arthur Georgi.

Goldschmidt, R. 1934. Lymantria. *Bibliographia Genetica* 11:1-180.

Heberer, G., ed. 1943. *Die Evolution der Organismen*. Jena: G. Fischer.

Kraepelin, K. 1919. *Einführung in die Biologie*, 4th ed. Leipzig and Berlin: B. G. Teubner.

Kuhn, O. 1947. *Die Deszendenztheorie*. Bamberg: Meisenbach.

Montgomery, W. M. 1974. Germany. In *The comparative reception of Darwinism*, ed. T. F. Glick. Austin and London: University of Texas Press, pp. 81-116.

Peter, K. 1920. *Die Zweckmässigkeit in der Entwicklungsgeschichte*. Berlin: Springer.

―――― 1957. Funktionelle Embryologie der Wirbeltiere. *Nova Acta Leopoldina* 133.

Weidenreich, F. 1929. Vererbungsexperiment und vergleichende Morphologie. *Palaeontologische Zeitschrift* 11:275-286.

Weismann, A. 1883. *Über die Vererbung*. Jena: G. Fischer.

―――― 1885. *Die Kontinuität des Keimplasmas*. Jena: G. Fischer.

Historical Development of the Present Synthetic Neo-Darwinism in Germany

Bernhard Rensch

Outlining the development of neo-Darwinistic explanations of evolution in Germany can be accomplished only provisionally. It is not possible to consider all relevant publications, particularly because the opinions of many authors are often mentioned only incidentally. I therefore restricted my review to more important books and articles including the main textbooks for students.

The chapters on evolution in those German biological textbooks that students normally used between 1912 and 1945 show that neo-Darwinistic explanations of speciation prevailed. But in most cases some skeptical remarks were added to point out that other mechanisms might also be possible, particularly in the origin of higher categories. (See the genetics textbooks of Haecker, 1921; Baur, 1922; Goldschmidt, 1923, 1928; Kühn, 1939; the books on general zoology of Kühn, 1922, 8th ed. 1944; Hartmann, 1927; Claus, Grobben, and Kühn [see Kühn, 1932]; Stempell, 1935; Buchner, 1938; Hesse and Doflein, 1943; and the Strasburger textbook of botany [Fitting et al.], 1947 edition.)

By the end of the 1930s, several authors thought that mutation, gene recombination, and selection were sufficient to explain the whole phylogenetic development of all organisms. The geneticist Timoféeff-Ressovsky (1939), for example, wrote: "No particular objections exist to extrapolate the mechanism of microevolution to macroevolution and its special problems (higher systematical categories, special adaptations, special development of new organs)."[1] He regarded selection as the main directing factor of evolution (1937; see also Bauer and Timoféeff-Ressovsky, 1943). The botanist Zimmermann emphasized that "no plausible objection against the selection theory exists" (1938). He also pointed out that macrophylogeny is composed of microphylogenetic sections and must have happened in the same manner. Ludwig (1938, 1939, 1940, 1943) held the opinion that evolution can generally be explained by the analyzed factors of speciation, although other mechanisms are imaginable. The geneticist von Wettstein (1942) and the zoologists Franz (1935) and Heberer (1943) also published neo-Darwinistic conceptions. I attempted to explain phylogenetic trends, "explosive" development, origin of new organs, higher categories, and phylogenetic progress by mutation, gene recombination, and selection (Rensch, 1939, 1943).

Limitations of and Objections against Neo-Darwinistic Explanations

Several German biologists and particularly paleontologists still defended Lamarckian explanations between 1912 and 1945. They believed that, at least in some cases, speciation could also come about by direct influence of the environment on the genes. In the fourth edition of his book on the principle of selection, Plate (1913) pointed out that phylogenetic alterations can be explained in a neo-Darwinistic manner, but he assumed that

---

1. I have translated all quotations of German texts as literally as possible.

in some cases a direct influence of the environment on the genes took place, and at a later date (1931) he still defended this Lamarckian view against the objections of Federley (1930). He appealed to the results of Fischer (1901), who seemingly had proved that alterations of the color of butterflies caused by coldness became heritable. The biologist O. Hertwig (1912) and the paleontologists Weidenreich (1921, 1929) and Hennig (1929) were also convinced that speciation could be explained in a Lamarckian manner. Until 1934 I also tried to defend a Lamarckian interpretation for the origin of those geographic races with variations that paralleled climatic conditions. Like all authors who held similar opinions I was well aware that great theoretical difficulties existed for such explanation. Haecker (1921, 1925), for example, pointed out that a causal chain $x, y, z, \ldots$, leading from the soma to the germ cells cannot alter the chromosomes, which would produce the same somatic characteristic in the next generation because quite another causal chain $m, n, o, \ldots$, leads from the genes to the embryonic and adult stages.

After 1929 several paleontologists continued to defend Lamarckian explanations, for example, Hennig (1932, 1944) and Beurlen (1937). The anatomist Böker (1935a and b, 1937) assumed that changes of anatomical structures (*Umkonstruktionen*) came about by operational stimuli (*Betriebsreize*) acting as formative stimuli. The zoologist Harms (1934, 1935) could prove that an increase of thyroxin causes alterations of many characteristics of fish species in such a manner that they become better able to live outside of water for some time. He interpreted these interesting results in a manner similar to Böker.

Several important statements of paleontologists also contributed to raise doubts about purely neo-Darwinistic explanations of evolution. Since the end of the last century, many evolutionary trends, so-called orthogeneses (Haacke, 1893), had been described in which the same direction of change was maintained in long lines of descent. Particularly the increase in body size and the relative increase of the face bones in mammals, the successive enlargement of the tusks in the forerunners of elephants, of the nosehorns of Titanotheria, or of the antlers of stags are well-known examples. Plate (1913) had already explained such trends by more or less constant conditions of selection, called "orthoselection." Abel (1929) spoke of a biological law of inertia (*Trägheitsgesetz*) causing the maintenance of an evolutionary direction, a law by which he perhaps wanted to dissipate the suspicion that he believed in some autonomous driving force. However, so-called overspecializations, as the successive phylogenetic development of ultimately nonadaptive, disadvantageous organs like the antlers of *Megaceros*, the more or less circular canines of *Babirussa*, and the excessive growth of mandibles, feelers, or legs in large beetles seemed to contradict an explanation by selection. As a result, the paleontologists Hennig (1929, 1944), Dacqué (1935), Beurlen (1937), von Huene (1941), and Kuhn (1947) assumed autonomous directing evolutionary forces (that Osborn in 1934 had called "aristogenesis"). Hennig wrote: "Not guided by the environment, but in spite of it life follows its way" (1929, p. 38). Von Huene (1941) spoke of a "directing principle"; Kuhn of "planning intellect" (*planende Vernunft*); the zoologist Woltereck of "self-raising" (*Selbststeigerung*, 1932, p. 474) a category of life that cannot be analyzed. None of these authors took into consideration that all special orthogeneses of single organs and all cases of overspecialization can be sufficiently understood only by paying attention to their correlations with other organs, particularly with body size. Selection could primarily influence such correlated characters.

Paleontologists found that they could prove that many lines of descent show a regular sequence of different stages of evolutionary differentiation. Haeckel (1866) already spoke of a "flourishing time" (*Blühezeit, acme*), in which a genus, family, order, or class of animals increases the number of types "due to manifold adaptations to different conditions of their existence," and in which they reach a quantitative as well as qualitative culmination. The paleontologist Jaekel (1902) could show that the evolution of Crinoidea and Cystoidea started with the development of a great number of new types, but this period was followed by a long phase of slower evolution during which several lines of descent died out. Walther (1908) coined the term "anastrophes" for the phases of rapid radiation of types; Wedekind (1920) called the explosive phases *Virenzperioden*. Osborn spoke of adaptive radiation.

Beurlen (1932, 1937) claimed that all higher taxa show such an "explosive" phase at the beginning, followed by a slowing down of structural radiation, finally leading to a senile phase of overspecialization and degeneration of structures. He concluded that these regularly occurring phases must be caused by an autonomous evolutionary driving force (*Entfaltungstrieb*), a "will" to create new forms (*Wille zur Gestaltung*), a teleological process governing even the most minute details. He regarded the three evolutionary phases as the result of a basic plan that existed since their beginning: "die Auseinanderlegung eines von Anfang an gegebenen Grundplanes" (1937, p. 131). However, Beurlen did not pay attention to the many exceptions of the supposed regular sequence of evolutionary phases. Several families and orders show a phase of increased radiation of new types only after a long period of slower phylogenetic development, which sometimes lasted many millions of years. And in some lines of descent explosive phases occur several times. Moreover, the lines of descent do not normally lead to overspecialization and degeneration; rather, in most cases they die out because of their inability to compete with newly appearing, better adapted, or superior species. And all recent species, families, orders, and classes of organisms are members of long lines of descent that underwent manifold changes and did not always show a clear sequence of evolutionary phases.

Discounting Beurlen's exaggerated generalizations, the existence of "explosive" phases and a subsequent slowing down of the rate of evolutionary change is nevertheless an important fact. Schindewolf also discussed relevant examples. I tried to explain these phases by a change in selection pressure (Rensch, 1939a and b, 1943; Chapter 4). Heberer (1943) also criticized Beurlen's theory of three evolutionary phases.

Many leading biologists and paleontologists doubted that the well-analyzed factors of speciation were sufficient for understanding the phylogenetic origin of new organs, such as sense organs, brains, wings of birds, instincts, or complicated types of flowers. At the end of his book on selection, Plate (1913) pointed out that the principle of selection could not yet explain how assimilation, respiration, reproduction, and sensation originated. Other biologists later expressed similar doubts about the development of new complicated adaptations, new organs, and totally new types of anatomical constructions that characterize new families, orders, and classes of organisms. Such remarks were made by plant geneticists (Baur, 1922; Schwanitz, 1943; and Fitting et al., 1947), and zoologists (Hartmann, 1927; Kühn, 1932, 1944). The paleontologist Hennig wrote: "Nobody can seriously claim that he could imagine that all such refinements had been developed by random mutations and extinction of those competing types which were not capable to reach the same level" (1944, p. 306).

Remane (1939, 1941) substantiated his skepticism by more definite statements. He repeated Weidenreich's criticism (1929) that we do not yet know mutations that cause those characters which are typical for higher categories. We do not know mutations that lead to the differentiation of sepals and petals, elytra and membranous wings in insects, or legs of vertebrates into wings or flippers. He felt that considerations of larger phylogenetic alterations should only be based on well-known mutations (*Realmutationen*). However, he failed to recognize that new types of anatomical constructions and new organs normally come about by a series of different mutations, or new gene combinations, and that each individual and each population is subjected to a series of manifold selection tests during the whole life cycle. Both facts together guarantee an always harmonious structure and make coadaptation understandable. For his objections he used examples of phylogenetic changes that cannot be supported by fossil material. In other cases, the lines of descent show successive alterations that lead to quite new types of construction, such as the phylogenetic development of ear ossicles of mammals or the transformation of feet with four toes into those with one toe in the horse family. Even where paleontological evidence is lacking, series of recent species often show all the intermediate stages between different types of construction. In such series of anatomical characters, the increasing advantage of the changes can often be demonstrated, which helps us to understand how natural selection could have caused the change. This advantage holds good for the conversion of the air bladders of fish into lungs of amphibians and the further successive improvement of the lungs of reptiles, birds, and mammals. In addition, the different types of eyes of coelenterates, chaetopods, or gastropods can be arranged so that they show the gradual improvement from eye grooves to vesicular eyes with retina, lens, and cornea. These improvements can definitely be explained by mutation and selection (Rensch, 1943; 1959; 1960, p. 72, fig. 17). As a result, I believe that Remane's remark on evolution was too pessimistic, when he wrote: "The question of the causes has been solved only to a very slight degree, we do not know the conditions which lead to the formation and alteration of functional systems" (1941, p. 121). But he correctly claimed that geneticists should search for mutations that could particularly contribute to the understanding of the phylogenetic development of new organs.

Several other authors, such as Weidenreich (1921), Philipchenko (1927), Woltereck (1932), and Beurlen (1937), also doubted that the origin of totally different types of anatomical constructions that characterize phyla, classes, and orders of animals and plants could be explained by the continuous effect of mutation and selection. Beurlen (1937, p. 131) believed in a "basic plan," and Woltereck even wrote: "Chance plus selection as creator of the diversity and the well-planned order of organisms . . . this idea will be recorded in the future as one of the strangest errors of the human propensity for causal explanations" (1932, p. 230). Among paleontologists, the skepticism was mainly based on the fact that new phyla and classes of animals nearly always appear suddenly in the fossil record and that intermediate fossils are lacking that would bridge the gap between the new type and its phylogenetic forerunners. I tried to explain all paleontological rules concerning longer lines of descent by mutation, gene recombination, and selection (Rensch, 1939).

The paleontologist Schindewolf (1929, 1936, 1944a and b) did not deny the possibility of tracing new types of construction to phylogenetic precursors. But he assumed that major leaps took place suddenly which altered the whole basic structure and that such leaps were caused by macromutations like those which were found in the moss *Marchantia* of *Antirrhinum*. According to Schindewolf, such happened in early ontogenetic stages, because they ...y plasticity "to melt the new complex of character-.. the inherited characters during further ontogenetic develop-ment, and form a new equilibrium." Selection should not have been ef-fective to a noticeable degree. It would therefore be vain to search for intermediate types "because these never existed . . . The first bird hatched from a reptile-egg" (1936, p. 59).

Schindewolf tried to substantiate his "law of the origin of types in early ontogeny" by paleontological findings. He could show that, in the oldest species of some well-known lines of descent, certain characters only ap-pear in the juvenile stage but in the following species in later stages and finally in the adult stage. The statement of such "proterogenesis" is doubtless very important. But it concerns only differences in species and genera and does not explain the origin of totally new types of construc-tion.

The authors who criticized Schindewolf's typostrophe theory pointed out particularly that macromutations which drastically alter the type of construction are normally lethal in animals or show a very reduced via-bility or fertility (compare Gross, 1943; Heberer, 1943; Rensch, 1943). Moreover, many phylogenetic gaps between phyla, classes, and orders of animals have been bridged or narrowed to an increasing degree because of intense paleontological research, as for instance the gaps between *Cros-sopterygia* and *Stegocephalia*, reptiles and mammals, apes and man. The difficulty of bridging other gaps is probably a result of the fact that these strong phylogenetic changes took place rather rapidly when environmen-tal conditions underwent equally rapid changes so that the selection pres-sure increased very much; for example, the change of life from water to land or vice versa. Moreover, the first populations of early groups of new phyla or classes were apparently relatively small. As a consequence, fossil remains are relatively rare and can only be detected in a restricted area. The paleontologist Stromer (1940), who doubted a sudden origin of totally new types of construction, also pointed out that several new phyla and classes which first appeared in the Cambrian probably derived from animals that did not yet have hard parts that could be fossilized.

The Tübingen Conference of 1929

In 1929 German paleontological and genetical societies met at Tübingen to discuss the differences between the Lamarckian and neo-Darwinian positions. Although the variant views were clearly evident, a better mutual understanding of the different arguments was made possible, and this conference was perhaps the first step toward the later synthesis.

The discussions were introduced by a lecture by Weidenreich (1929) on genetic experiments and comparative morphology. He pointed out that the views of Lamarckians and geneticists were not as incompatible as assumed, although paleontological investigations had led to some findings that geneticists could not yet explain. Weidenreich said that it could clearly be proved that phylogenetic alterations run parallel to alterations of the environment. Furthermore, all major alterations do not affect merely single characteristics, but the whole structural system. When the forerunners of *Homo* adopted upright posture, the bones of legs, feet, pelvis, neck, and head became transformed, along with changes in muscles, intestines, and brain.

Geneticists claim that mutations are independent of the environment although they admit that X rays and chemical substances can produce a higher rate of mutations. The decisive question seems to be whether or not a series of mutations can produce successive changes in the same direction. One has to take into account that mutations do not alter characters of the adult organism, but only the genetic basis, the *Anlage*—that is, the possibility of developing in certain directions. They only shift the norm of reaction (*Reaktionsnorm*). Changes of the arrangement of genes could perhaps be sufficient to produce such shifts. The aftereffects that remain constant in a series of generations (*Dauermodifikationen*) could become inherited in time.

Federley's lecture, which asked why genetics refuses the assumption of inheritance of acquired characters, emphasized that genes have pleiotropic effects and many structures are determined by a series of genes. These facts already make a somatic induction of genes very improbable. Genes are stable and do not react to stimuli. Some well-known mutations alter characters that can be decisive for the development of new structures which characterize higher categories; for example, mutations that cause hairlessness or reduction of legs.

Federley also criticized Plate's hypothesis that species are based on stable genes and that only certain unstable genes are subjected to mutations which lead to minor alterations (*Erbstock-Hypothese*). Treatment of the butterfly *Arctia caja* with coldness does not prove aftereffects, because Fischer did not work with pure strains. Geneticists, however, must admit that they cannot yet explain the origin of functional adaptations.

In the discussion of both lectures Hartmann emphasized that aftereffects in *Protozoa* (*Dauermodifikationen*) did not lead to heritability after four or five thousand generations. Zimmermann pointed out that most mutations are disadvantageous, whereas adaptations are advantageous. Selection can probably be regarded as the directing factor in evolution. Weissermel said that the parallelism between alterations of structures and changes of the environment are evident in many cases. Hennig insisted that mutations cannot explain orthogenetic trends and the coadaptation of different structures. Weidenreich doubted that the aftereffects in butterflies treated with coldness are disproved sufficiently. So far no mutations have been found that characterize species. Possibly also the cytoplasm of the germ-cells may transmit inherited characters. Besides, no paleontologist would deny that selection plays an important role by wiping out unfavorable characters.

In his final remarks Federley stressed that species are composed of individuals with different gene combinations. Coadaptations cannot yet be explained by geneticists. He closed the discussion by the rather disheartening remark that Lamarckians do not understand genetics and discussions seemed therefore to be useless.

The conference did not lead to an integration of the statements of paleontology and genetics. In my opinion it was a mistake that selection and all already known relevant experiments were not discussed. However, after the meeting both groups of scientists began to consider the statements of their scientific opponents more carefully. During the next decade the conceptions of many paleontologists remained unchanged, however. Schindewolf, for example, who certainly paid very much attention to genetics, wrote: "Although one should expect that the results of both sides very well complement one another and flow together into a complete sphere of ideas, just the opposite is the case, the way of thinking of both branches of science developed in increasingly different directions" (1936, p. iii). Apparently he was already too much committed to his typostrophe theory.

The geneticist von Wettstein, the zoologist Marinelli, and the paleontologist Ehrenberg also published articles in *Palaeobiologia* in 1942 to harmonize the results of each branch of science with regard to evolution. However, objections to opposing views were not sufficiently discussed, and no synthesis occurred.

The Relatively Slow Development of the Neo-Darwinist Synthesis

Many discoveries of paleontologists did not seem to fit the explanation of evolution by random mutation and selection. Evolutionary trends, parallel development of different lines of descent, iterations, irreversibility, explosive development at the phylogenetic beginning of higher taxa, the origin of totally new organs and types of anatomical constructions, and particularly the origin of man seemed to indicate that some factors of evolution cannot be analyzed through the study of speciation. On the other hand, some experiments of geneticists seemed to indicate that after-effects could possibly be regarded as a first stage of somatogenic induction. It was difficult to explain coadaptations before the far-reaching manner in which single genes could cause pleiotropic effects was understood. It also took some time before evolutionists began to take into account the various correlations of structures and organs in the wholeness of the body (mainly correlations by allometrical growth and by compensation of material during ontogeny). As a result, German biologists' warnings of premature generalizations seemed justified.

Despite all understandable skepticism, the neo-Darwinist synthesis in its present version could have made its way more rapidly. The synthesis happened only between 1939 and 1945, apparently because the geneticists and other leading biologists were not sufficiently familiar with the paleontological discoveries. Paleontology was not yet included in the standard biological instruction in nearly all German universities. Paleontologists, on the other hand, had only limited knowledge of genetics because they did not have adequate access to the relevant literature. A clear picture of the process of evolution, however, must be based on genetic, ecological, morphological, embryological, and paleontological evidence. Such a synthesis became particularly difficult in a period of rapidly increasingly specialization that led to one-sided premature generalizations.

Sometimes the skepticism was reinforced by religious convictions. Dacqué (1935), for instance, wrote that the "chance theory" is "atheistic in its proper essence," and the paleontologist Kuhn (1947) traced the structure of organisms back to a "logical, absolute cause (God) . . . The primary cause of nature is to be found in God." And finally, it was widely held that the evolution of man should not be the result of undirected mutations, accidental mistakes of genetic transmission, and the accidental condition of natural selection.

My Own Ideas on the Mechanisms of Evolution

A few biographical data may help to clarify the development of my own ideas.[2] I was born January 21, 1900, at Thale, Germany. At the University of Halle I studied zoology, botany, chemistry, and philosophy, receiving my Ph.D. in 1922 under the geneticist Haecker, on the cause of dwarfism and giantism in the domestic fowl. From 1925 on I was assis-

2. Since this chapter was originally written (1974), Rensch has published an autobiography (1979)—Eds.

tant at the Zoological Museum, University of Berlin; after 1937 reader; and from 1947-1968 professor of zoology at Münster. As a student of Haecker I was well informed about the process of speciation and the course of phylogeny. Haecker emphasized that Lamarckian explanations meet with strong theoretical objections. However, the possibility of somatogenic induction could not yet totally be denied because of the experiments of Fischer (1901) and some other authors. Even though I regarded him very highly, I had a number of scientific disagreements with Haecker because he did not accept the species status of the sibling species of *Parus* and *Certhia*. He also thought that the geographic color differences of birds and mammals—more brownish in Western and more grayish in Eastern Europe—were nongenetic modifications, whereas I was convinced of their genetic nature.

In my student days and during my tenure in the Zoological Museum of the University of Berlin, I mainly worked on the formation of geographic races and species of birds and mollusks, inspired by the investigations of Kleinschmidt (1909-1928), Hartert (1910-1923), and Stresemann (1926) on birds; Sumner (1923-1924) on *Peromyscus*; Schmidt (1917) on fish; and Goldschmidt, Seiler, and Poppelbaum (1924) on *Lymantria*. I soon recognized the great theoretical importance of the new systematics. My own studies led me to the conviction that subspecies and closely related species that represent one another geographically could be combined into large complexes, which I termed *Rassenkreise*, in most classes of animals and plants. I presented this thesis at the International Ornithological Congress at Copenhagen in 1926 (Rensch, 1929a), when I also emphasized that most geographic races are the precursors of new species. In some cases very different races of the same *Rassenkreis* do not hybridize and behave like "good species" when they come together—the reason I did not call such large complexes "species." I found that borderline cases between geographic races and species are very common among birds. For complexes of closely related young species that replace one another geographically I proposed the term *Artenkreis* (Rensch, 1928, 1929b).

I was particularly interested in studying the dependence of body size, proportions of legs, wings, and bill, and colors of the plumage on climatic conditions. An expedition to the Lesser Sunda Islands in 1927 enlarged my experience with speciation. I summarized my results in my first book *Das Prinzip geographischer Rassenkreise und das Problem der Artbildung* (1929b). I held the opinion that the climatic parallelism of race formation, which could be formulated in such rules as Bergmann's rule and the rules that I termed Allen's and Gloger's rules, could be explained by a direct influence of the environment on inherited characters. I tried to base such a Lamarckian explanation, which was widely accepted in these years by other biologists, on the following facts. Nearly all the mutations of birds analyzed by Stresemann, Chapman, and others concerned larger differences. The gliding differences between neighboring races, running parallel to climatic changes, were on the contrary rather insignificant. They concerned, for example, only shades of brownish gray and gray, or differences of size of a few millimeters. I then thought it improbable that climatic selection could produce such insignificant differences. In my opinion, slight color differences that ran parallel in birds and mammals could not be caused by selection because the enemies of both classes were different. Nearly nothing was known about the pleiotropic effect of genes in these years and we did not consider that selection could be effective by way of correlated characters. Moreover, noninherited differences of size and color that correspond to those of geographic races could be produced experimentally by similar exposure to temperature and moisture. Aftereffects of such experimental alterations in the next generations (*Dauermodifikationen*, Jollos, 1939) were often discussed then, and somatogenic induction leading to heredity seemed to be possible in certain cases.

On the other hand, I was able to prove experimentally the effect of natural selection in other cases. To find out why the color of the eggs of the European cuckoo normally resembles that of the egg of its hosts, I painted either stronger or weaker brown, gray, or dark spots on a single egg in the clutch of host birds. When the painted egg differed very much from the other eggs, the host birds removed it in nearly all cases, but they incubated it when the difference was only slight (Rensch, 1924, 1925). I concluded that the color adaptation of the cuckoo's eggs must therefore have been developed through selection by the host birds.

At the Berlin Museum I had many discussions with Stresemann. I hardly ever went to Berlin-Dahlem and had no close contact with anyone there. On the other hand, I did have contacts with the Berlin-Buch group. I had met Timoféeff-Ressovsky at the Berlin Museum and he invited me to test my ideas with *Drosophila* in his laboratory. As a result I spent most afternoons of the better part of the years 1931-1933 at the Brain Research Institute at Buch testing the aftereffects of different temperature regimes during ontogeny, at that time a burning question because of Jollos' claims. Even though I continued my experiments through many generations the results were entirely negative and I published nothing. However, I learned the *Drosophila* technique and became acquainted with *Drosophila* mutations. I also had many stimulating conversations with Timoféeff-Ressovsky, Klaus Zimmermann, Zarapkin, and occasionally with Oskar Vogt, the director of the institute.

There was a strong paleontological group at the Naturkunde Museum at Berlin (including Pompecky, the younger Quenstedt, and Janensch), but they limited themselves to a descriptive treatment of their material. I presume they had largely Darwinian interpretation, with the usual reservations, but I do not recall active discussions with any member of this group.

My thinking on the nature of the evolutionary mechanisms changed in the early 1930s. When experiments by geneticists showed that nearly all genes have pleiotropic effects and that selection can become effective during some thousands of generations even when the advantage of a new allele is only 1 to 2 percent, I gave up all Lamarckian explanations. I now explained climatic parallelism of race differences in size, proportions, shape of wings, and number of eggs per clutch in a much more satisfying manner by natural selection (Rensch, 1936, 1938).

At the same time I began to study the rules that govern the evolution in longer lines of descent, such as Cope's rule, orthogenetic development of organs, irreversibility, and orderly sequence of different rates and intensities of phylogenetic radiation. I had the impression that the explanations of such paleontologists as Osborn, Hennig, Beurlen, Kuhn, and Schindewolf, who postulated an autonomous directing evolutionary factor, were wrong, because they did not take into consideration the correlations between structures and organs in the wholeness of the body. I was convinced that the regularities of longer lines of descent, including seeming evolutionary progress, could be explained by the same factors that account for speciation—that is, random mutation, gene recombination, and selection. My conviction was strengthened by studies of literature of comparative anatomy, developmental physiology, and particularly by the important books of Sewertzoff, *Morphologische Gesetzmässigkeiten der Evolution* (1931); Huxley, *Problems of Relative Growth* (1932); and Dobzhansky, *Genetics and the Origin of Species* (1937). A few other books had greatly impressed me: D'Arcy Thompson, *Growth and Form* (1917); Haecker, *Entwicklungsgeschichtliche Eigenschaftsanalyse* (1918); Goldschmidt, *Physiologische Theorie der Vererbung* (1927); and Ziehen, *Erkenntnistheorie* (1913, 1939).

In 1938 I began to prepare a book on problems of evolution. Because I had been invited to write an article for the *Biological Reviews*, I gave there a brief, although rather imperfect, outline of my conception. I treated the different modes of speciation, parallel phylogenetic development, loss of genes without selection (which Reinig, 1939, had termed "elimination"), discussed examples of compensation of material, and pointed out that phenomena such as evolutionary trends (orthogenesis), "explosive" phases of evolution, followed by decreasing radiation of types, dying out of lines of descent, and evolutionary progress, can be explained by selection (Rensch, 1939a and b).

The outbreak of the war in 1939 and my being called up for military service prevented further scientific work. Only in 1942, when I was released from military service owing to a myocarditis, could I take up my studies of evolution again. In a lengthy article (Rensch, 1943) I treated the paleontological rules of evolution to show that one is not at all obliged to postulate autonomous directing forces as Osborn, Dacqué, Beurlen, and Schindewolf had claimed. I pointed out that no physiological basis for such directing forces can be detected. Instead, it is very probable that these rules came about because the evolutionary changes in question were advantageous and therefore subject to positive selection. I discussed Cope's rule of phylogenetic increase of body size in more detail and mentioned that some lines of mammals showed an increase in body size corresponding to the gradually lowering of the temperature in the late Tertiary, so that these alterations can possibly be explained in the same manner as Bergmann's rule by climatic selection. Moreover, I tried to prove that larger animals have a more efficient metabolism and more neurons in their absolutely larger brain so that they are more versatile. In poikilothermous animals, larger species normally produce many more offspring than related smaller species, which is also advantageous. Like Huxley (1932), I explained by positive or negative allometrical growth the special orthogenesis of single organs, excessive structures in large species, overspecializations, and, on the other hand, the development of vestigial organs in lines of descent that became gradually larger or smaller. I also illustrated the relative enlargement of the facial bones of larger mammals by drawings of the skulls of smaller and larger rodents, mustelides, and cats.

In my opinion, the explosive phases at the phylogenetic beginning of higher categories came about by intense selection when animals penetrated into a different habitat where competing weaker species were lacking or could be driven away or wiped out. The radiation of new species, genera, and families then gradually decreased in the same measure as all appropriate habitats were occupied and the possibilities of adaptation to special ecological conditions were exhausted. Lines of descent do not die out because of autonomously determined degeneration, as Beurlen tried to show, but by an inability to compete with superior species. Parallel development in related lines of descent, which I illustrated by marine genera of snails, I attributed to homologous genes or gene combinations as well as to similar conditions of selection.

In 1943 I also tried to show how the origin of new organs and new types of anatomical constructions that characterize higher categories can be explained by mutation and selection. I assumed that the same holds good for evolutionary progress in many lines of descent and in the phylogenetic tree as a whole. I pointed out that an increase in complication and more rational structures and functions, particularly increasing centralization of functions, could arise on all phylogenetic levels and that increasing rationalization was supported by selection.

In the summary I stated: "All known evolutionary rules can be explained by mutation and selection. The assumption of autonomous creative principles or driving forces is inappropriate. Only alterations of environmental factors are decisive for the formation of new species and higher categories" (p. 52).

I have abstracted my article in *Biologia Generalis* (1943) rather completely because here I treated nearly all the problems that I discussed in much more detail in my book *Neuere Probleme der Abstammungslehre: die transspezifische Evolution*, which I wrote at the end of the war. Because of the breakdown of the German economy after the war it could not be published until 1947. Here I also treated the absolute rate of evolution, histological, cytological, and physiological consequences of allometric growth, phylogenetic lower and upper size limits, effects of phylogenetic alterations in various stages of the individual cycle, the evolution of life on earth, bionomogenesis, and the evolution of conscious phenomena and relevant epistemological questions.

After the proof sheets of this book were corrected, I obtained information about Huxley's *Evolution, the New Synthesis* (1942), Mayr's *Systematics and the Origin of Species* (1942), and Simpson's *Tempo and Mode in Evolution* (1944), which the authors kindly sent me. I was able to mention them in a short postscript. I was surprised, but also gratified, that synthetic neo-Darwinism was argued in these works in a similar manner as my own attempts. I could take into account the results of these authors as well as many other publications and the results of our own investigations of the anatomical, histological, and functional effects of differences of body size in the second edition of my book (1954), which later appeared in English translations (1959, 1960).

It is difficult to establish how long it took before the neo-Darwinian interpretation was universally adopted in Germany because of the disruptions caused by the war and the postwar dislocations. I think it is correct to say that all the younger evolutionary biologists have now adopted the selectionist interpretation, including, for example, Schindewolf's students. To what extent my book and my other writings have been influential is hard to say. I know that my versions of the species concept and of speciation were adopted by such systematists as Eisentraut, Pax, Arndt,

Ramme, and Mertens. Various zoologists confirmed an influence of my book on their thinking, such as R. Hesse, B. Klatt, C. Zimmer, and O. von Wettstein. I had some favorable reactions also among paleontologists, as from K. Ehrenberg and O. Abel in Vienna. The paleontologist Walter Gross, although he opposed (1943) the saltationism of Schindewolf and Beurlen, did not particularly modify his views.

References

Abel, O. 1929. *Paläobiologie und Stammesgeschichte*. Jena: G. Fischer.

Bauer, H., and N. W. Timoféeff-Ressovsky. 1943. Genetik und Evolutionsforschung bei Tieren. In *Die Evolution der Organismen*, ed. G. Heberer. Jena: G. Fischer, pp. 335-429.

Baur, E. 1922. *Einführung in die experimentelle Vererbungslehre*, 5th ed. Berlin: Bornträger.

——— 1924. *Einführung in die experimentelle Vererbungslehre*, 6th ed. Berlin: Bornträger.

Beurlen, K. 1932. Funktion und Form in der organischen Entwicklung. *Naturwissenschaften* 20:73-80.

——— 1937. *Die stammesgeschichtlichen Grundlagen der Abstammungslehre*. Jena: G. Fischer.

Böker, H. 1935a. Artumwandlung durch Umkonstruktion, Umkonstruktion durch aktives Reagieren der Organismen. *Acta Biotheoretica* A 1:17-34.

——— 1935b, 1937. *Einführung in die vergleichende biologische Anatomie der Wirbeltiere*, vols. 1 and 2. Jena: G. Fischer.

Buchner, P. 1938. *Allgemeine Zoologie*. Leipzig: Quelle und Meyer.

Dacqué, E. 1935. *Organische Morphologie und Paläontologie*. Berlin: Bornträger.

Dobzhansky, Th. 1937. *Genetics and the origin of species*. New York: Columbia University Press.

Ehrenberg, K. 1942. Paläozoologie, Stammesgeschichte und Abstammungslehre. *Palaeobiologica* 7:196-211.

Federley, H. 1930. Weshalb lehnt die Genetik die Annahme einer Vererbung erworbener Eigenschaften ab? *Zeitschrift für Induktive Abstammungs- und Vererbungslehre* 54:20-50.

Fischer, E. 1901. Experimentelle Untersuchungen über die Vererbung erworbener Eigenschaften. *Zeitschrift für Entomologie* 6:49, 363, 377.

Fitting, H., R. Harder, W. Schumacher, and F. Firbas, eds. 1947. *Lehrbuch der Botanik für Hochschulen*. Jena: G. Fischer.

Franz, V. 1935. Der biologische Fortschritt. *Die Theorie der organismengeschichtlichen Vervollkommnung*. Jena: G. Fischer.

Goldschmidt, R. 1923, 1928. *Einführung in die Vererbungswissenschaft*, 4th ed., 5th ed. Berlin: Springer.

———, I. Seiler, and H. Poppelbaum. 1924. Untersuchungen zur Genetik geographischer Variation. *Archiv für mikroskopische Anatomie und Entwicklungsmechanik* 101:92-337.

Gross, W. 1943. Paläontologische Hypothesen zur Faktorenfrage der Deszendenzlehre: Über die Typen- und Phasenlehren von Schindewolf und Beurlen. *Naturwissenschaften* 31:237-245.

Haacke, W. 1893. *Gestaltung und Vererbung.* Leipzig: Weigel.

Haeckel, E. 1866. *Generelle Morphologie der Organismen.* Berlin: Reimer.

Haecker, V. 1911. *Allgemeine Vererbungslehre,* 1st ed. Braunschweig: Vieweg.

—— 1921. *Allgemeine Vererbungslehre,* 3d ed. Braunschweig: Vieweg.

—— 1925. *Pluripotenzerscheinungen. Synthetische Beiträge zur Vererbungs- und Abstammungslehre.* Jena: G. Fischer.

Harms, J. W. 1934. *Wandlungen des Artgefüges.* Tübingen: Heine.

—— 1935. Die Plastizität der Tiere. *Revue Suisse de Zoologie* 42:461-476.

Hartert, E. 1901-1923. *Die Vögel der paläarktischen Fauna,* vols. 1-3. Berlin: Friedländer.

Hartmann, M. 1927. *Allgemeine Biologie: Eine Einführung in die Lehre vom Leben.* Jena: G. Fischer.

Heberer, G. 1943. Das Typenproblem in der Stammesgeschichte. In *Die Evolution der Organismen,* ed. G. Heberer. Jena: G. Fischer, pp. 545-585.

Hennig, E. 1929. *Von Zwangsablauf und Geschmeidigkeit in organischer Entfaltung. Reden bei der Rektoratsübergabe am 25. April 1925 Tübingen.* Tübingen: Mohr, pp. 13-39.

—— 1932. *Wesen und Wege der Paläontologie.* Berlin: Bornträger.

—— 1944. Organisches Werden paläontologisch gesehen. *Paläontologische Zeitschrift* 23:281-316.

Hertwig, O. 1906. *Allgemeine Biologie,* 2d ed. Jena: G. Fischer.

—— 1912. *Allgemeine Biologie,* 4th ed. Jena: G. Fischer.

Hesse, R., and F. Doflein. 1943. *Tierbau und Tierleben in ihrem Zusammenhang betrachtet,* 2d ed., vol. 2. Jena: G. Fischer.

Huene, F. von. 1941. Die stammesgeschichtliche Gestalt der Wirbeltiere—ein Lebensablauf. *Paläontologische Zeitschrift* 22:55-62.

Huxley, J. 1932. *Problems of relative growth.* London: Methuen.

Jaekel, O. 1902. *Über verschiedene Wege phylogenetischer Entwicklung.* Jena: G. Fischer.

Jollos, V. 1939. Grundbegriffe der Vererbungslehre, insbesondere Mutation, Dauermodifikation, Modifikation. In *Handbuch der Vererbungswissenschaften,* ed. E. Baur and M. Hartmann, vol. 1. Berlin: Bornträger.

Kleinschmidt, O. 1909-1928. *Corvus Nucifraga; Corvus Perisoreus, Falco Palumbarius; Parus Salicarius; Falco Peregrinus. Berajah, Zoographia infinita.* Halle.

Kühn, A. 1932. Allgemeine Zoologie. In *Lehrbuch der Zoologie,* ed. C. Claus, K. Grobben, and A. Kühn, 10th ed. Berlin, Vienna: Springer.

—— 1939. *Grundriss der Vererbungslehre.* Leipzig: Quelle und Meyer.

—— 1944. *Grundriss der allgemeinen Zoologie,* 8th ed. Leipzig: Thieme.

Kuhn, O. 1947. *Die Deszendenz-Theorie.* Bamberg: Meisenbach.

Ludwig, W. 1938. Beitrag zur Frage nach den Ursachen der Evolution auf theoretischer und experimenteller Basis. *Verhandlungen der Deutschen Zoologischen Gesellschaft, 1938. Zoologischer Anzeiger Supplementband* 11:182-193.

———— 1939. Der Begriff "Selektionsvorteil" und die Schnelligkeit der Selektion. *Zoologischer Anzeiger* 126:209-222.

———— 1940. Selektion und Stammesentwicklung. *Naturwissenschaften* 28:689-705.

———— 1943. Die Selektionstheorie. In *Die Evolution der Organismen*, ed. G. Heberer. Jena: G. Fischer.

Osborn, H. F. 1934. Aristogenesis, the creative principle in the origin of species. *American Naturalist* 68:193-235.

Philipchenko, J. 1927. *Variabilität und Variation*. Berlin: Bornträger.

Plate, L. 1913. *Selektionsprinzip und Probleme der Artbildung*, 4th ed. Leipzig, Berlin: Engelmann.

———— 1931. Warum muss der Vererbungsforscher an der Annahme einer Vererbung erworbener Eigenschaften festhalten? *Zeitschrift für Induktive Abstammungs- und Vererbungslehre* 58:266-292.

Reinig, W. F. 1939. Die Evolutionsmechanismen, erläutert an den Hummeln. *Verhandlungen der Deutschen Zoologischen Gesellschaft, 1939. Zoologischer Anzeiger Supplementband* 12:170-206.

Remane, A. 1939. Der Geltungsbereich der Mutationstheorie. *Verhandlungen der Deutschen Zoologischen Gesellschaft, 1939. Zoologischer Anzeiger Supplementband* 12:206-220.

———— 1941. Die Abstammungslehre im gegenwärtigen Meinungskampf. *Archiv für Rassen- und Gesellschaftsbiologie* 35:89-122.

Rensch, B. 1924. Zur Entstehung der Mimikry der Kuckuckseier. *Journal für Ornithologie* 72:461-472.

———— 1925. Verhalten von Singvögeln bei Änderung des Geleges. *Ornithologische Monatsberichte* 33:169-173.

———— 1928. Grenzfälle von Rasse und Art. *Journal für Ornithologie* 76:222-231.

———— 1929a. Die Berechtigung der ornithologischen systematischen Prinzipien in der Gesamtzoologie. *Proceedings of the Sixth International Ornithological Congress*. Copenhagen (1926), pp. 228-242.

———— 1929b. *Das Prinzip geographischer Rassenkreise und das Problem der Artbildung*. Berlin: Bornträger.

———— 1936. Studien über klimatische Parallelität der Merkmalsausprägung bei Vögeln und Säugern. *Archiv für Naturgeschichte* N.F. 5:317-363.

———— 1938a. Einwirkung des Klimas bei der Ausprägung von Vogelrassen, mit besonderer Berücksichtigung der Flügelform und der Eizahl. *Proceedings of the Eighth International Ornithological Congress*. Oxford (1934), pp. 285-311.

———— 1938b. Bestehen die Regeln klimatischer Parallelität bei der Merkmalsausprägung von homöothermen Tieren zu Recht? *Archiv für Naturgeschichte* N.F. 7:364-389.

———— 1939a. Klimatische Auslese von Grössenvarianten. *Archiv für Naturgeschichte* N.F. 8:89-129.

———— 1939b. Typen der Artbildung. *Biological Reviews* 14:180-222.

———— 1943. Die paläontologischen Evolutionsregeln in zoologischer Betrachtung. *Biologia Generalis* 17:1-55.

———— 1947. *Neuere Probleme der Abstammungslehre.* Stuttgart: Enke.

———— 1954. *Neuere Probleme der Abstammungslehre,* 2d ed. Stuttgart: Enke.

———— 1959. *Evolution above the species level.* London: Methuen.

———— 1960. *Evolution above the species level.* New York: Columbia University Press.

———— 1979. *Lebensweg eines Biologen in einem turbulenten Jahrhundert.* Stuttgart: G. Fischer.

Schindewolf, O. H. 1929. Ontogenie und Phylogenie. *Paläontologische Zeitschrift* 11:54-67.

———— 1936. *Paläontologie, Entwicklungslehre und Genetik. Kritik und Synthese.* Berlin: Bornträger.

———— 1944a. *Grundlagen und Methoden paläontologischer Chronologie.* Berlin-Zehlendorf: Bornträger.

———— 1944b. Der Kampf um die Gestaltung der Abstammungslehre. *Naturwissenschaften:* 269-282.

———— 1947. *Fragen der Abstammungslehre.* Frankfurt: Kramer.

Schmidt, J. 1917. Zoarces viviparus L. and local races of the same. *Compte Rendu des Travaux du Laboratoire de Carlsberg* 13:279-397.

Schwanitz, F. 1943. Genetik und Evolutionsforschung bei Pflanzen. In *Die Evolution der Organismen,* ed. G. Heberer. Jena: G. Fischer, pp. 430-478.

Sewertzoff, A. N. 1931. *Morphologische Gesetzmässigkeiten der Evolution.* Jena: G. Fischer.

Stempell, W. 1935. *Zoologie im Grundriss.* Berlin: Bornträger.

Stresemann, E. 1926. Übersicht über die Mutationsstudien I-XXIV und ihre wichtigsten Ergebnisse. *Journal für Ornithologie* 74:377-385.

Stromer, E. 1940. Kritische Betrachtungen. 3. Die Lückenhaftigkeit der Fossilüberlieferung und unsere derzeitigen Kenntnisse sowie die Folgerungen daraus. *Zentralblatt für Mineralogie, Geologie, und Paläontologie* B:262-276.

Sumner, F. B. 1923-1924. Results of experiments in hybridizing subspecies of *Peromyscus. Journal of Experimental Zoology* 38:245-292.

Timoféeff-Ressovsky, N. W. 1937. *Experimentelle Mutationsforschung in der Vererbungslehre.* Dresden, Leipzig: T. Steinkopff.

———— 1939. Genetik und Evolutionsforschung. *Verhandlungen der Deutschen Zoologischen Gesellschaft, 1939. Zoologischer Anzeiger Supplementband.*

Walther, J. 1908. *Geschichte der Erde und des Lebens.* Leipzig: Veit.

Wedekind, R. 1920. Über Virenzperioden (Blüteperioden). *Sitzungsberichte der Gesellschaft zur Beförderung der gesamten Naturwissenschaften zu Marburg.*

Weidenreich, F. 1921. *Das Evolutionsproblem und der individuelle Gestaltungsanteil am Entwicklungsgeschehen.* Berlin: Springer.

———— 1929. Vererbungsexperiment und vergleichende Morphologie. *Paläontologische Zeitschrift* 11:275-286.

Weismann, A. 1902. *Vorträge über Deszendenztheorie.* Jena: G. Fischer.
———— 1913. *Vorträge über Deszendenztheorie.* 3d ed. Jena: G. Fischer.
Wettstein, F. von. 1942. Botanik, Paläobotanik, Vererbungsforschung und Abstammungslehre. *Paläobiologica* 7:154-168.
Woltereck, R. 1932. *Grundzüge der allgemeinen Biologie.* Stuttgart: Enke.
———— 1940. *Die Ontologie des Lebendigen.* Stuttgart: Enke.
Zimmermann, W. 1938. *Vererbung erworbener Eigenschaften und Auslese.* Jena: G. Fischer.

Evolutionary Theory in Germany: A Comment

Viktor Hamburger

In Germany, in the twenties and thirties, the problem of adaptation was in the minds of most biologists much more than finer points of evolutionary theory. There was a widespread skepticism that straightforward selectionism could be the ultimate explanation of all complex phenomena of adaptation. Underlying the scientific discussion of these problems was also the inclination of the Germans for a metaphysical underpinning and the often subconscious need to combine their scientific thinking with a metaphysical *Weltanschauung.* Their metaphysics came, for the ones I knew, from *Naturphilosophie* of the early nineteenth century—Goethe, Carus—and from Kant and not, as in some cases in this country, from a religious, Christian background.

Let us focus on the problem of adaptation. Most scientists, at all times, are very reluctant to give away their private personal creed, in print or even in talks; that is why I consider the speech of Boveri of 1906, although it antedates the period we are interested in, as a singularly interesting document, because here he really spoke off the cuff (see chapter 3). As I understood from Baltzer, his student and biographer who knew him very well, Boveri regretted it later and never came back to this topic. But his essay on organisms as historical beings is really a unique document, and it was not repeated among my acquaintances until Spemann, shortly before his death in 1941, committed to print some similar confessions in his autobiography that I'm sure nobody has read.

The biologists I knew, including Spemann and Kühn and many other nonembryologists and nongeneticists, did not feel the need for a unitary explanation that would be valid for all adaptations. Boveri and Spemann made it very clear that they wanted to believe that different options were open to them. I think that was very important. None of them were strongly opposed to selectionism; they all conceded considerable ground to selectionism but they did not consider it a cure-all. Whenever needed they would let Lamarckian thinking slip in and some (Pauly, Spemann) would even admit psychic factors, but always in conjunction with everything else. Only Spemann had an overall psychic idea—a vitalistic view that was taken over from Pauly; Boveri never went that far.

Why this hesitation to accept selectionism as the one and only explanation? I think first and foremost it was the complexity of adaptations, such as concealing or other coloration patterns. For example, the display of the tail feathers of the peacock forms a total *Gestalt* to which each feather and each part of each feather contributes a fragment; these fragments must fit into a total pattern when all feathers come together in this display. To make things more complicated, the hormonally controlled behavior, the state of excitation, must be linked up with the display. Everything must fit, from behavior to the deposition of pigment granules by melanophores and other pigment cells during the development of the feather. Feather development poses many such complexities that challenge the imagination of the embryologist. Each feather develops from a small epidermal elevation that grows to form the shaft and its side-branches, the barbs, and barbules. The pigment cells feed pigment granules into the barbules, and each eye spot of the completed feather, for example, with its dark center and colored rings around it, is "painted" across many barbs. But it is much more like weaving, where the dyes (granules) are arranged in a sequence of different colors in each thread (barb), each at a definite length; and the adjacent thread (barb) has the same sequence but in proportionately longer or shorter lengths, and so forth, until the deposition of black central pigment and then of the colors of the inner, then of the outer, rings is discontinued. And an eyespot is only a small fraction of the total *Gestalt*. Then the pigment-producing cells die, leaving the dead pigment pattern encased in the dead keratin of the barbs and barbules—and natural selection operates on these "dead" patterns. Even if we make allowance for epigenetic interactions between pigment-producing cells and other embryonic regulatory devices, the embryologist is faced with interpreting even minor changes of the total *Gestalt* in terms of small mutational steps. This example could be multiplied hundredfold, in protective coloration alone.

We were always confronted with arguments against selectionism, but no one made an effort, except perhaps the Kühn-Henke school, to really get at one specific example of a complex adaptive color pattern and analyze it in terms of developmental mechanisms.

Süffert, who was a colleague and friend in Freiburg, analyzed these complex adaptations structurally and behaviorally. His own position was very interesting. In Spemann's laboratory we had almost daily tea sessions in the afternoon that sometimes extended for hours, centering often through Süffert's presence on the origin of such complex patterns of adaptation. (Evolutionary problems were hardly discussed; we were not aware of population genetics.) Süffert studied adaptation in the wings of butterflies and moths and he showed how, for example, the pupa of a particular species has a pattern that, on the principle of countershading (which obliterates three-dimensionality) is effective only if the caterpillar selects for pupation the shaded side of a tree. But the hatched moth must move to the other side of the bark of the same tree to make its concealing pattern effective. The switch in behavior has evolved parallel with the change from concealing pupal to imaginal coloration. Süffert was a strict selectionist; he tried to explain to Spemann that all these extremes could be explained in principle by selectionism, but Spemann remained adamant. That dialogue was carried on over many years.

The same scientists found it difficult to accept pure selectionism for other reasons, as well. For example, the individual is always completely integrated and it seemed to Boveri, Goldschmidt, and others that a major evolutionary change would require macroevolutionary "mutations" that would lead immediately to a new organization. "It is not the evolutionary changes of organisms per se that stimulate our curiosities so powerfully, but that the changes are teleological (*zweckmässig*) measured by human standards, or more concretely: not the small modifications are important for us whereby a new species can be distinguished, but those big steps call for an explanation according to which the water animals become land animals, land animals again water animals, crawling animals become flyers, nonvisual ones become visual, and instinctive drive becomes reasoned action" (Boveri, 1906, p.16). Then Boveri asked the loaded question: "Which forces could be able to effect this?" And from there he got into psychic explanations.

It was also argued that it is hard to believe that use itself should not have any formative effects on adaptations because they can actually be seen in modifications. The living organism can individually adapt itself by developing strong muscles or keratin pads on the skin—in other words, the Lamarckian argument was revived. In summary, in the minds of many biologists who were evolutionists there were two or three very powerful phenomena that acted strongly against acceptance of selectionism as the one and only solution.

Although Spemann's ideas were not unique, he was articulate, and his way of thinking came out very well in personal discussions and also in his book. Spemann came from comparative anatomy, from Gegenbaur in Heidelberg, but also from Goethe's idealistic morphology and the principle of interconnectedness of St. Hilaire. This tradition was very much alive; Goethe was still widely read; after all, he had coined the term "morphology," so there was some legitimacy in these connections. Of course, Spemann himself, sustained by his discoveries of epigenetic mechanisms in development, always retained the holistic-interconnection picture. On the other hand he was, as he says in his autobiography, very strongly inclined toward causal analysis and he found his way into experimental embryology through Weismann, who inspired the constriction experiments whose interpretation in turn led to everything else that Spemann did. Spemann is a good example for the observation that a broad, synthetic vitalistic *Weltanschauung* is entirely compatible with a keen analytical acumen, and in his case energized it.

Spemann's only excursion into phylogeny was an article on homology (1915). I remember that Spemann got himself into deep water because he couldn't easily adjust the old concept of homology with data of experimental embryology. For example, he got into this conceptual dilemma: whether a lens that in the embryo is formed by invagination from the epidermis as a result of induction by the optic vesicle is homologous to the lens that in salamander lens regeneration originates from the upper iris. Obviously it is, but that means then that developmental mechanisms cannot be used for definition of homologies.

I did not find in Spemann a burning interest in Mendelian or Morgan-school genetics, in contrast to Baltzer who was also in Freiburg when I was (1918-1920); both were Boveri's students. Baltzer inherited this interest in genetics and actually lectured on genetics. At that time he was working on species hybrids and the hybridization method was one way of looking at evolution. I suspect that the "atomistic" aspect of the gene theory was uncongenial to Spemann although I don't remember ever having heard him say so.

The strange psycho-Lamarckian convictions that Spemann developed became stronger as he got older. Pauly's influence, which dated back to Spemann's student days, persisted throughout his life. (For more details see Hamburger, 1969.) Boveri was about fifteen years older than Spemann, and Boveri was the first one who established contact in Munich with Pauly. Pauly was a very artistic, violent-tempered biologist in Munich who developed a strong psycho-Lamarckian concept of evolution and was very intolerant of anything and anyone who contradicted or slightly disagreed with him. As a matter of fact, Boveri's *Antrittsvorlesung* of 1906 was a direct answer to Pauly's *Darwinismus und Lamarckismus* (1905). Pauly had a strange, persuasive influence on two of the most strong-minded persons of this period, Boveri and Spemann. Both were very critical of Pauly's dogmatism that went to absurd unscientific extremes. In Baltzer's book on Boveri (1962) it is told that Boveri wrote to Pauly: "I'm sure you won't be quite satisfied with my lecture," because Boveri obviously did not agree with all of Pauly's ideas. A strong letter came back from Pauly, the content of which is unknown; but it must have been hard on Boveri. He wrote back: "It isn't that bad, what I did." Nevertheless both Boveri and Spemann were attracted to Pauly by the widespread German tendency to have a unified *Weltanschauung*, a more general overview of metaphysical and scientific creeds than just having *Weltanschauung* and scientific work compartmentalized side by side. Many churchgoing scientists also have a similar compartmentalization. I believe that Spemann never quite resolved this conflict. Whereas Pauly had an all-pervasive psychic principle to explain evolutionary changes, for Spemann and Boveri it was only the last resort. They took recourse to this principle only when everything else—Lamarckism or selectionism— failed; but it was recognized as a third option.

Spemann, toward the end of his life in 1940, wrote in his autobiography a kind of confession along these lines that indicates that there were obviously conflicts in his thinking; it was not a monolithic, uniform philosophy of science. On the one hand, he said, he used psychic agents only for the specific cases that he could not explain otherwise; on the other hand, he was vitalistic on general principle. He wrote: "The encounter with Pauly was for me of decisive significance. It reinforced my basic conviction so consonant with my earlier thinking that the organism in all its living parts is *beseelt* (endowed with psychic properties), no less although in a different way than the organ from which we know this function in ourselves, the brain." Then he went on: "Today I am more firmly convinced than ever of this basic kinship of all life processes, since I have come to know through my own experimental work that the same cell group which seemed to be destined to form skin can also become brain if transplanted in early development into the region of the future brain. Hence we are standing and walking with parts of our body which we could have used for thinking if they had been developed in another position in the embryo . . . This for me is not an assumption or a supposition but an irrefutable certainty" (1943, p. 167).

Two points must be considered in deciding whether Spemann's metaphysical tendency inhibited or rather stimulated his research. The causal analytical kind of thinking in which Spemann was extremely talented was not influenced by his metaphysical thinking. On the other hand, I have pointed out that his selection of experiments from among a large number of possibilities was dictated by his holistic, organismic thinking (Hamburger, 1969). For example, very early he raised the question of the chemical nature of the organizer. He pushed it back in his mind, and only much later, in 1931, performed experiments in which organizer tissue deprived of its structure was tested for its inductive capacity; he was not successful. A year later Holtfreter proved the chemical nature of the organizer by bold, large-scale experiments. In a relatively early paper, in 1918, Spemann mentioned that completely isolating a part of the gastrula in tissue culture would be an ideal way of getting at the problem of determination, but he never performed the experiment—which indicates that the option was there but he chose not to take it. He chose instead to do other experiments in which the embryo was left intact; for example, he tried to make an organizer out of a piece of belly ectoderm by transplanting it in the organizer region. Everything that could be done with the living embryo was preferable to using dead embryos, dead tissues, and explanted, isolated tissues. In that respect, I think, his basic philosophy entered subconsciously into his scientific work.

References

Baltzer, F. 1962. *Theodor Boveri, Leben und Werk eines grossen Biologen*. Stuttgart: Wissenschaftliche Verlagsgesellschaft. (*Theodor Boveri*, trans. D. Rudnick. Berkeley: University of California Press, 1967.)

Boveri, T. 1906. *Die Organismen als historische Wesen: Festrede*. Würzburg: H. Stürtz.

Hamburger, V. 1969. Hans Spemann and the Organizer Concept. *Experientia* 25:1121-1128.

Pauly, A. 1905. *Darwinismus und Lamarckismus*. Munich: E. Reinhardt Verlag.

Spemann, H. 1915. Zur Geschichte und Kritik des Begriffs der Homologie. *Die Kultur der Gegenwart*. III. Teil, IV. Abteilung, vol. 1. Allgemeine Biologie, ed. C. Chun and W. Johannsen. Leipzig and Berlin: B. G. Teubner, pp. 63-86.

——— 1943. *Forschung und Leben*, ed. F. W. Spemann. Stuttgart: J. Engelhorn.

10 France

France is the only major scientific nation that did not contribute significantly to the evolutionary synthesis. In the absence of a French architect of the synthesis, we invited Ernest Boesiger to report on the situation in France, where neo-Lamarckian ideas have flourished almost up to the present. Boesiger had been deeply interested in the recent history of evolutionary biology; unfortunately he died on August 30, 1975, before he was able to prepare the transcript of his conference presentation for publication. This was done by the editors, who are responsible for the manuscript in its present form. They are grateful to Michel Solignac, who helped in preparation of the references.   E.M.

Evolutionary Biology in France at the Time of the Evolutionary Synthesis

Ernest Boesiger

A study of what has been going on in France in the domain of evolutionary biology provides very important insights. It permits an examination of phenomena that in other countries are a matter of history. France today (1974) is a kind of living fossil in the rejection of modern evolutionary theories: about 95 percent of all biologists and philosophers are more or less opposed to Darwinism.

Only a few people assisted, at least to some extent, in the creation of the synthetic theory in the 1930s and 1940s in France. Certainly what happened in the 1930s in France might be called a saltation. Two people, Georges Teissier and Philippe L'Héritier, decided not to follow the general way of French biologists, but to create (for France) a completely new approach. Teissier said in the publication he submitted as a candidate for entrance into the Academy of Sciences in 1958: "In the years of my studies at the university, Darwinism was usually considered an obsolete theory and no biologist in France really believed in natural selection. But in England, first under the influence of Haldane and then of Fisher, and then in the United States, under that of Wright, a new tendency of thinking was created, which led to the magnificent flourishing of a renovated Darwinism." As a result, he wanted to study problems of natural selection in collaboration with his younger colleague, L'Héritier (see the note by Ernst Mayr in the next selection). One point in particular probably persuaded Teissier and L'Héritier to enter evolutionary biology in this way: both had exhibited a tendency to treat biological problems mathematically, which was a revolutionary standpoint then in France. Teissier, to continue the quotation, went on to say: "For the experiments, one must realize, we had to create a completely new technique, with apparatus, breeding procedures, counting techniques and determination of the genetical structure of the studied populations. This technique, which is now very much utilized also in foreign countries, has contributed more than any other to the actual development of the experimental research in population genetics."

This new technology of which Teissier spoke was of course the population cage, which permitted Teissier and L'Héritier to make findings that fit quite well with the synthetic theory. In their second publication (1934) of results using the population cage technique in studying *Drosophila melanogaster*, L'Héritier and Teissier showed that there was a strong selective advantage of the wild type over the mutant type *white*. They deduced, without having the proof of a completed experiment, that the wild type would eliminate the mutant *white*. Soon afterward they performed another experiment of competition between the mutant *Bar* and wild type. They gave figures (not a graph) that showed a substantial selective pressure against *Bar*, but the selective pressure decreased when the frequency of *Bar* became low. *Bar* was not eliminated from the population in this first experiment. Nevertheless, Teissier and L'Héritier stated that they had supplied proof that natural selection had eliminated an allele in favor of the wild type. Soon afterward, in a second paper, they reported that in reality the mutant *Bar* had not been entirely eliminated— an equilibrium had been established between wild type and *Bar*. They concluded in the case of *Bar* that, in conflict with usually accepted theory, the mutant gene is not eliminated, but maintained for a long time at a stable level of equilibrium.

Later L'Héritier and Teissier presented new results, in this case on competition between the mutant *ebony* and wild type. They found an equilibrium frequency of 15 percent for *ebony* and concluded that "natural selection has not only the effect of the maintenance of the uniformity of the species by eliminating monstrosities. It seems that she can, in certain cases, maintain indefinitely a genetical variability in a population of a species" (1937). This first experimental proof of balanced polymorphism was, I think, a significant contribution to the modern synthesis.

In work done with the mutant *curly*, Teissier showed that even a lethal gene can be maintained in a population cage for quite a long time at a high equilibrium frequency. The next question Teissier studied was why lethal genes can be maintained in a population. He found that the fecundity of heterozygote females was higher than that of homozygote wild-type females and stated clearly that the maintenance of *curly* at an equilibrium frequency was the result of this heterozygote advantage.

In 1943 Teissier reported on the behavior of an apparently spontaneous mutation for *sepia* in a population cage. The frequency of *sepia* increased in this population up to an equilibrium level of 22 percent. He concluded: "These facts are remarkable since they show that a new gene can be installed in a population where it appeared by mutation . . . As a general rule mutant genes are rapidly eliminated by the normal alleles and one has even deduced from this situation that natural selection plays exclusively a role of conservation of the normal type of the species. But the modern statistical genetics postulates on the contrary that natural selection may play a role of renovator." Later he proposed two terms to describe these two possibilities of natural selection; he spoke of conservative natural selection and of creative (innovative) natural selection. His words were "selection conservatrice et selection novatrice."

Another very interesting phenomenon was discovered in Teissier's laboratory in frequency selection. He and one of his students observed that when a gene becomes rare in a population, its selective advantage may increase. Teissier later stated in lectures that his findings of changes in gene frequencies in the population cages might explain the maintenance of polymorphism in natural populations. But in 1955 Teissier said in a lecture: "The phenomena of selection which we have observed are nevertheless exceptions, and as a general rule natural selection has a tendency to eliminate all other alleles of a gene, with the exception of the normal allele. This elimination is more or less rapid according to the deleterious effect of the mutation, but it is never immediate and a disadvantageous allele may persist indefinitely in a population, if it reappears constantly by mutation." Because this statement from an older Teissier reduces drastically what he had stated in earlier years, Teissier did not categorically establish in France the modern synthesis.

Very few other contributions to the synthetic theory were made at that time. L'Héritier published a book, *Génétique et Evolution* (1934), but this small book did not contain the work he and Teissier had done. He simply presented in a French book the ideas of Fisher and Wright. Another French colleague, Vandel, made a contribution to the synthetic theory, but he certainly cannot be considered a Darwinian. In 1938 Vandel presented a first paper on terrestrial isopods (woodlice) and later several others, especially on the family of Trichoniscidae. Many species of this family, and especially of the genus he studied, live in caves, while others live outside caves under normal terrestrial conditions. Vandel stated that very often Lamarckian interpretations have been given to explain the specific traits of these cave animals, such as losing the eyes, becoming albinos, having a different size. He performed many experiments. In the laboratory he bred animals coming from caves and others coming from the normal terrestrial situation and produced albinos or individuals with reduced eyes and other changes in both groups. And he said: "The study of the evolutive lines permits postulation that cave forms are a result of a series of mutations oriented in the same sense . . . which are the sign of a long divergent evolution. The appearance of albino mutations in laboratory stocks where all individuals are maintained under identical conditions makes the Lamarckian interpretation, that means the direct intervention of the environment, improbable." He added that some cave forms have eye and pigment reduction and some surface forms are more or less albinos. Only the forms having by preadaptation (in the sense of Cuénot) the traits characteristic for cave forms have penetrated into caves where their evolution continued in the direction typical for cave animals. Vandel concluded that life in caves is a consequence of the depigmentation and not its cause.

He published in 1940 another interesting study in the same vein. He studied triploid races in isopods and showed that these triploids have a greater resistance to low temperatures and aridity than the normal diploid races. He concluded quite correctly that chromosomal mutations can be a factor in evolution and can determine the areas of distribution of a species.

Horvasse, who was professor of zoology in Clermont-Ferrand, did not really work in this field but in 1943 wrote a book on adaptation and evolution. In this book he adopts a more or less Darwinian viewpoint: "Even if a majority of naturalists seem now to believe in the evolution of organisms, evolution which Caullery (1931) and others no longer consider as a theory but as a fact, there are still obstinate opponents to this doctrine." He adds: "To be Darwinist or Lamarckian became a question of nationality, age, temper, specialization and one might even write, of fashion." Horvasse also states that the crisis of transformism (as he calls evolution) is more profound than it ever was. He hopes that the Mendelians will give a direct demonstration of evolution, an ultimate Darwinian solution of the crisis, by studying laboratory and natural populations. Horvasse concludes by saying that Darwin showed the way to go, excluding finalism. Certainly his work has to be adapted to our actual knowledge by accepting the importance of the environmental factors—and by incorporating mutationism. We must understand, he says, that selection is not only conservative (by elimination), but, as Cuénot postulated, that selection also has a constructive role by the action of mutations with a small effect, by preadaptation, and by adjustment to changes in the environment. These conclusions were not based on actual research, and later in his life Horvasse made statements in which he reverted to a more or less skeptical viewpoint.

The majority of the French biologists has often been called a majority of neo-Lamarckians; however, no well-articulated theory existed in France that could be called neo-Lamarckism. Several of the people I describe have strongly protested against the label "neo-Lamarckian." Others can perhaps be considered to have been more or less neo-Lamarckians. Some traits are characteristic of all these people who formed the great majority of opposition to the synthetic theory. It is misleading to call these scientists neo-Lamarckians because all of them rejected the most essential ideas of Lamarck, with the exception of the hereditary transmission of acquired traits.

Let me give a short list of Lamarck's ideas not accepted by the so-called neo-Lamarckians:

(1) Opposition to vitalism: Lamarck was not a vitalist.
(2) Opposition to finalism: Lamarck was not a finalist.
(3) Opposition to dualism: Lamarck was not a dualist. For him feelings, morals, spirit, and mind are functions and expressions of matter.
(4) Actualism: Lamarck was an actualist, exactly as Lyell somewhat later. He said that what we can see today is sufficient to explain the mechanisms of past evolution.
(5) Lamarck made no distinction between microevolution and macroevolution. He stated clearly that all higher taxa are, as he said, "les parties de l'art"; that means created by man. Evolution concerns individuals, but does not create something else. Differentiation occurs, but no specific mechanisms are needed to explain macroevolution.
(6) Lamarck clearly opposed the typological approach in taxonomy, a great achievement that has usually been completely overlooked. He proposed a very new approach for taxonomy.
(7) For Lamarck man evolved exactly like all other animals and plants; no new specific mechanisms were required for human evolution.

The situation in France at the moment of the creation of the synthetic theory can be illustrated by Paul Lemoine, a former director of the Musée de l'Histoire Naturelle in Paris, who wrote the introduction and the concluding statement of a big volume on animals and plants in the French encyclopedia published in 1937, the same year in which Dobzhansky's *Genetics and the Origin of the Species* appeared. Lemoine wrote: "The reading of this volume shows that the theory of evolution will very soon be abandoned. No biological fact favors the theory of evolution . . . Natural selection plays no role . . . The data of genetics furnish no argument in favor of evolution—to the contrary . . . The data of paleontology show that there was no evolution . . . This presentation demonstrates that the theory of evolution is impossible . . . Evolution is a kind of dogma in which the priests no longer believe but which they maintain for their people."

The authors who had written articles for this volume met in 1938 to protest Lemoine's statements in the introduction and conclusion which they had not seen. Guyénot, Caullery, Cuénot, Rostand, Grassé, Arambourg, and Piveteau issued a resolution protesting these two chapters.

Vialleton had only a slightly less ignorant attitude than Lemoine. He published a book on evolution (1929) subtitled *The Illusion of Transformism*, which allows that a special orthogenetic evolution had formed some new genera, but evolution created no types of organization.

Rabaud said in his book (1937) that Morgan's theory was an usurpation of Mendel. There were no facts to support Morgan's theory. The hypothesis of crossing-over was false. The conception of genes was an error. Rabaud also denied the existence of natural selection, the heredity of acquired traits, preadaptation, adaptation, and orthogenesis. Modifications of organisms occur at random and under the direct influence of the environment; most are teratological.

Before one can fully understand the intellectual milieu in France at that period, it is necessary to say a few words about Bergson, who was not a biologist but a philosopher. He wrote a whole book (based on his lectures at the Collège de France) about creative evolution. All French biologists were very heavily and directly influenced by Bergson. Arguments derived directly from Bergson occur in the writings of Grassé, Cuénot, Vandel, and even Teilhard de Chardin.[1]

Confirming Bergson's importance, Monod said in his famous bestseller, *Chance and Necessity* (1971): "When I was young, there was no hope to get the bachelor's degree if one had not read Bergson's *Creative Evolution*." The main point of Bergson's theory is to explain evolution by a mysterious factor, which is of course not observable. He says it would be nonsensical to try to find this factor. He calls it *élan vital*, which can

---

1. Bergson's influence is still strong; see Grassé, *L'évolution du vivant* (1973)—Eds.

be translated as "vital drive." Bergson heavily criticized the mechanistic theorists. He said it was unthinkable that a series of accidents, maintained by selection if they were advantageous, could produce in two different evolutionary lines the construction of identical structures.

Bergson actually declared that the Darwinian idea of adaptation through the automatic elimination of the unadapted was simple and clear. But after having said that the Darwinian idea was brilliant, Bergson declared that the directing agents of evolution were exterior to the organism, were negative, and could not explain the progressive and orthogenetic development of complex structures. How could natural selection explain identity of structure of extremely complicated organs—such as the eye of higher vertebrates—in divergent evolutionary lines? Bergson used this example; so did Cuénot, Caullery, Grassé, and many others.

Vital competition, said Bergson, and natural selection are of no help. We are not interested in what has disappeared, but very much interested in what has been conserved. How could accidental causes, occurring in a random order, give the same result in different evolving lines? A mysterious principle must be postulated that comes to the aid of function. But then this phenomenon would no longer be accidental variation.

Bergson thought that neo-Lamarckism, of all actual forms of evolutionism, was the only one providing an internal and psychological principle of development. It is also the only one that explains the formation of identical complex organs in independent lines of evolution. In the Lamarckian view, it is conceivable that under the same circumstances the same effort would lead to the same result. Bergson was a very successful, liberal philosopher. He was against Darwinism, and he was even against Lamarckism. As a philosopher he had to find an explanation of evolution that was acceptable in France and that did not introduce God. His extremely elegant and very successful solution was to say living matter had an *élan vital*. Bergson said that *élan vital* was a fact even if its exact nature could not be specified. Life had this property from the very beginning. It was a property of living matter to be able to go further and further and to produce evolution.

One of the important biologists of the period was Caullery. In *The Problem of Evolution* (1931) he chose a very characteristic epigram: "The fact of evolution is obvious, only its mechanism remains uncertain." And he presented a certain number of typical arguments that almost exactly duplicate Bergson's.

Although Jean Rostand, son of the great writer Edmond Rostand, was not really a biologist, he wrote in his 1928 book about chromosomes (that he wrote a book about chromosomes shows that he believed they existed and were important): "The evolution of machines which are so complex and well adapted, as most plants and animals are, cannot be explained by the accumulation of random variations . . . Darwinian selection might be considered as an agent of evolutionary progress. But natural selection has only a destructive role of elimination . . . Natural selection maintains the level of a species and cannot improve it . . . If mutations had made evolution, they must have been of a very different type, since those we know actually produce only monsters. There are only regressive, destructive, and no progressive, constructive mutants." Some very curious statements appear in Rostand's book: "Is the evolutive force of the chromosomes exhausted? Did biologists arrive too late?" Lamarck's theory is quite seductive, Rostand claimed, and it explains adaptation. However, it is based on the heredity of acquired traits, which is not proved. Rostand concluded: "We have never been as sure as today that evolution of living forms occurs, but we have never been less sure about the mechanisms of evolution."

Cuénot was a very intelligent man and the most influential of the opponents of the synthetic theory in France. He was also the man who introduced genetics in France. In 1941, in *Invention et Finalité en Biologie*, he says in the very first lines that he fears to be treated as a finalist by mechanists and vice versa because he cannot believe that complex organs and especially "coaptations" have been built up simply by chance. Coaptation (not coadaptations, but a completely different phenomenon that has nothing to do with coadaptation of chromosomes) refers to the perfect fit of two structures; for example, parts of two folded wings, or any two structural elements that must fit perfectly with each other. The upper and lower jaws and the teeth on these jaws are another example. Cuénot said mutations cannot create coaptations, which are not chance effects. On the other hand, he thought mechanistic explanations had very often replaced former erroneous finalistic interpretations. So he saw quite clearly that finalism had to be eliminated.

He hoped to show in his 1941 book that the capacity of invention is one of the important traits of life. Finality in nature is not a theoretical interpretation, but one of the most obvious facts. Accepting finality as a fact does not at all mean accepting a metaphysical principle of finalism. It is simply an observation. Cuénot understood that a biologist was not supposed to be interested in any philosophical consideration of biological problems. But restriction implied a lack of curiosity, and he preferred to have this philosophical curiosity. Cuénot did not deny that natural selection has efficiency; he gave a number of examples of adaptations that can easily and strictly be attributed to natural selection. He thought Darwinism an admirable and very attractive theory, but a theory whose range of legitimate application is rather limited, a theory unable to account for numerous morphological adaptations, and insufficient to explain how in a fluctuating environment trends manifest in these adaptations and especially in coaptations have been sustained.

The logical chain of arguments of Darwinism would break down when it became obvious that death of animals was not really a sorting out of fit and unfit. He gave a very curious, concrete example: Take a pair of green frogs. They produce 5,000 eggs a year, but only two adults remain to replace the parents. Evidently, said Cuénot, this survival is not a consequence of anatomical or physiological advantage, but only of pure chance, particularly because massive destruction affects particularly the young stages. Even if natural selection did exist, it would not explain many evolutionary phenomena, such as the gradual genesis of organs like wings, which are not yet functional as rudimentary organs and cannot be selected for at that incipient stage.

He presented then his requirements for explaining evolution. Living matter is certainly regulated by physicochemical laws—but it also requires for its evolution an anti-chance factor. The Cartesian machine needs an inventor, a conductor, some "obscure" profound and unknown cause of the biological finality. In his conclusion he said that he adheres to a mitigated finalism, somewhat restricted, which is more satisfactory to him than a radical mechanism. He was aware that this conclusion would lead to embarrassing questions. Who had the idea? Who made the plan? Who stimulated the execution of the plan? These questions forced him to admit a transcendental force, a will, an intelligence that guided nature. He said that he always thought that life and man and the whole universe should not be imprisoned in a rigid, definitely fixed, blind determinism without any goal.

Vandel is also a great man who did some excellent work. His evolutionary ideas underwent an interesting development. One of Vandel's first books about the evolution of man was extremely skeptical. But he wrote in his late years (1968) an account of the genesis of the living matter that was much closer to the synthetic theory than the earlier book. However, his very latest publications are again very much opposed to the synthetic theory. In his relatively less skeptical book, *The Genesis of Living Matter* (1968), he made the following statements:

(1) Heredity and evolution must be distinguished, and are actually opposed to each other. Because heredity assures stability and evolution assures change, there is no connection between heredity and evolution.

(2) It is doubtful that mutations produce the evolutionary variations. Mutations are variations in individuals. Evolutionary variations must concern a whole group.

(3) *Drosophila* has not changed since the Tertiary period despite the high number of mutations; as a result, mutations have no evolutionary effect.

(4) Adaptation is always a specialization and thus a way without issue in which evolution stops.

(5) Evolutionary variations responsible for macroevolution can only appear in the egg or in the first stages of ontogeny; only macromutations ("ontomutations," as he calls them) can give suddenly a new constitutional type very different from that of the original stock. (Vandel obviously takes only sympatric speciation into account and ignores allopatric speciation.)

(6) Evolution is not conducted by a supreme being; but it is also not simply the result of lucky chance events, as Epicurus believed and as neo-Darwinists still believe.

(7) In bacteria the sequence of mutation and selection may explain evolution. But unfortunately bacteria did not evolve much. The sieve model no longer applies to metazoans. In metazoans, cytoplasm is as important as the genome. The cytoplasm is responsible for regulations, including the activation of genes.

Surprisingly, French biologists conspicuously neglected the English language literature. Actually, these anti-Darwinians were aware of the writings of Morgan, Dobzhansky (1937), Huxley, and other Darwinians, but they were so unconvinced by their arguments that in most cases they did not even mention them in their writings. This striking and quite peculiar isolationism demands an explanation. But I am not a sociologist or philosopher, and it appears that philosophical and sociological arguments or causes are even more important than scientific issues. I realize that these arguments or causes are of a very different order from those that I encounter in my scientific work.

The first cause that seems significant is that philosophy was extremely important for all French intellectuals since about the seventeenth century, which may partially explain why French scientists like Vandel, Cuénot, or Grassé are unhappy with simple facts; they also want a philosophy. Because France is a Latin country, it is very strongly under the influence of typological thinking as it is taught in high schools, lyceums, and the universities, starting with Aristotle and Plato. This typological thinking is still very important and explains some of the arguments that are given against nontypological populational thinking. In France the philosophy of Descartes has certainly had an influence by reinforcing this typological thinking. And probably the French people, being very deeply influenced by the Cartesian philosophy, have real difficulties understanding Darwinism simply because they are unprepared for it.

Lamarck's theory is a kind of typological or Cartesian theory, but not extremely so. It is evident that Darwin's theory and the modern synthesis fit less well with a Cartesian scheme than Lamarckism. The belated influence of Lamarckism is quite curious because Lamarck had absolutely no success in his own lifetime. He was completely rejected by the powerful Cuvier. However, when the theory of Wallace and Darwin became famous in a very short time, it was no longer possible for French biologists to continue rejecting the idea of evolution. Some did, of course, and I have provided examples. The more reasonable biologists, however, admitted the fact of evolution; but they would not accept the Anglo-Saxon Darwin. Because they needed a Frenchman, they rediscovered Lamarck and made him into something very different from what his ideas actually convey.

When France was defeated in 1871 by the Germans, the chauvinism in France increased. People had to insist on all that might be French against all that was coming from abroad. There has always been discord between France and England. Essentially the same reason applied after World War Two for opposing the American synthetic theory, because it came to France from the United States.

Perhaps even more important were sociological reasons that are a consequence of the structure of French universities and research centers. In France and in some other countries the whole career of a research worker is determined either by consultative commissions at the university or by consultative commissions of a research organization like the French National Research Center in Paris. These commissions are of course very powerful, and their decisions are heavily influenced by individual professors. The chiefs of the committees (*Comité Consultatif*) have enormous power. How can anyone find a job in France if he opposes a boss? Even when big bosses disagreed with each other they usually nominated their favorite candidates by trading off before the official meeting. Because the great bosses were unanimously, with one exception, against the synthetic theory, it is quite natural that the inbreeding of theories continued from Giard to Caullery to Grassé. The exception was my chief, Teissier. In 1945 when Teissier became professor at the Sorbonne, the situation in France was rather special. He received his professorship not as the favorite candidate of a powerful professor, but as the reward for his activities in the resistance movement—an opportunity that might not have been open in 1948. Of course, he could indulge in exactly the same political games. He introduced students agreeing with his own views, but certainly no neo-Lamarckians would have had a chance. So the system remained exactly the same.

Finally, the presentation of neo-Darwinism in France has not always been very happy. It was mostly presented as a simple sieve model: there are chance mutations, and then selection. Only the two-allele system was considered. As a result, the important new achievements of the modern synthesis did not come through in France. For example, the main ideas presented by Dobzhansky in 1937 (heterogeneity of natural populations, the advantage of heterozygotes, coadaptation of chromosomes, interaction of genes, mechanisms of reproductions, isolation being important for the speciation, and the feedback between environment and organisms) were simply not taken into account in most cases even by the few neo-Darwinians.

For all these reasons it is not so astonishing that the situation in France is what it is.

References

Bergson, H. 1907. *L'évolution créatrice*. Paris: F. Alcan et Guillaumin.

Caullery, M. 1931. *Le problème de l'évolution*. Paris: Payot.

Cuénot, L. 1941. *Invention et finalité en biologie*. Paris: Flammarion.

Grassé, P.-P. 1973. *L'évolution du vivant*. Paris: Albin Michel.

Horvasse, R. 1943. *De l'adaptation à l'évolution par la sélection*. Paris: Hermann et Cie. (*Actualités Scientifiques et Industrielles* 941).

L'Héritier, P. 1934. *Génétique et évolution. Analyse de quelques études mathématiques sur la sélection naturelle*. Paris: Hermann et Cie. (*Actualités Scientifiques et Industrielles* 158).

———— and G. Teissier. 1937. *Comptes rendus de la société biologique* 124:880, 882.

Ostoya, P. 1951. *Les théories de l'évolution*. Paris: Payot.

Rabaud, E. 1931. *Le transformisme*. Paris: Presses Universitaires de France.

———— 1937. *La matière vivante et l'hérédité*. Paris: Les éditions rationalistes.

———— 1953. *Le hasard et la vie des espèces*. Paris: Flammarion.

Rostand, J. 1928. *Les chromosomes artisans de l'hérédité et du sexe*. Paris: Hachette.

Teissier, G. 1943. Génétique apparition et fixation d'un gène mutant dans un population stationnaire de *Drosophiles. Comptes rendus de l'academie des sciences* 216:88-90.

Vandel, A. 1938. Contribution à la génétique des Isopodes du genre *Trichoniscus*. I. Les mutations *alba* et *pallida* de *Trichoniscus* (*Spiloniscus*) *elisabethae* Herold, et l'origine des formes cavernicoles. *Bulletin biologique de la France et de la Belgique* 73(2):121-146.

———— 1940. La Parthénogenèse géographique. *IV*. Polyploidie et distribution géographique. *Bulletin biologique de la France et de la Belgique* 73(1):94-100.

———— 1968. *La genèse du vivant*. Paris: Masson et Cie, 7 (*Les grands problèmes de la biologie* 6).

Vialleton, L. 1929. *L'origine des êtres vivants. L'illusion transformiste*. Paris: Plon.

The Arrival of Neo-Darwinism in France

Ernst Mayr

Boesiger's account of the situation in France raises one tantalizing question: How did two young Frenchmen (Teissier and L'Héritier) suddenly become Darwinians in a universally Lamarckian milieu? Boesiger leaves this question unanswered. During a recent visit to France (April 1978) I met the survivor of this team, Philippe L'Héritier, and from conversations with him I was able to reconstruct the following account.

L'Héritier grew up in central France (near Clermont-Ferrand), passionately interested in natural history (particularly birds) and gifted in mathematics. As a result of performing extremely well in his entrance examinations, he was accepted by both the Ecole Polytechnique and the Ecole Normale Supérieure in Paris. He chose the latter to study mathematics. After completing his course he remained there on the staff. During his studies he became a friend of Teissier, who was seven years his senior. Teissier was equally a mathematician with certain biological interests, only in the case of Teissier it was in biometry, particularly the study of allometry. A senior professor, André Mayer of the Collège de France, who was interested in L'Héritier, told him that with his interest in both biology and mathematics he should explore the new field of mathematical genetics (presumably he had Haldane and R. A. Fisher in mind). When L'Héritier showed interest in this suggestion, Mayer procured a Rockefeller Fellowship for him, and in 1931 L'Héritier went to Ames, Iowa, to study genetics. He studied under the corn geneticist Lindstrom, although Jay Lush taught animal breeding at Ames as well. At the end of the academic year 1931-32 L'Héritier, like the other Europeans studying in the United States, made a "grand tour," including a pilgrimage to Morgan's laboratory at Cal Tech in Pasadena where he met Dobzhansky among others. His long conversations with Dobzhansky might have given him the idea of populations. On his way back to France, L'Héritier visited Woods Hole, another traditional stopping point for European biologists. There, he remembers vividly, while walking on the beach he suddenly had the idea that one might be able to experiment with populations by constructing population boxes. With Teissier's help after his return to Paris, he translated this idea into reality and soon the first papers on the behavior of genes in artificial populations were published.

In 1938 L'Héritier was offered a position at the University of Strasbourg, which ended his teamwork with Teissier. They agreed that transporting the heavy population cages to Strasbourg was impractical; Teissier continued the population work, while L'Héritier concentrated on the cytoplasmic factor (CO_2) they had discovered. When war threatened, the University of Strasbourg was transferred to Clermont-Ferrand where L'Héritier continued his work as much as the war allowed.

L'Héritier told me that he and Teissier had no difficulty in accepting the Darwinian interpretation, particularly the noninheritance of acquired characters and the principle of natural selection, because as members of a mathematics department they had not been exposed to the prevailing French ideas and natural selection was an eminently reasonable process for a mathematician.

Boesiger's thesis that the French system of authoritarian institute directors made converting the younger staff members to modern evolutionary thinking quite impossible is indirectly corroborated by L'Héritier's account. The new ideas had to be brought in from the outside and mathematicians, who were not familiar with the biological objections raised by the Lamarckians against selection, were well qualified to serve as the entering outside wedge. Many details are still lacking in this sketchy account, particularly concerning the role of André Mayer. But the major puzzle seems now to have been resolved.

A Second Glance at Evolutionary Biology in France

Camille Limoges

My brief remarks will focus on three points that seem to me of particular interest: the place and significance of neo-Lamarckism in French evolutionary thinking, the reluctant acceptance of Mendelian genetics and of its relevance for evolution theory, and some institutional peculiarities of French biology.

Neo-Lamarckism is not a French invention, but rather an American-born movement represented by biologists like Cope, Hyatt, and Packard before it started to take shape in France in the very late 1880s. The first generation of French evolutionists, including Giard and Perrier, did not see at first any antagonism between Lamarck and Darwin, but complementarity, and considered themselves as Darwinians. Like Romanes in England, they believed that true Darwinism embodied not only natural selection but also so-called Lamarckian factors. This syncretic attitude was maintained up to the time when a neo-Darwinian viewpoint was defined by people like Weismann, Lankester, Wallace, Poulton, and Meldola, who advocated the "all-sufficiency" of natural selection and excluded all Lamarckian remnants. About 1888-1890, in reaction against neo-Darwinism, by refusing to accept total rejection of Lamarck, French evolutionary biology becomes clearly neo-Lamarckian, leaving very little room for the efficiency of natural selection.

Giard was the preeminent advocate of French neo-Lamarckism, but the movement was never unified. Though some of his students, such as Le Dantec and Caullery, followed more or less closely his teachings, other major French biologists of the time, including Perrier and Delage, remained aloof though they also strongly advocated the inheritance of acquired characteristics as the core of the explanation of evolutionary phenomena. These rivalries seemed to have been, on the whole, more personal and institutional than theoretical.

In a way, the presuppositions of Lamarckian evolutionary biology were accepted by the majority of French biologists until the 1960s; some may still adhere to these views. However, the movement never succeeded after the 1910s in presenting an organized and consistent theoretical body of knowledge. The impossibility of proving the occurrence of any inheritance of acquired characteristics led some French biologists with neo-Lamarckian inclinations to propose that evolution has come to an end, that truly evolutionary processes are no longer observable in nature. The idea, of course, was that when evolution did occur, it had conformed to Lamarckian mechanisms. This position was, for instance, repeatedly expressed by Caullery and Rostand.

This peculiar conclusion led to an attitude of "theoretical agnosticism." Highly skeptical about the very possibility of building a true (Lamarckian) evolutionary theory, most French biologists of the 1920-1950 period were thoroughly critical of all theorizing and were remarkably reluctant to accept any of the major theoretical breakthroughs of biology in this century: Mendelism, the chromosomal theory of heredity, population genetics, and the evolutionary synthesis. Their rejection may account for the heavily descriptive character of French biology and ecology up to the 1960s. It may also account for the fact that all books discussing evolutionary theory in France have been more or less of a popularizing nature. Indeed, advocates of theoretical agnosticism in evolutionary matters are not likely to write theoretical literature for the specialist. As professionals they will, by force, have to adopt a rather descriptive approach to biological problems. Only reluctantly will they write on great theoretical issues and then merely to caution against any exaggerated hopes, hopes that are more likely to be those of the laymen than of the professionals.

The few exceptions to that generalization are very interesting, particularly Cuénot. With Cuénot, France had made a good start in Mendelian genetics. In the very first years of the century, at about the same time as Bateson and Punnett, Cuénot had verified the validity of Mendelian ratios for animals, described multiple allelomorphs, done work relating to the discovery of lethal genes, and speculated early on the possible relationship between enzymes and genes. However, he alone in France was engaged in experimental Mendelian genetics; he presented his results in a climate of considerable opposition or at best skepticism. In the 1890s he had been the only French biologist to have taken side with the neo-Darwinian camp, by resolutely rejecting all aspects of the neo-Lamarckian approach. (To understand Cuénot's work and the French evolutionary biology of the time, see Limoges, 1976.) By 1903, however, he had rejected neo-Darwinism and adopted a mutationist viewpoint, for a time close to that of de Vries. From that time on, there seems to have been no Darwinian biologist in France, until Teissier and L'Héritier started to publish their population genetics researches in 1933. However, they did not get chairs until after World War Two.

The reluctance in France to accept Mendelian genetics and the Morgan school theory was deeply rooted in some of the presuppositions of the neo-Lamarckian approach. Central to these was the conception of the organism as a whole of tightly correlated parts such that any particulate theory seemed wrong and naive. People like Le Dantec and Delage emphasized, for instance, the idea of the cell as a whole of such a nature that any autonomy of the nucleus from the cytoplasm, which they saw somewhat as the "environment" of the nucleus, seemed very farfetched. Le Dantec said, for example, that the characters on which Mendelian genetics was based were of a superficial and trivial nature and had no more to do with the deep nature of the organism than the pieces of clothes that the circus clown removes one after another leaving at the end a no less complete man.

Caullery, who later became the main popularizer of Mendelian-Morganian genetics in France, held similar, if less extreme, views. In 1913, in his book *Les Problèmes de la Sexualité*, Caullery still saw the chromosomal interpretation of Mendelian genetics as an unacceptable offshoot of Weismannism. (Weismann at that time in France was the scapegoat for everything that supposedly went wrong in the development of evolutionary theory since the 1880s, which can be attributed to both the reaction against neo-Darwinism and the traces of anti-German nationalism in the biological literature.) In 1913 Caullery still argued against the chromosomes as a permanent component of the nucleus, rejected the idea that they are the bearers of the hereditary material, and believed that, had it not been for the fact that chromosomes are easy to dye, there would be no such fuss about them. Mendelism is seen mainly as the building of a symbolism to describe regularities, the causes of which are not understood, and the approach of the Morgan school, for instance, is seen as a sort of naive—or crass—realism that reifies symbols into things. The idea that sex may be determined by the "x factor" is denounced as a clear instance of that reification of symbolism. The idea of the determination by a single factor is predicated, Caullery said, on the assumption that sexuality is an element superimposed on the organism, whereas any good biologist knows that sexuality pervades the organism. No atomistic partitioning of the organism could do.

Caullery's position seems to be representative of the thinking of the French biological community of the time. For example, at the Fourth International Conference on Genetics held in Paris in 1911 all the French speakers were opposed or at best skeptical toward the new developments. Cuénot, the only significant French contributor to the development of Mendelian genetics, did not attend but stayed in Nancy. The leading French presentation was made by two students of Caullery, Delcourt and Guyénot, who argued that by not keeping the environment controlled in their *Drosophila* experiments, the biologists of the Morgan school had placed themselves in a very weak position. From an evolutionary standpoint, nothing could be concluded from their work, especially as far as the origin of the variations was concerned. In other words, they repudiated the chromosome theory. Delcourt and Guyénot had painstakingly worked out a set of techniques by which such physical variables as temperature and moisture were controlled, the environment kept sterile, and the composition of the food given to the flies analyzed and constant. Their research program was clearly Lamarckian, with the expectation that studying the effects of the fluctuation of single variables of the environment would enable them to attribute the source of hereditary variations to environmental influences. They had few results in 1911, but their expectations ran high. However, when Guyénot defended his doctoral dissertation in 1917, none of the anticipated results had been obtained; he essentially showed only the development of sophisticated techniques of breeding *Drosophila* in an aseptic environment. The next year Guyénot had become a convinced Morganian and published articles pleading for a mutationist, non-Lamarckian viewpoint. In 1924 he published the first textbook on classical genetics in French.

After Caullery visited Morgan in 1916 he became an advocate and popularizer of the chromosome theory of heredity. But he never stopped being a Lamarckian, remaining convinced that Mendelism explains nothing beyond the boundaries of the species and has nothing to do with evolution, a process seemingly ended. As late as 1936 Caullery had to defend himself against Rabaud, one of his colleagues at the Sorbonne who still rejected the chromosome theory of heredity and the concept of the gene, which indicates that Caullery in context was a progressive. However, he did not contribute anything to the development of genetics; Guyénot did, but he was a Swiss teaching in Geneva. By the mid-1920s Cuénot, who had stopped his original contributions to genetics many years before, was becoming more and more of a finalist and rather dissatisfied with the mutationist viewpoint. Except for Teissier and L'Héritier, who were attentive to the works of Wright and Fisher, and Boris Ephrussi working at the time mainly outside France, the French contribution to genetics was rather bleak.

The generation of students trained after the war by Teissier and L'Héritier began the integration of genetics and evolutionary thinking. As for the spread of the concepts of the synthetic theory in France—though a complex study is needed for confirmation—L'Héritier seems to have played a more consistent role than Teissier who, as a Communist, was placed in a very difficult position at the time of the Lysenko affair. (Among all the European Communist parties, the French one played the leading role in Lysenkoist propaganda.)

I do not entirely agree that Descartes and Bergson were the villains whose influence caused a detrimental effect on the minds of French biologists. However, I agree that philosophical influences must be taken into account to understand the French biological literature of the period. The spiritualist overtones of even the most rationalist of French philosophers created an intellectual environment uncongenial to a Darwinian approach. The language barrier and the university system may help to explain the slow advance of the synthetic theory in France. The French university was extremely provincial in language matters; unlike American universities, which expected its science graduate students to know one or two foreign languages, many French biologists during that period had to rely mainly on French sources owing to a lack of a working knowledge of English. The related fact that French students often had to rely entirely on rather grotesque presentations of what Darwinism is about should also be taken into account. The peculiarities of the institutional system also influenced French thought. For example, the system is closed; no one can hold a chair in a French university who is not French and who does not have a French doctorate. It is an isolating factor of significance; it means that France is largely deprived of direct foreign inputs. The history of twentieth-century American evolution biology might have been quite different if all faculty positions had been closed to foreigners.

Another aspect is the centralization of the system, now largely diminished as a consequence of the transformation of the French universities, triggered by the events of 1968. That centralization, providing for career control throughout France in the hands of a few senior professors in Paris, tended to induce intellectual conformity among biologists and dampen theoretical audaciousness. Conversely, the independent-minded Cuénot remained all his life in Nancy; Guyénot had a chair in Geneva; Ephrussi, an associate professor at Johns Hopkins from 1941 to 1944, had previously held no chair in France; and Lwoff, Monod, and Jacob, of Nobel fame, were based in a private institution, the Institut Pasteur.

The top of the system, the Paris Faculty of Science, until the 1960s was quite aloof from evolutionary developments. Its major chair devoted to evolutionary biology, created for Giard in 1887, shows a striking pattern of continuity: Giard was succeeded by his student Caullery in 1908, who was himself succeeded by Grassé until his retirement in 1965. Only then with Bocquet, the first Darwinian ever to occupy the chair, did the second wind of evolutionary thinking, first heralded in France by Teissier and L'Héritier, reach the institutional summit.

References

Caullery, M. 1913. *Les problèmes de la sexualité*. Paris: Flammarion.
———— 1931. *Le problème de l'évolution*. Paris: Payot.
———— 1935. *Les conceptions modernes de l'hérédité*. Paris: Flammarion.
———— 1937. A propos des commentaires de M. E. Rabaud sur l'hérédité. *Bulletin biologique de la France et de la Belgique* 71:1-9.
Cuénot, L. 1902a. La loi de Mendel et l'hérédité de la pigmentation chez les souris. *Compte rendu hebdomadaire des séances et mémoires de la Société de Biologie* 54:395-397.
———— 1902b. Sur quelques applications de la loi de Mendel. *Compte rendu hebdomadaire des séances et mémoires de la Société de Biologie* 54:397-398.
———— 1903a. L'hérédité de la pigmentation chez les souris noires. *Compte rendu hebdomadaire des séances et mémoires de la Société de Biologie* 55:298-299.
———— 1903b. Transmission héréditaire de pigmentation par les souris albinos. *Compte rendu hebdomadaire des séances et mémoires de la Société de Biologie* 55:299-301.
———— 1904a. L'hérédité de la pigmentation chez les souris (3ième note). *Ar-*

chives de zoologie expérimental et générale 4 (3):123-132.

———— 1904b. Un paradoxe héréditaire chez les souris. *Compte rendu hebdoma-daire des séances et mémoires de la Société de Biologie* 56:1050-52.

———— 1905. Les races pures et leurs combinaisons chez les souris (4e note). *Archives de zoologie expérimental et générale* 4 (3):123-132.

———— 1907. L'hérédité de la pigmentation chez les souris (5e note). *Archives de zoologie expérimental et générale* 4 (6):1-13.

———— 1908. Les idées nouvelles sur l'origine des espèces par mutation. *Revue générale des sciences pures et appliquées* 6:860-871.

———— 1909. Les peuplements des places vides dans la nature et l'origine des adaptations. *Revue générale des sciences pures et appliquées* 20:8-14.

———— 1911. *La genèse des espèces animales.* Paris: Alcan.

———— 1936. *L'espèce.* Paris: Doin.

Delage, Y. 1895. *La structure du protoplasma et les théories de l'hérédité.* Paris: Reinwald.

———— and M. Goldsmith. 1919. Le Mendélisme et le mécanisme cytologique de l'hérédité. *Revue scientifique* 57:97-109, 130-135.

Delcourt, A., and E. Guyénot. 1911. Génétique et milieu. Nécessité de la déter-mination des conditions. Sa possibilité chez les Drosophiles. *Bulletin scienti-fique de la France et de la Belgique* 45:249-332.

———— 1911, 1913. Variation et milieu. Lignées de Drosophiles en milieu stérile et défini. *Proceedings of the Fourth International Congress of Genetics* (Paris), pp. 478-487.

Giard, A. 1904. *Controverses transformistes.* Paris: Naud.

———— 1911. *Oeuvres diverses.* Paris: Laboratoire d'évolution des êtres orga-nisés.

Guyénot, E. 1917. Recherches expérimentales sur la vie aseptique et le développe-ment d'un organisme (*Drosophila Ampelophila*) en fonction du milieu (the-sis).

———— 1921. Lamarckisme ou Mutationnisme. *Revue générale des sciences pures et appliquées* 32:598-606.

———— 1924. *L'hérédité.* Paris: Doin.

———— 1930. *La variation et l'évolution.* Paris: Doin.

Le Dantec, F. 1904. L'hérédité des diathèses ou hérédité mendélienne. *Revue sci-entifique* 41:513-517.

———— 1913. *Evolution individuelle et hérédité,* 2d ed. Paris: Alcan.

Limoges, C. 1976. Natural selection, phagocytosis, and preadaption: Lucien Cuénot, 1887-1914. *Journal of the History of Medicine and Allied Sciences* 31:176-214.

Rabaud, E. 1936. Commentaires sur l'hérédité. *Bulletin biologique de la France et de la Belgique* 70:409-455.

Rostand, J. 1928. *Les chromosomes, artisans de l'hérédité et du sexe.* Paris: Hachette.

———— 1932. *L'évolution des espèces.* Paris: Hachette.

11 England

The course of the evolutionary synthesis in England, as in each of the other countries, was quite distinctive. England was the land of Charles Darwin, who was not only the most famous evolutionist but also the most comprehensive in his outlook. Darwin had attempted to synthesize all the available facts from many diverse fields: animal and plant breeding, geology, paleontology, geographical distribution, systematics, embryology, anthropology, psychology, and sociology. He believed that he had created a truly synthetic view of evolution.

Although T. H. Huxley and Romanes had died in the 1890s, many old-guard Darwinians were still alive at the turn of the century and at the rediscovery of Mendelism. Wallace, Hooker, Lankester, Seward, and Shipley had all known Darwin. Darwin's sons Leonard, Francis, and George were all active in promoting their father's ideas. The outpouring of Darwinian literature in England in 1909, the centenary of Darwin's birth, and the fiftieth anniversary of the publication of *On the Origin of Species* indicate that Darwin's ideas still lived. Cambridge University Press alone published Darwin's essays of 1842 and 1844 (in *The Foundations of the Origin of Species*), a huge volume of essays (*Darwin and Modern Science*), and a stunning guide to Darwin exhibits at Christ's College, Cambridge, on display in 1909 (*The Portraits, Prints, and Writings of Charles Robert Darwin*). Poulton published *Darwin and the Origin*, and many other books and innumerable articles also appeared in England.

But was Darwinism really so healthy in England? The third generation of Darwinians, which included people like Poulton and Goodrich, was much smaller than the second generation. And the grand Darwinian synthetic view of the evolutionary process had become narrower and more fragmented with time. The second-generation Darwinians had already lost some of the diverse expertise Darwin had brought to evolutionary problems. By 1909 Poulton was clearly an entomologist and Goodrich a morphologist, both specialists rather than generalists. The problem was greater than lack of education. Each of the fields Darwin had synthesized had grown during the last third of the nineteenth century, and as they grew, the glue binding Darwin's synthesis came apart. Although the grand Darwinian synthesis remained an ideal, by the early 1900s much work in special fields seemed necessary before the synthesis could be spelled out in any detail. Bateson expressed this idea clearly several times when he argued that before we could understand evolution, we had to know more about heredity. And he threw himself into the study of heredity after 1900.

The naturalist-systematist Darwinians in England had little understanding or use for the new science of genetics that emphasized large single mutations. Naturalists rarely found such mutations in natural populations. The biometricians Pearson and Weldon, both Darwinians, fought furiously with the Mendelians, led by Bateson. Paleontologists and morphologists continued on their own specialized routes. Consequently, just as Darwin was being celebrated in 1909, his grand synthesis had in practice been largely fragmented. The fragmentation was, perhaps surprisingly, more severe in England than elsewhere. By the early 1910s experimental geneticists who were Darwinians could be found in the United States (Castle, East, Jennings) and Germany (Baur, Goldschmidt). But England had no experimental geneticist who was a Darwinian, and no Darwinian naturalist who incorporated the new science of heredity into his work. Furthermore, as Richard Burkhardt shows in this chapter, a substantial and vociferous group of neo-Lamarckians also flourished in England.

Just how little the naturalist Darwinians appreciated the contributions of genetics to the study of evolution in nature is exemplified by Poulton. In his *Essays on Evolution* (1908), Poulton had argued that Mendelism was unimportant in the study of evolution. Thirty years later one of Poulton's last papers was titled "Insect Adaptation as Evidence of Evolution by Natural Selection" (1938). One might expect Poulton at least to mention the important contributions of Fisher to Darwinian mimicry theory (1927, and 1930), or Haldane's analysis of the spread of the *carbonaria* gene in *Biston betularia* (1924), or the work of Ford on insect adaptation (1931). Poulton knew the work of all three and had even communicated Fisher's 1927 paper on mimicry to the Entomological Society of London, but he mentioned no geneticist in his 1938 paper, which was intended "to bring together certain observations on insect life which appeared to demand a Darwinian and exclude a Lamarckian interpretation." He concluded that "the same conclusions, based on the same kind of evidence, had been fully reached by Henry Walter Bates only two years after the publication of the Origin of Species" (1938, p. 1). In Poulton's interpretation of the evolutionary process, the rise of genetics was of negligible importance.

The problem of analyzing the rise of the evolutionary synthesis in England is especially interesting because of the deep gulfs between the fields, later synthesized into one comprehensive view. Just how the synthesis occurred in England is a controversial subject. One conclusion is certain, however. J. S. Huxley, grandson of T. H. Huxley, played a central role in creating a neo-Darwinian synthetic view of evolution in England. The weight of the Darwinian tradition fell squarely upon Huxley's shoulders. His grandfather thought Julian would be a fine biologist; Julian himself dearly wished to continue the tradition of his grandfather and Darwin.

While a student at Oxford he worked with the comparative anatomist Goodrich, who was a Darwinian. More importantly, he studied with two rising young Darwinians, Jenkinson and Smith (who served as Huxley's zoological tutor). Both Jenkinson and Smith were killed in World War One. Huxley also studied at the famous Naples Marine Biological Station where he met many of Europe's finest biologists, and in Germany with Warburg and Richard Hertwig. He visited the United States and met at Columbia with Morgan, Wilson, and their students; he also visited many other institutions. Huxley was much impressed by the work of the Morgan school (his 1942 book, *Evolution: The Modern Synthesis*, was dedicated to Morgan), and when he taught at Rice Institute he invited Muller to come as his assistant, "one of the most sensible things I ever did" (1970, p. 91). By the time Huxley returned to England in 1920 to teach at Oxford University, he was a fine field naturalist, well acquainted with results in experimental genetics, and a generalist in biology as a whole. Ford, de Beer, Elton, and Baker were among his students at this time (Baker, 1976).

Huxley unquestionably exerted a strong Darwinian influence in England with his teaching, writing, and organizational activities. He brought Ford and Fisher together, communicated with most geneticists and evolutionists in England, read voraciously and widely, and wrote voluminously. His first explicitly Darwinian book explaining evolution was a little volume in the "Things to Know" series, titled *The Stream of Life* (1927). Here Huxley emphasized the primacy of gradual natural selection as the prevailing mechanism of evolution in nature. This book is only 63 pages long, but it contains a surprisingly sophisticated and complete statement of the neo-Darwinian view of evolution that became widely accepted in the 1930s and 1940s.

In the same year Huxley accepted an offer from H. G. Wells to join him and his son G. P. Wells in writing *The Science of Life* (Wells, Huxley, and Wells, 1930), the biology sequel to Wells's famous *Outline of History*. Huxley resigned his professorship at King's College in London to write full time for more than two years, producing more than half a 1,514-page book. He wrote a total of 630 pages in the published version devoted exclusively to processes of evolution. Set in the context of the entire sweep of biology, Huxley's discussion of evolution was the single most encompassing presentation of a neo-Darwinian viewpoint available in 1930. Huxley explicitly downplayed the importance of the inheritance of acquired characters, orthogenesis, and evolution by discontinuous leaps, while emphasizing the importance of natural selection operating on small heritable differences. Moreover, Huxley incorporated the most recent ideas from genetics, selection theory, gene control of development, systematics, and animal behavior in his conscious attempt to portray the neo-Darwinian synthesis. The influence of *The Science of Life* on scientists and the educated public deserves careful study by historians. Evolutionist Thomas Eisner of Cornell University has, for example, recently stated that reading *The Science of Life* was decisive in fixing his view of evolution in the neo-Darwinian framework.

Huxley made original contributions to evolutionary theory in the areas of animal behavior, problems of relative growth, sexual selection, gene control of development, and embryology in general. But far more important was his overall synthetic view of the evolutionary process. Following his popular presentation in *The Science of Life*, Huxley attempted to keep abreast of new developments in evolutionary theory and to make a fully documented modern synthesis directed toward biologists rather than the general public. Three major publications followed in rapid succession: his presidential address to the Zoology Section of the British Association (1936); his edited volume, *The New Systematics* (1940); and *Evolution: The Modern Synthesis* (1942). All these publications relied heavily on the recent research by Dobzhansky, Muller, Goldschmidt, Mayr, Sumner, Waddington, Ford, Darlington, Turrill, and especially Fisher, Haldane, Wright, and Hogben.

Evolution: The Modern Synthesis was more comprehensive in subject matter and documentation than the other major works of the evolutionary synthesis period, including Haldane's *Causes of Evolution* (1932), Dobzhansky's *Genetics and the Origin of Species* (1937), Mayr's *Systematics and the Origin of Species* (1942), and Simpson's *Tempo and Mode in Evolution* (1944). Haldane's and Dobzhansky's books were written primarily for geneticists, Mayr's for systematists, and Simpson's for paleontologists. Perhaps because of the greater appeal of these books for specialists, they together exerted a greater impetus toward the evolutionary synthesis than did Huxley's book with its broader approach. Had Huxley put his book together in 1936 (for the most part, he could have), it might well have become the dominant force of the evolutionary synthesis.

In constructing his new synthesis, Huxley made extensive use of the more specialized researches of English evolutionists whose work, as Huxley said, was crucial for the synthesis. Some English evolutionists have suggested that Fisher's *Genetical Theory of Natural Selection* (1930) was the beginning of the evolutionary synthesis in England; others have chosen his 1918 paper, "The Correlation between Relatives on the Supposition of Mendelian Inheritance." Ford's comments in this chapter single out Fisher's 1927 paper on the evolution of mimicry as the best candidate. In any case, Fisher undoubtedly played an important role in England. Haldane and Hogben also contributed substantially to the theoretical underpinnings of the synthesis in England. Darlington's ideas about cytology in relation to evolution, especially the evolution of genetic systems, had a wide influence beginning in the early 1930s (see chapter 2). Mather's work on quantitative inheritance marked the arrival in England of a strong genetic analysis of continuous variability. And finally came the important fieldwork on genetics and evolution in natural populations initiated by Diver, Ford, and Ford's students. Once begun, the synthesis occurred quite rapidly in England, almost in the decade 1930-1940.

E. B. Ford, M.A., D.Sc., F.R.S., entered Wadham College, Oxford, in 1920. Even before going up to the university, many of his lifelong interests were formed, including classical studies, archaeology, traveling to Italy and Sicily, and field studies of Lepidoptera.

Ford's presentation at our 1974 conference was not recorded; the tape machine malfunctioned. Because of other heavy commitments, Ford was unable to re-create his remarks. The loss is severe. Ford is a central figure in the evolutionary synthesis in England, and he described at length his family background and his contacts with the great evolutionary thinkers of the previous generation. At the conference Ford spoke without notes, giving a delightful presentation that everyone enjoyed greatly.

We could not simply leave blank Ford's contribution to the synthesis or his evaluations of the contributions of others. By rearranging his detailed answers to the original questionnaire, and his answers to the many questions that Ernst Mayr and I posed to him by letter, I have put together the following selection, which helps to fill the gap left by the malfunctioning tape recorder. Ford read and edited this final version.

Alexander Weinstein has contributed a note on Tower's neo-Lamarck-ism, which follows Burkhardt's selection on neo-Lamarckism in Britain and the United States.   w.b.p.

References

Baker, J. R. 1976. Julian Sorrel Huxley. *Biographical Memoirs of the Fellows of the Royal Society* 22:207-238.

Fisher, R. A. 1918. The correlation between relatives on the supposition of Mendelian inheritance. *Transactions of the Royal Society of Edinburgh* 52:399-433.

———— 1927. On some objections to mimicry theory: statistical and genetic. *Transactions of the Entomological Society of London* 75:269-278.

———— 1930. *The genetical theory of natural selection.* Oxford: Oxford University Press.

Ford, E. B. 1931. *Mendelism and evolution.* London: Methuen.

Haldane, J. B. S. 1924. A mathematical theory of natural and artificial selection. *Transactions of the Cambridge Philosophical Society* 23:19-41.

Huxley, J. S. 1927. *The stream of life.* New York and London: Harper.

———— 1936. Natural selection and evolutionary progress. *Reports to the British Association for the Advancement of Science* 106:81-100.

———— ed. 1940. *The new systematics.* Oxford: Oxford University Press.

———— 1942. *Evolution: the modern synthesis.* London: Allen and Unwin.

———— 1970. *Memories.* London: Allen and Unwin.

Poulton, E. B. 1908. *Essays on evolution.* Oxford: Oxford University Press.

———— 1938. Insect adaptation as evidence of evolution by natural selection. In *Evolution: essays on aspects of evolutionary biology,* ed. G. R. de Beer. Oxford: Oxford University Press, pp. 1-10.

Wells, H. G., J. S. Huxley, and G. P. Wells. 1930. *The science of life.* New York: Doran.

Some Recollections Pertaining to the Evolutionary Synthesis
E. B. Ford

My father's interest and companionship were a constant stimulus and inspiration to me. He was a most interesting and widely educated man (he was, of course, an Oxford graduate). He was very well read, an outstandingly good public speaker, and very much a figure in society. He was not a naturalist and had never collected Lepidoptera. When I started to do so, on July 27, 1912, he almost at once was delighted to join me, and we gradually developed our entomological studies together: each starting without previous knowledge.

We together made the now well-known observations on the Marsh Fritillary, *Melitaea aurinia*, from 1917 on, and we wrote a joint paper (Ford and Ford, 1930), the first time that a relationship between numerical fluctuations and variation, and its impact on evolution, especially human evolution, was suggested. Although confirmed since on a number of occasions, it has never been developed, in the way it should, as a major evolutionary device.

I fancy that my appreciation of the value of experimental work came via archaeology. We lived in one of the family houses until I was ten. The long front was of the William and Mary period (about 1690-1700), though parts of the house were much older. It was built on the views of the Roman fort on the hill above, and Roman coins and pottery were constantly turning up in the garden, which had been cultivated for hundreds of years, the better pieces removed and the rest thrown back. But even the latter were sufficient to fire the interest of a small boy. Just, therefore, as I came to think of archaeology a good deal from the point of view of fieldwork, so I did of evolution. Working on archaeology induced me to cultivate the habit of deduction from my own observations, a habit I began to apply in a scientific direction through collecting Lepidoptera. As a result I had begun evolutionary studies before going up to Oxford. Though a classical scholar, I was a convinced Darwinian before going to the university. I had read *On the Origin of Species* in bits, as a boy, but was indirectly more than directly influenced by it.

The fieldwork with my father on *Melitaea aurinia* was very important for my development as an evolutionist. This *Melitaea* work made me think I could contribute something to evolution and caused me to change from philosophy and the classics to zoology. I wanted to know more about the variation (which *M. aurinia* sometimes did, and sometimes did not show) and I found Darbishire's book, *Breeding and the Mendelian Discovery* (1911), in the little library of the department of zoology. Then Julian Huxley advised me in regard to more extended genetic literature. The *Melitaea* work also gave me a clue about the rate of evolutionary change, a subject that later became very influential in my researches. When my father and I first looked in 1917 for *M. aurinia* in a neighboring locality where it had once been very common, we were disappointed to find it rare there. We caught some specimens each year and looked for varieties, but there seemed very few. Suddenly the numbers and variability began to increase greatly. When Leonard Darwin told me later that his father (Charles Darwin) thought it would take about fifty years to detect and study evolution in an annual species I, remembering *M. aurinia*, felt sure he was wrong.

The Honours School of Zoology at Oxford, which I read, covered the whole field of zoology, but with special reference to comparative anatomy and embryology. There were some lectures on genetics given by Julian Huxley, but genetics was the subject least dealt with. There was little in the lectures on it that I was not getting out of the literature. I myself was concerned with the study of evolution by means of observation and experiment, an interest that derived from our work on *Melitaea*. No such idea was considered in the course at Oxford. I was the only undergraduate who researched on genetics. We were given the amphipod crustacean *Gammarus chevreuxi* to show us segregation. I went on studying it because I realized that the genes for eye color controlled developmental rates.

As an undergraduate, I read quite a considerable portion of the existing books and articles on genetics, but it was not too large a field to cover in the early 1920s. I thought Darbishire's book the best. It was obvious even then that Punnett writing on mimicry had not properly looked at the butterflies he was writing about, and Poulton agreed. Punnett's book on poultry was trivial. I thought Morgan's book, *The Physical Basis of Heredity*, one of the worst written books I had ever encountered and I think so still. Having, in my classical studies, read books by people who could write, I was horrified. How dreadful are the blots on the pages that are called figures. If anything could have stopped me from taking an interest in genetics, it would have been that book of monumental dullness and incompetent presentation.

The morphologist and embryologist Goodrich was the principal lecturer in zoology at Oxford. He was a genuine Darwinian, though he never himself contemplated the impact of genetics on evolutionary studies. Doubtless he was a Darwinian because he was Lankester's favorite pupil. But I doubt if Goodrich was much involved in the selectionist trend of Oxford. In that matter his interest was quite passive. As a pupil of Lankester, he believed that selection was the controlling force in evolution. Having accepted the primacy of natural selection, he felt free to develop his real interest in comparative anatomy. He had no interest at all in studying the mechanism of evolution; nor could he, because he did not know enough genetics.

The lecturer whose interests most closely reflected mine was Julian Huxley. I owe him a great debt, especially for inspiration. While I was still an undergraduate, he and I combined, as equal partners, to research on the amphipod *G. chevreuxi* and showed that specific genes may control the time of onset and rate of processes in the body. We were always alive to the evolutionary importance of that discovery (see Ford and Huxley, 1927; Ford, 1938, pp. 149-154, 243-245).

Even though Julian Huxley was only (except as an undergraduate) at Oxford from 1919 to 1927, he was the most powerful force in developing the selectionist attitude there. He took this view largely from Lankester, though he did not know him personally.

I also became friends with a number of older evolutionists, such as Leonard Darwin, Lankester, and Poulton. I frequently asked them what Charles Darwin said to them on various topics. It was through John Scott Haldane (father of J. B. S. Haldane) that my own family was brought in touch with the Darwin family. I was particularly close to Leonard Darwin, that most unusual of Darwins—a military man. I had met him several times when I was a child. He came to Oxford and lectured on eugenics in about 1924-25, and we naturally talked together. He got me to join the London Eugenics Society in which he was interested. But a number of people, including he and I, disapproved of views expressed there and resigned. Darwin and I made common cause over the matter. He had become interested in my ideas on evolution and genetics, and by the later 1920s we were intimate friends.

Poulton, directly under the influence of Lankester, whom he knew well, was a most dedicated and vociferous selectionist. But he did not carry as much weight as he might have because he knew absolutely no genetics at all. About 1930 he said that Mendelism was not of any general application; for if it were, people would have seen 3:1 ratios long ago, whenever a European married a Negro.

Poulton was not really anti-Mendelian. He thought Mendelism interesting and, at a superficial level, important. He was quite prepared to believe that the switch control of mimetic polymorphism is Mendelian. He did not think that the fundamental qualities of an organism were determined on Mendelian lines, but in a way not yet discovered. I should add that he was never a Lamarckian. I have often heard him say in discussing some feature that "it is certainly inherited and probably Mendelian." It would be a major matter to bring together any published account of his views on heredity. They must be scattered as odd sentences here and there among his two hundred or so papers.

Mayr has suggested that Poulton's writings on evolution and selection seem very powerful and much closer to modern evolutionary views than Bateson's, yet Bateson had a much greater reputation than Poulton. Poulton actually became famous as an entomologist. Bateson is better known because he immediately took notice of Mendelism on its "rediscovery" and was the first in England to do so. Also, of course, his names (genetics, and some of the basic terms such as heterozygote and homozygote) did much to bring him to the attention of scientists. Having known them both, I would be inclined to think that Poulton was the cleverer man.

I met Lankester through Poulton. He was already an old man but with an absolutely clear memory and mind. He talked to me a good deal about Darwin and Pasteur, both of whom he knew. Lankester was a personal friend and great admirer of Darwin; he stayed with Darwin twice at Down. Lankester's father (the first public analyst) was a close personal friend of Darwin's and so deeply impressed by Darwin that he was determined that one of his sons should become a great biologist. He named all three of them suitably: Forbes, Ray, and Owen!

I was not really able to influence Lankester's views on evolution because his interest was by then firmly fixed on comparative anatomy. In his younger days he had been more alive to other aspects of zoology. For example, he was the first to use the spectroscope in zoological studies. Lankester once threw a great auk's egg at me.

Fisher was interested in natural history from childhood. He was in doubt whether to take up mathematics or biology. On going into some museum he saw the skull of a fish with all the bones labeled: this decided him in favor of mathematics. But it was this joint interest that turned his attention to applying mathematics to scientific and, especially, biological problems.

He caught butterflies as a child but any he may have kept disappeared before he was grown up. Yet he retained some interest in them as biological material. In later life somewhere in Wyoming he found a patch of surviving prairie, where he caught a few butterflies. Someone "set" them for him, and for a time he would show them as examples of prairie life. They were not, if I remember, more than a dozen. Eventually they disappeared. He never formed a collection of butterflies.

Fisher and I were intimate friends from 1923 to his death in 1962. His encouragement was magnificent. The true start of the modern evolutionary synthesis of which we are speaking was, in my opinion, his 1927 paper, "On Some Objections to Mimicry Theory: Statistical and Genetic." This paper (supported by a consequence of it, his dominance theory in 1928) in 1928 led me to plan my book *Ecological Genetics*. His 1927 paper opened up the possibility that I had had in mind for some years of taking genetics into the field. Before that, my own work on *Melitaea aurinia* indicated powerful selection in nature. When Fisher's mimicry paper drew attention to the selective adjustment of the effects of genes, I thought this extremely important for wild populations. Later work showed up powerful selection in the field, especially in view of the polymorphism concept. It would not have been possible to work on the synthesis of genetics and ecology without Fisher's *Statistical Methods for Research Workers*. His *Genetical Theory of Natural Selection* was its great early landmark.

I met Fisher, through Julian Huxley, when I was still an undergraduate. Julian told him about me, and he decided to travel to Oxford (from Cambridge, where he was then a Fellow of Caius College) to see me. It did not occur to him to let me know he was coming, so when he arrived at Wadham College (I suppose Julian had given him the address), I was out. Accordingly, he settled down in my rooms to wait for me to return. Fisher often visited me at Oxford, and I was constantly going to see him at Harpenden or at Cambridge. He never visited anyone at Oxford but me.

Fisher's mathematical demonstrations provided essential proofs that very small selective advantages could be important in evolution, genes of neutral survival value must be very rare, and when they do occur, they spread too slowly to preserve their neutrality. As the evidence developed, Fisher became fully aware of powerful selection in nature and wholly accepted my polymorphism concept. By the time I propounded my formal definition of genetic polymorphism (Ford, 1940), Fisher was accustomed to high selection values in nature, far higher than he would have believed possible in 1930, partly because of accumulating evidence as in the cases of *Panaxia* and the heterostyle-homostyle mechanism in *Primula*, on which we carried out detailed fieldwork together. He was very glad to have a clear definition of what he easily recognized to be a distinct form of variation. He was delighted at each stage as I indicated that the human blood groups are subject to powerful selection: that they are polymorphisms, and that they must be associated with liability to specific diseases and infections, as well as with large physiological effects. Fisher was most impressed by the fact that polymorphism concealed powerful selective forces. The selection pressures in, for example, polymorphic mimicry are balanced, and therefore often neutral as a whole, or very slight in total effect. But the balance is between opposed powerful selection pressures, which allows rapid and large adjustments when necessary—quite different from what would happen if the balanced selective forces themselves were small.

Fisher took an intense interest in my fieldwork, especially that on *Panaxia* and *Maniola*. It confirmed the importance of developing statistical methods appropriate to actual field research. He was always glad to visit me and see my fieldwork in progress, as he often did. Perhaps without these observations he would not have placed so much emphasis on field studies; but he would always have been interested in them and have tried to some extent to take part in them himself, as far as his bad sight would allow. The suggestion that he and I should study wild populations of *Primula* together came from him.

Fisher influenced my fieldwork in the mathematical deductions to be drawn from my studies on the moth *Panaxia dominula* (Fisher and Ford, 1947, pp. 143-174) and in the mathematical treatment of marking release and recapture as a method of estimating population numbers and survival rates in nature (Fisher and Ford, 1947; Dowdeswell, Fisher, and Ford, 1949, pp. 67-84). Practically all of the statistical methods that made my accurate fieldwork possible came directly or indirectly from Fisher.

Fisher always held that the mathematical approach in biology is one way, but not the only way, of studying the subject. He would specifically stress this view, citing Darwin's work, which Fisher greatly admired even though Darwin was quite unmathematical.

As an undergraduate, I did not know of any naturalists who were geneticists, and I think there were none. I wanted to study the genetics of wild populations, a study that seemed quite natural. Those who had worked on the genetics of mimicry were using wild material, but they did not think of the ecological aspects of the matter as well. When I heard of Gerould's work on blue and green larvae in *Colias*, I thought that he and I might be working along the same lines; but he never seemed to take it further. I was troubled because though zoology seemed oriented toward evolution, evolution did not seem to be studied experimentally. I could not find naturalists who were geneticists (or the reverse) and so, subconsciously, I suppose I was aware of the importance of taking that step.

Most of the earlier Darwinian zoologists (including Wallace, Lankester, and Poulton) were extremely unmathematical, and they felt Mendelism could be an intrusion of mathematics into biology. I once spent part of an afternoon trying to explain $p^2 : 2pq : q^2$ to William Bateson. Not only could he not understand it but he could see no possible point in it. The Darwinian naturalists were opposed (rightly) to the mutation theory of evolution put forward by de Vries and thought Mendelism was an aspect of it, believing Mendelism to be concerned with "sports" and abnormal forms appearing under domestication. So necessary did it seem to push the Mendelian aspect of evolution that I called my 1931 book *Mendelism and Evolution*. The right title today would be *Genetics and Evolution*. It was not the right title then.

One must remember the climate of thought in the 1920s. It is probably not realized today how obsessed were those zoologists who had experimental leanings toward experimental embryology, as developed by Spemann and Harrison. I remember more than one person speaking in almost the same terms, saying: "Surely you are not now going on with genetics when we are seeing what can be done in experimental embryology?" Yet if ever there were a "damp squib," experimental embryology, as conceived then, was one.

One must remember, too, that those who had made a great name in zoology then (such as Goodrich) were almost wholly nonexperimental, being chiefly concerned with comparative anatomy in its most observational aspects. They did not even do what, indeed, I have done rather extensively: apply chemistry to taxonomy.

Six steps were fundamental to the evolutionary synthesis of the 1930s and 1940s, which I place in no particular order.
(1) The study of evolution in progress, by a combination of ecology and laboratory genetics, with the rest of the relevant techniques involved (such as the founding of artificial colonies in nature). Without that combination, the power of selection in wild populations (similar to Darwin's own views) could not have been realized. Pure ecologists know nothing of the genetics of the variation they encounter. Those who sit in their laboratories and decide mathematically how evolution ought to work, could never have found, and did not find, how high are the selective forces for advantageous qualities in nature.
(2) The polymorphism concept, showing also that polymorphism must necessarily be immensely common and maintained by selection (which has great medical as well as evolutionary importance).
(3) The recognition that since particulate inheritance requires a combination of great heritable variability and great heritable stability in organisms, mutation *must* be very rare.
(4) Fisher's analysis showing that genes of neutral survival value cannot contribute to the evolution of any but the smallest populations, and even then their effect will soon be overcome.
(5) Fisher's recognition that the effects of single genes are subject to selective adjustment in wild populations.
(6) The supergene concept, as developed by Darlington.
My most important contribution to the evolutionary synthesis was to develop the subject of ecological genetics from 1928 on, even though my book by that title was not published until 1964. This work gave rise to the discovery from 1940 on (though inherent in our work on fluctuation in numbers during the early 1920s) that selection for advantageous qualities is often immensely more powerful than had previously been supposed. In *The Genetical Theory of Natural Selection* (1930) Fisher suggested that selection for advantageous qualities in nature might rise to as much as 1 percent. As a result of the approach by means of ecological genetics, it is now known that values of 20 to 60 percent and well over

are common and usual—a fact of wide evolutionary significance. Incidentally, it further indicates how unimportant, though not nonexistent, is random drift. In 1924, when Haldane calculated a high selective value for the *carbonaria* gene of the moth *Biston betularia* during its spread in Manchester, his result was regarded as entirely exceptional, being a response to the abnormal situation in an industrial area.

Industrial melanism, principally but not entirely in the Lepidoptera, has provided information of special importance for ecological genetics and the work on it has proved to be a perfect example of its methods. In earlier years I studied the subject myself, but it was taken much further by the brilliant researches of Kettlewell. Clarke and Sheppard have also contributed greatly to our knowledge of such melanism, as have Creed and Lees.

References

Darbishire, A. D. 1911. *Breeding and the Mendelian discovery*. London: Cassell.

Dowdeswell, W. H., R. A. Fisher, and E. B. Ford. 1949. The quantitative study of population in the Lepidoptera, 2. *Maniola jurtina L. Heredity* 3:67-84.

Fisher, R. A. 1925. *Statistical methods for research workers*. Edinburgh: Oliver and Boyd.

———— 1927. On some objections to mimicry theory: statistical and genetic. *Transactions of the Entomological Society of London* 75:269-278.

———— 1930. *The genetical theory of natural selection*. Oxford: Oxford University Press.

———— and E. B. Ford. 1947. The spread of a gene in natural conditions in a colony of the moth *Panaxia dominula L. Heredity* 1:143-174.

Ford, E. B. 1931. *Mendelism and evolution*. London: Methuen.

———— 1938. *The study of heredity*. London: Thornton Butterworth.

———— 1940. Polymorphism and taxonomy. In *The new systematics*, ed. J. Huxley. Oxford: Oxford University Press, pp. 493-513.

———— 1964. *Ecological genetics*. London: Methuen.

———— and J. S. Huxley. 1927. Mendelian genes and rates of development in *Gammarus chevreuxi. British Journal of Experimental Biology* 5:112-134.

Ford, H. D., and E. B. Ford. 1930. Fluctuation in numbers and its influence on variation in *Melitaea aurinia. Transactions of the Royal Entomological Society of London* 78:345-51.

Gerould, J. H. 1927. Studies in the general physiology and genetics of butterflies. *Quarterly Review of Biology* 2:58-78.

Morgan, T. H. 1919. *The physical basis of heredity*. Philadelphia: Lippincott.

Lamarckism in Britain and the United States

Richard W. Burkhardt, Jr.

Although the deaths of the Americans Cope, Hyatt, and Packard between 1897 and 1905 deprived the English-speaking world of its only well-defined school of neo-Lamarckian biologists, sympathies for a Lamarckian interpretation of evolution continued to exist among British and American biologists well into the twentieth century. The sources of these sympathies and the occasions for their expression are too numerous and diverse to allow an adequate description of them in a short paper. An introduction to this topic may be attempted, however, by sampling the ideas of one of the leading British Lamarckians and by setting forth some general observations on Lamarckism in Britain and the United States in the twentieth century.

Wheeler once remarked that expressing a belief in Lamarckism amounted to committing "the ninth mortal sin" (Wheeler, 1921). Viewed in the light of this remark, E. W. MacBride, professor of zoology at the Imperial College of Science and Technology in London from 1913 to 1934, appears as one of the most impenitent sinners of the first third of the twentieth century. Early in his career MacBride's study of embryology and his basic acceptance of the doctrine of recapitulation led him to endorse the idea of the inheritance of acquired characters (MacBride, 1895). Nothing shook his faith in Lamarckism in the remaining years of his life. The evidence from embryology, systematic zoology, and paleontology seemed to him to provide a clear case for the inherited effects of habit. In the 1920s he was making the additional claim that "the doctrine of Lamarck has been submitted to the crucial test of experiment and proved to be true" (MacBride, 1925a). This claim was based primarily on the work of Kammerer, though MacBride also cited studies by Dürken at the University of Breslau and Heslop-Harrison at the University of Durham as experimental proof of Lamarckism's validity.

Kammerer's suicide in 1926 did not dampen MacBride's enthusiasm for Kammerer's experimental studies, at least not immediately. Nor did it enhance MacBride's view of rival evolutionary theories. In 1927, reviewing Kammerer's book on variation in lizards of the Adriatic Islands, MacBride could not resist a shot at the Mendelians: passing from the writings of the Mendelians to the "clear and beautiful arguments" of Kammerer's work, he wrote, was like passing "from the babbling of the nursery to the reasoned debate of the forum." When de Beer attempted to "dethrone the theory of recapitulation" in his book *Embryology and Evolution* of 1930, MacBride saw the attempt as the kind of "lost cause" or "errand of hopeless knight-errantry" that one had come to expect from "brilliant Oxford men" (1930a). Contrary to de Beer's view, MacBride asserted, "genes have played no part in evolution." "Mutations and adaptations," he explained later, "have nothing to do with one another" (1930b). In 1932 he characterized Haldane's views on natural selection as "evidences of an archaic type of thought," and he noted: "Selection as an effective cause of anything is a superstition which dies hard."

MacBride believed that the key to understanding evolution was not held by the geneticists. Regarding himself to be quite familiar with modern research in genetics, he wrote in 1931: "I have watched all the work going on [at the John Innes Institution], and the more I see of it the more I am convinced that Mendelism has nothing to do with evolution." Later he referred caustically to the "chairs of 'Drosophily' " that had been set up in North America "devoted to the exclusive study of this type-animal" (1937). He seems to have drawn two major conclusions from the geneticists' work. The first, based on the pure-line studies of Johannsen, Jennings, and Agar, was that natural selection is unable to operate on small, fluctuating variations. The second, based in large measure on his reading of the work of the Morgan school, was that mutations are characteristically deleterious (in general, MacBride seems to have thought of mutations in terms of large-scale de Vriesian jumps). He cited with approval Johannsen's doubts about the way the theory of the gene had developed, and he promoted Tornier's conclusion from studies on goldfish that in mutations "what is inherited is . . . not a factor or gene . . . *but a certain grade of germ-weakness which in each succeeding generation produces the same morphological effects*" (1925). Given this view of mutation, it was obvious to MacBride that "mutations can have played no part whatever in evolution. Since they are the outward and visible signs of a weakened constitution, they are in a state of Nature ruthlessly weeded out by natural selection." He maintained that once "small individual differences and larger occasional mutations as the raw material of evolution" had been rejected, only one alternative remained: the inherited effects of use and disuse (1920).

As Ernst Mayr indicated in the prologue, there were more Lamarckians besides MacBride in the 1920s and 1930s. No one in this period, however, was more outspoken than MacBride in Lamarckism's defense. MacBride carried the Lamarckian banner in the pages of *Nature*, debating the mechanisms of evolution with Lankester (1925b), with Huxley (1925c, 1926), with Haldane (1931a, 1932), and with others. Furthermore, although a fair number of biologists sympathized with the idea that acquired characters had *some* role to play in the evolutionary process,

MacBride seems to have been distinctive in the 1930s in his bald assertion that "habit is the driving factor in evolution" (1931b).

Several general observations seem important for understanding the history of neo-Lamarckism. First is the common observation that the neo-Lamarckians of the late nineteenth and early twentieth centuries held views quite different from Lamarck's a century earlier. Neo-Lamarckism arose less as a continuation of Lamarck's own thinking than as a legitimate response to problems left unanswered in Darwin's *On the Origin of Species*. The idea of the inheritance of acquired characters, the defining characteristic of neo-Lamarckian thought, was indeed central to Lamarck's thinking. But it was never an issue for Lamarck himself; he and the vast majority of his contemporaries simply took the inheritance of acquired characters for granted. Of more fundamental importance for Lamarck's own theory of evolution were his ideas on the tendency toward increased complexity exhibited by the animal scale. The idea of the inheritance of acquired characters came to be known as the "Lamarckian" mechanism through something of a historical accident.

Second, in the years immediately after the publication of *On the Origin of Species*, the question about acquired characters was not whether they could be inherited, which was widely assumed to be true, but instead their importance in the evolutionary process relative to natural selection, isolation, and other factors. Darwin did not provide an explanation of the causes of heritable variation. Because the effects of use and disuse appeared to be a likely source of heritable variation, many biologists turned to the inheritance of acquired characters as either an aid to natural selection or a primary factor of evolution in its own right. I would like to restrict the label "Lamarckian" or "neo-Lamarckian" to those scientists who regarded the inheritance of acquired characters as *primary* in the evolutionary process. (I do not think that it is appropriate to call figures such as Romanes (1895) or Plate (1903) Lamarckians; each regarded himself as a true Darwinian who, like Darwin, supposed natural selection to be the primary mechanism of organic change but still believed that the inheritance of acquired characters was possible.) The neo-Lamarckians did not promote the inheritance of acquired characters to the exclusion of all other factors. The American entomologist Packard (1894), for example, considered the fundamental cause of variation to be the direct and indirect effects of the environment on the organism, but he also acknowledged the evolutionary significance of geographical isolation, "geological extinction, natural and sexual selection, and hybridity." The American paleontologist Cope, in his early writings at least (see, for example, 1887), allowed that natural selection might well be responsible for the adaptive features that provided the distinguishing characters of species, though the distinguishing characters of genera and higher categories had been developed by other means.

Third, the idea of the inheritance of acquired characters had great breadth of appeal in the last decades of the nineteenth century and the early decades of the twentieth century. This appeal cut across both national and disciplinary boundaries, and it drew support from philosophical and social considerations as well as scientific ones. Philosophically, Lamarckism seemed to relieve man of the chilling thought that the organic world was the result of "nothing but" chance (the apparent "chanciness" of Darwin's theory of natural selection, in the eyes of many, was the most unpalatable thing about it). In the social realm, Lamarckism held out the hope that man's attempts at social engineering might have more than short-term effects (under the assumption that improving the environment would have a positive effect on the genetics of the human race). The Lamarckian position was supported in England, France, Germany, Austria, Switzerland, Italy, Russia, and the United States by embryologists, paleontologists, physiologists, bacteriologists, and plant geographers. It seemed to fit well with the embryologist's assumption that ontogeny recapitulates phylogeny, with the paleontologist's fossil sequences that seemed to display the accumulated effects of use and disuse, with the physiologist's interest in causal rather than statistical relationships, with the bacteriologist's understanding of the bacterium's adaptation to environmental change, and with the plant geographer's data on the geographic variation of forms. In a more straightforward manner than Darwinism, Lamarckism also seemed capable of explaining the degeneration of useless organs, correlated variation, and the origin of various kinds of instinctive behavior.

Fourth, as a result of the great diversity of sources from which Lamarckism drew its support, the Lamarckians at the turn of the century did not constitute a unified group. The Americans were recognized as having a distinctive school of neo-Lamarckians led by the paleontologists Cope and Hyatt. But the Lamarckians in general remained fragmented and never achieved anything like a synthesis of their own. There seems to have been a significant lack of communication between the Lamarckians from different fields. The Lamarckian botanists, zoologists, physiologists, paleontologists, and microbiologists of the time did not seem inclined to seek support for their evolutionary position outside of their own respective disciplines. Communication among the Lamarckians was evidently cut down by language as well as disciplinary barriers. For example, while MacBride was well acquainted with the German sources, he drew relatively little on French studies, although the French strongly supported a more or less Lamarckian position at the beginning of the century.

Fifth, experimentation played a role in the decline of Lamarckism in the twentieth century, but not as straightforward a role as is often claimed. It was naive of biologists like Hertwig (1909) and Morgan (1903) to suggest that the issue of the inheritance of acquired characters could be resolved in the laboratory. Complaining about the failure of the neo-Lamarckians to offer experimental evidence in support of their views, Morgan wrote: "One of the chief virtues of the Lamarckian theory is that it is capable of experimental verification or contradiction, and who can be expected to furnish such proof if not the Neo-Lamarckians?" In fact, designing an experiment to provide an unequivocal answer to the question of the reality of the Lamarckian mechanism proved to be no easy matter. The Lamarckians were faced with the grave difficulty of providing experimental evidence that could be interpreted only by the assumption of the inheritance of acquired characters. The anti-Lamarckians, on the other hand, were faced with the ultimate impossibility of proving a universal negative: they could report particular cases in which acquired characters were apparently not passed on to the next generation, but they could not prove that such transmissions never took place. Weismann's experiment with mouse tails is well known, but its significance for the history of biology has been much overrated. Interestingly enough, Weismann himself never supposed that the deciding factor in his debate with the Lamarckians would be experimentation. He supposed, to the contrary, that the Lamarckian principle would be rejected when it was realized that "the observed phenomena of transformation" could be explained without it (1891).

In the last analysis, Weismann's assessment of the way Lamarckism would fall from favor appears to have been correct. Lamarckism was not falsified by experiment so much as it was rejected as an unnecessary hypothesis. But if the role of experimentation in Lamarckism's downfall must not be overestimated, so too must it not be underestimated. In the early years of the twentieth century, as increasing stress was put on the importance of experimentation for the development of biology, the debate on the inheritance of acquired characters was drawn more and more into the experimental arena. And though decisive evidence for or against the Lamarckian hypothesis was not forthcoming from the work of the experimentalists, the inability of the Lamarckians to provide themselves with unassailable experimental support did come to be regarded as a serious liability for their position.

Many experiments were conducted to test the Lamarckian principle. The best-known experiments in the nineteenth century, prior to Weismann's, were those of Brown-Séquard on induced epilepsy in guinea pigs (1859-1860). Brown-Séquard's results were difficult to interpret, however, as were the results reported by Standfuss (1898), Fischer (1901, 1902), and Schröder (1903) from studies of the effects of abnormal temperatures on the genetics of Lepidoptera. (Schröder's results were eventually understood as cases of parallel induction rather than as somatic changes transmitted to the germ-plasm; Detlefsen, 1925; Sokoloff, 1966). In certain other experiments the nontransmission of acquired characters appeared clearer, for example, in the study by Castle and Phillips (1911) involving the transplantation of ovaries from two black guinea pigs to a white albino female. But the negative results of individual cases could not overthrow the whole Lamarckian position. When Kammerer began reporting positive instances of the inheritance of acquired characters in the prestigious *Archiv für Entwicklungsmechanik der Organismen*, a great deal of attention was directed at his claims. Kammerer's studies, like those of others before him, proved open to more than a single interpretation (to say nothing of being difficult to verify). Kammerer's suicide, which followed shortly after the revelation that his only remaining specimen of *Alytes obstetricans* had been doctored, was widely construed as an admission of scientific fraud. The Lamarckians who had come to hope that experimentation would confirm their views suffered a severe setback.

A number of pro-Lamarckian observations and experiments had been reported in the early 1920s, the most notable of which were the experiments of Guyer and Smith (1918). These experimental results failed to be confirmed, however, and after Kammerer's suicide experimental activity directed toward testing the inheritance of acquired characters slackened considerably, though it did not die out entirely (for example, work by McDougall, 1927, 1930; and Rhine and McDougall, 1933; and studies cited by Thorpe, 1930). The Lamarckian cause was not helped by suspicions that other experimentalists besides Kammerer had made fraudulent claims about environmentally induced genetic changes (see the next selection).

What did the failure to demonstrate the Lamarckian principle experimentally mean to biologists in general? Many naturalists had never supposed that the mechanisms of evolution would be revealed in the laboratory. Nonetheless, in the United States, at least, the failure to demonstrate the Lamarckian principle experimentally seems to have been taken quite seriously. I doubt that it is simply a coincidence that the country in which the new science of genetics flourished most conspicuously was the country in which Lamarckism seems to have faded fastest (Rosenberg, 1967). The deaths of Cope, Hyatt, and Packard brought an end to the American school of neo-Lamarckism, even if some American biologists such as McClung continued to harbor Lamarckian sympathies. After the second decade of the twentieth century, the message conveyed by the typical American biology text to the beginning student was that the Lamarckian position had to be regarded with considerable skepticism because there was no experimental proof of that position's validity (Hegner, 1912, 1942; Abbott, 1914; Calkins, 1914; Galloway, 1915; Newman, 1929; Woodruff, 1932; Kinsey, 1938; Fitzpatrick and Horton, 1940). Some texts suggested that even if acquired characters were sometimes passed on to succeeding generations, it happened only rarely and could not constitute a major factor in the evolutionary process (Brooks, 1907; Burlingame and others, 1922).

The failure to confirm the Lamarckian principle experimentally did not, of course, occur in an intellectual vacuum. At the same time there was a continuing sense of the theoretical inadequacies of Lamarckism and an increasing appreciation of how much natural selection operating on small, Mendelian differences might actually accomplish. Simpson barely addressed Lamarckism in *Tempo and Mode in Evolution* (1944). In a few words he pointed to both the experimental and the theoretical deficiencies of the Lamarckian position. "Experiments in the present century, . . . not only have failed to corroborate that there is such a process [as Lamarckian evolution] but also have shown that it is highly improbable, if not impossible." Indicating then that he did not "propose to consider the theory further in the present study," he simply noted that Lamarckism provided "neither a necessary nor a sufficient hypothesis to explain the pertinent facts of evolution as seen by the paleontologist." Improbable, unnecessary, and insufficient—this was the final assessment of Lamarckism. Experimentalism had weakened the Lamarckian cause in the eyes of many. In the last analysis, however, Lamarckism was not so much disproved as discarded. In his *Systematics and the Origin of Species* (1942), Mayr did not even bother to reject Lamarckism. He simply did not refer to it.

References

Abbott, J. F. 1914. *The elementary principles of general biology.* New York: Macmillan, p. 321.

Brooks, W. K. 1907. *The foundations of zoology,* 2d ed. New York: Macmillan, p. 123.

Brown-Séquard, C.-E. 1859-1860. Hereditary transmission of an epileptiform affection accidentally produced. *Proceedings of the Royal Society of London* 10:297-298.

———— 1871-1872. Quelques faits nouveaux relatifs à l'épilepsie qu'on observe à la suite de divers lésions du système nerveux, chez les cobayes. *Archives de physiologie normale et pathologique* 4:116-120.

———— 1882. Faits nouveaux établissant l'extrême fréquence de la transmission, par hérédité, d'états organiques morbides, produits accidentellement chez des ascendants. *Comptes Rendus de l'Académie des Sciences, Paris* 94:697-700.

Burlingame, L. L., H. Heath, E. G. Martin, and G. J. Peirce. 1922. *General biology*. New York: Holt, p. 429.

Calkins, G. N. 1914. *Biology*. New York: Holt, pp. 201-203, 227-230.

Castle, W. E., and J. C. Phillips. 1911. On the germinal transplantation in vertebrates. Publication 144. Washington, D.C.: Carnegie Institution.

Cope, E. D. 1887. On the origin of genera. In *The origin of the fittest*. New York: Appleton, pp. 110-111, 122-123.

Detlefsen, J. A. 1923. Are the effects of long-continued rotation in rats inherited? *Proceedings of the American Philosophical Society* 62:292-300.

———— 1925. The inheritance of acquired characters. *Physiological Reviews* 5:244-278.

Duerden, J. E. 1920. Inheritance of callosities in the ostrich. *American Naturalist* 54:289-312.

Fischer, E. 1901. Experimentelle Untersuchungen über die Vererbung erworbener Eigenschaften. *Allgemeine Zeitschrift für Entomologie* 6:49-51, 363-365, 377-381.

———— 1902. Weitere Untersuchungen über die Vererbung erworbener Eigenschaften. *Allgemeine Zeitschrift für Entomologie* 7:129-134, 161-167, 201-205, 241-246, 266-272, 301-306, 452-456, 476-483, 506-514, 521-525.

Fitzpatrick, F. L., and R. E. Horton. 1940. *Biology*. Boston: Houghton Mifflin, p. 403.

Galloway, T. W. 1915. *Zoology: a textbook for universities, colleges and normal schools*, 3d ed. Philadelphia: Blakiston, p. 496.

Guyer, M. F., and E. A. Smith. 1918. Studies on cytolysins. I. Some prenatal effects of lens antibodies. *Journal of Experimental Zoology* 26:65-82.

———— 1920. Studies on cytolysins. II. Transmission of induced eye-defects. *Journal of Experimental Zoology* 31:171-233.

Hegner, R. W. 1912. *An introduction to zoology*. New York: Macmillan, pp. 227-228.

———— 1942. *College zoology*, 5th ed. New York: Macmillan, p. 429.

Hertwig, R. 1909. *A manual of zoology*, 2d U.S. ed. (from 5th German ed.) New York: Holt, pp. 54-55.

Kinsey, A. C. 1938. *New introduction to biology*, rev. ed. Chicago: Lippincott, pp. 407-408.

MacBride, E. W. 1895. Sedgwick's theory of the embryonic phase of ontogeny as an aid to phylogenetic theory. *Quarterly Journal of Microscopical Science* 37:325-342.

———— 1920. The method of evolution. *Scientia* 28:23-33.

———— 1925a. The theory of evolution since Darwin. *Nature* 115:89-92.

———— 1925b. The blindness of cave animals. *Nature* 116:818.

———— 1925c, 1926. Genes and linkage groups in genetics. *Nature* 116:938; 117: 232, 340-341.

———— 1927. Variety and environment in lizards. *Nature* 120:71-74.

———— 1930a. Embryology and recapitulation. *Nature* 125:883-885.

———— 1930b. Embryology and evolution. *Nature* 126:918-919.

———— 1931a. Embryology and evolution. *Nature* 127:55-56.

———— 1931b. Habit: the driving factor in evolution. *Nature* 127:933-944.

———— 1932. The inheritance of acquired characters. *Nature* 129:900-901; 130: 128-129.

———— 1937. Mendel, Morgan, and genetics. *Nature* 140: 348-350.

McDougall, W. 1927. An experiment for the testing of the hypothesis of Lamarck. *British Journal of Psychology* 17:267-304.

———— 1930. Second report on a Lamarckian experiment. *British Journal of Psychology* 20:201-218.

Mayr, E. 1942. *Systematics and the origin of species.* New York: Columbia University Press.

Morgan, T. H. 1903. *Evolution and adaptation.* New York: Macmillan, p. 260.

Newman, H. H. 1929. *Outlines of general zoology,* rev. ed. New York: Macmillan, p. 485.

Packard, A. 1894. On the inheritance of acquired characters in animals with a complete metamorphosis. *Proceedings of the American Academy of Arts and Sciences* 21:368-369.

Plate, L. 1903. *Über die Bedeutung des Darwinschen Selectionsprinzips und Probleme der Artbildung.* Leipzig: W. Engelmann.

Rhine, J. B., and W. McDougall. 1933. Third report on a Lamarckian experiment. *British Journal of Psychology* 24:213-235.

Romanes, G. J. 1895. *Darwin and after Darwin. An exposition of the Darwinian theory and a discussion of post-Darwinian questions,* vol. 2. Chicago: Open Court, pp. 1-156.

Rosenberg, C. E. 1967. Factors in the development of genetics in the United States: some suggestions. *Journal of the History of Medicine* 22:27-46.

Schröder, C. 1903. Die Zeichnungsvariabilität von *Abraxas grossulariata*. *Allgemeine Zeitschrift für Entomologie* 8:105-119, 145-157, 177-192, 228-234.

Sokoloff, A. 1966. Morphological variation in natural and experimental populations of *Drosophila pseudoobscura* and *D. persimilis*. *Evolution* 20:49-71.

Sommer, M. 1900. Die Brown-Séquard'sche Meerschweinchenepilepsie und ihre erbliche Uebertragung auf die Nachkommen. *Beiträge zur pathologischen Anatomie und zur allgemeinen Pathologie* 27:289-330.

Standfuss, M. 1898. Experimentelle zoologische Studien mit Lepidopteren. *Neue Denkschriften der allgemeinen schweizerischen Gesellschaft für die Gesammten Naturwissenschaften* 36:5-40.

Thorpe, W. H. 1930. Biological races in insects and allied groups. *Biological Reviews of the Cambridge Philosophical Society* 5:177-212.

Weismann, A. 1891. *Essays upon heredity and kindred biological problems,* ed. E. B. Poulton, S. Schönland, and A. E. Shipley, vol. 1, 2d ed. Oxford: Oxford University Press, p. 461.

Wheeler, W. M. 1921. On instincts. *Journal of Abnormal Psychology* 15:295-318.

Woodruff, L. L. 1932. *Animal biology.* New York: Macmillan, p. 374.

A Note on W. L. Tower's *Leptinotarsa* Work

Alexander Weinstein

W. L. Tower, who reported he had produced mutations in the potato beetle *Leptinotarsa* by changes in temperature and humidity (1906, 1910a and b), may have been influenced by Morgan's *Evolution and Adaptation* (1903) to undertake the experiments.

There is a critical discussion of Tower's work in Bateson's *Problems of Genetics*. This book contains the Silliman lectures that Bateson delivered at Yale in 1907, brought up to date and published in 1913. Bateson said that although Tower's claim to have produced true-breeding genetic changes by altering temperature and humidity was widely accepted, he himself, after seeing some of Tower's experiments in progress in 1907 and reading his papers carefully, remained unconvinced.

Tower reported in 1906 that normal individuals of the beetle *Leptinotarsa*, when subjected to altered conditions, produced some normal offspring and some variants and that each type bred true. Bateson pointed out that unless each type of germ cell united only with its own type, there must have been heterozygotes among the F_1's; but Tower reported that all the normal-appearing individuals (dominants) bred true and he seemed to think that this result was in harmony with expectation.

In 1910 Tower reported two experiments which, he said, confirmed each other; but T. D. A. Cockerell (1910) pointed out that they were actually in disagreement. Tower (1910b) admitted this and said that one of the experiments had been included by mistake; he wished to withdraw it and substitute another. With this substitution the discrepancy disappeared. But Tower said that the experiment he was withdrawing was nevertheless correct, that he knew why it did not agree with the others, but that he did not care at the time to make any statement about it. Bateson's conclusion was that under the circumstances, discussion of Tower's results must be postponed until there was "careful independent" confirmation of them.

Bateson also called attention to the fact that Tower's second large report (1910a) made no mention of the first report, published in 1906.

Finally, Bateson said his hesitation about Tower's publications had been increased by a destructive criticism by R. A. Gortner (1911). Gortner, a professor of biochemistry at the University of Minnesota, showed that Tower's discussion of the chemistry of pigments in *Leptinotarsa* exhibited such ignorance both of the special subject and of chemistry in general that it could not be taken seriously.

Portions of Bateson's book were included in the reading list for Morgan's course in experimental zoology when I took it in 1913-14. Morgan agreed with Bateson's criticisms of Tower's work. But Sturtevant told me that Morgan had found Tower interesting personally, and that at a society meeting the two men had stayed up all one night talking to each other.

I was later told by Jennings that among the circumstances that had generated doubt about Tower's work was that there had been a fire in Tower's greenhouse, and that Tower claimed that records of his experiments had been destroyed in it.

References

Bateson, W. 1913. *Problems of genetics*. New Haven: Yale University Press.

Cockerell, T. D. A. 1910. The modification of Mendelian inheritance by extreme conditions. *American Naturalist* 44:747-749.

Gortner, R. A. 1911. Studies on Melanin IV. The origin of the pigment and the color pattern in the elytra of the Colorado potato beetle (Leptinotarsa decemlineata Say). *American Naturalist* 45:743-755.

Morgan, T. H. 1903. *Evolution and adaptation*. New York: Macmillan.

Tower, W. L. 1906. An investigation of evolution in Chrysomelid beetles of the genus Leptinotarsa. Washington, D.C.: Carnegie Institution of Washington, publication no. 48.

―――― 1910a. The determination of dominance and the modification of behavior in alternative (Mendelian) inheritance by conditions surrounding or incident upon the germ cells at fertilization. *Biological Bulletin* 18:285-352.

―――― 1910b. The determination of dominance and the modification of behavior, etc. A correction and addendum. *Biological Bulletin* 20:67-69.

12   United States

The evolutionary synthesis occurred primarily in the United States. No other country had so many centers of research in genetics, systematics, cytology, and paleontology. In genetics alone, more than ten major centers of genetics research existed before 1925, and this number increased rapidly during the 1930s. At Columbia, Morgan and Wilson produced over a dozen students with doctorates in genetics between 1910 and 1930. During these same years, East and Castle turned out some twenty research geneticists at the Bussey Institution of Harvard. A very high proportion of these students achieved prominence in the field of genetics; they found jobs in the expanding biology departments of universities or in research centers such as state agricultural experiment stations.

Many of the most influential works of evolutionary synthesis were written and published by researchers in the United States. Particularly important were the four books commissioned as Jesup Lectures at Columbia University and published in the prestigious Columbia Biological Series: Dobzhansky's *Genetics and the Origin of Species* (1937), Mayr's *Systematics and the Origin of Species* (1942), Simpson's *Tempo and Mode in Evolution* (1944), and Stebbins' *Variation and Evolution in Plants* (1950).

In the United States the high level and wide extent of research in the various fields involved in the synthesis made a synthetic view more possible than elsewhere. Yet the mere existence of an advanced level of research in the separate fields did not guarantee the emergence of a synthetic view. Within each specialty, an idea or piece of research work might be common knowledge; yet the same knowledge might be almost unknown or completely neglected in a related but different research specialty.

Most geneticists in 1920 were well aware of Castle's selection experiments on hooded rats, East's work on multiple-factor inheritance, and the genetic analysis of *Drosophila* carried out by Morgan and his students. Every genetics textbook contained descriptions of this work.

354

Geneticists knew that selection of even small differences between individuals could lead to measurable changes in laboratory populations of rats, *Drosophila*, and maize. Yet systematists remained largely ignorant of these genetics results. On the other hand, geneticists were for the most part unaware of the facts about variability within and between natural populations as reported by systematists like Jordan, Osgood, and Grinnell.

Sumner, who tried to synthesize the fields of genetics and systematics in the 1910s and the 1920s, complained constantly that his systematist colleagues were ignorant of genetics and his geneticist colleagues were ignorant of systematics. Sumner had begun his research on natural populations of deer mice (genus *Peromyscus*) with neo-Lamarckian assumptions about the inheritance of acquired characters. Sources that funded genetics research refused to support Sumner's work until the late 1920s, about fifteen years after he began the project, and only after he had modified his neo-Lamarckian ideas. Sumner's synthesis of genetics and systematics was already complete before 1930, and indicates that, in theory, his synthesis could have been made about fifteen years earlier, except for the specialization and isolation of fields.

Even within a single research group, significantly different interpretative views often flourished. Morgan, who had strongly criticized Darwin's views of evolution by gradual natural selection (1903), never gave up all of his antiselectionist tendencies. Many of Morgan's students, especially Muller, Sturtevant, and Weinstein, were ardent selectionists. At the conference, Weinstein told how Morgan allowed his students to rewrite the sections on natural selection in *Critique of the Theory of Evolution* (1916) because he could not bring himself to give so much weight to gradual natural selection.

The extraordinary diversity within and between research fields makes the development of the evolutionary synthesis in the United States difficult to analyze. Complete coverage of developments in the United States is impossible in one brief chapter. Allen and Dobzhansky concentrate on the contributions of Morgan and his students, while Carson reveals a different school of thought in cytogenetics with McClung at the University of Pennsylvania. The course of the evolutionary synthesis in the United States is also covered in many of the chapters dealing with the specific fields involved in the synthesis and in Mayr's prologue.　　　W.B.P.

The Evolutionary Synthesis:
Morgan and Natural Selection Revisited

Garland E. Allen

The work of Thomas Hunt Morgan (1866-1945) occupies a unique position in the evolutionary synthesis. Morgan's name is most generally associated not so much with evolution, but with genetics. He received the 1933 Nobel Prize in physiology and medicine for his work in elucidating the structure of the chromosomes of higher organisms and their genetic role in the transmission of hereditary traits. In fact, Morgan's name is probably the one most frequently associated with the founding of the classical school of Mendelism in the early decades of the twentieth century. It is much less generally known that Morgan not only was strongly interested throughout his life in the Darwinian theory, but wrote four major books and numerous articles on evolution between 1900 and 1935. Although, strictly speaking, none of these works was a major contribution to evolutionary theory, Morgan's name deserves a more prominent place in the history of twentieth-century evolutionary theory than it is usually afforded. His interest in evolution was more than casual or passing.

But what is the exact nature of Morgan's interest in, and views about, Darwinian theory? Ernst Mayr's view is that "Morgan's opinions . . . impeded the eventual synthesis" of genetic and evolutionary theory (personal communication, December 15, 1975), thus suggesting that his place in the history of this synthesis was largely negative. But I do not think that view is historically very fair or accurate. Although Morgan's views on natural selection encompassed a number of very serious misunderstandings and in that sense he failed to comprehend some basic aspects of evolutionary thinking, he struggled to integrate his findings in genetics with evolutionary theory—and made several significant modifications on his views on Darwinian theory between 1903 and 1932. These changes indicate some aspects of the new synthetic thinking that Morgan and many of his contemporaries were developing in the 1900-1932 period, the period when Mendelian genetics as a science came into its own and its bearing on evolutionary theory began to become most apparent.

Morgan's evolutionary misunderstandings were shared by a number of other biologists during the latter part of the nineteenth and early decades of the twentieth centuries. These misunderstandings indicate some of the problems that many biologists brought to presynthetic concepts of Darwinian theory. Morgan's struggle with evolutionary issues between 1903 and 1932 indicates some of the real ways in which the development of

Mendelian genetics both positively and negatively influenced the eventual synthesis.

Morgan was by nature a very skeptical person who enjoyed playing the devil's advocate. If an idea was too popular, too well accepted, Morgan almost automatically began to find aspects of the idea that he could attack. He often went into the lion's den and challenged the most partisan supporters of a respected theory. Because Morgan wrote voluminously, some of his works demonstrate more care than others. No one could, on occasion, turn out a less well written paper than Morgan (though no paper he wrote lacks some new ideas and some challenging thoughts). He did not always present his arguments in a manner that was clearly and well thought out. As a result, Morgan may have seemed more confused about evolution than he actually was.

On the other hand, Morgan was careless about what he wrote in some situations and very careful in others. He often said things that he later retracted. Someone once asked him: "Professor Morgan, why is it that you so often retract what you say in earlier papers?" Morgan is reported to have replied: "Well, that way I publish twice as many papers."

Probably more important than these personal idiosyncrasies—though they are significant—is Morgan's background and his training. He was born in 1866, in Lexington, Kentucky, of a prominent southern family. He attended the State University of Kentucky in Lexington, from which he received his B.A. in 1886. From there he went to Johns Hopkins University where he obtained his doctoral degree in 1891. While at Hopkins his major adviser was William Keith Brooks, who was then a chief exponent of the morphological tradition in the United States. Brooks had been a student of Louis Agassiz and later Alexander Agassiz. A descriptive biologist of considerable merit, Brooks carried out some very fine morphological studies of a variety of marine invertebrates including the Chesapeake Bay oyster. He was an outspoken champion of the use and study of marine organisms for the elucidation of biological principles. Like all morphologists, Brooks's major and abiding interest lay in discovering phylogenetic relationships. He wrote voluminously; he had a kind of mystical streak that would come out periodically when he would lie in a hammock at his home and discuss philosophy with a few of his students.

Brooks transmitted to a number of his students not so much important facts, but rather a way of viewing biological processes—not as answers to a puzzle, but as problems to be investigated. He often saw in small topics problems of cosmic importance. As Bateson said of his summer experiences with Brooks, he came to see heredity as an unsolved problem that related to every area of study of the living organism. Brooks combined the morphological approach—itself so detailed—with a more general, philosophical interest that broadened his concern beyond the details themselves. Brooks's other students included Conklin, Harrison, and Wilson.

Morgan wrote his thesis on the phylogeny of the pycnogonids (sea spiders), testing Dohrn's theory that the group was derived from the annelids rather than, as was most generally believed, from the arachnids or crustaceans. Morgan carried out the same kind of microscopical and morphological analysis of early development that characterized Brooks's work. Morgan showed, in fact, that on about eight different criteria the embryogeny of the pycnogonids related them more closely to the arachnids than to the crustaceans, thus completely rejecting Dohrn's annelid theory. Morgan proved he could carry out exacting morphological work if he wanted to. The problem was that he did not want to.

Morgan's first job was at Bryn Mawr in 1891, where he remained until 1904, when he accepted a position at Columbia. During the first five years at Bryn Mawr several influences began to shift Morgan away from morphological research toward more experimental work. Jacques Loeb also joined the Bryn Mawr faculty in 1891. As the only members of a two-person department, Morgan and Loeb became close friends—a friendship that lasted throughout their lives. Loeb was a chief exponent of the experimental and mechanistic philosophy, a view he had developed fully while at Würzburg, where he had encountered the great experimental plant physiologist, Julius Sachs. Sachs was a strong proponent of experimental, as opposed to descriptive and especially morphological, biology. Morgan was initially impressed with Loeb's emphasis on experimentation.

Another influence on Morgan came in the summer of 1892 and again during the academic year 1894-95 when he went to the Zoological Station at Naples, where he worked directly with Driesch. At Naples Morgan completed his transformation from a descriptive to an experimental embryologist with a very distinct bias toward the *Entwicklungsmechanik* tradition. Morgan had learned about the Naples lab from his close friend Wilson, who had gone there some ten years earlier. Wilson had brought back such tidings of praise about the station, its tremendous atmosphere of inquiry, excitement, and impressive constellation of people passing through that Morgan felt he had to see what it was like. Naples represented a break with the descriptive, morphological tradition, and thus had come to be a kind of mecca for younger biologists in that day.

The *Entwicklungsmechanik* tradition brought Morgan into direct contact with the problem of regeneration, resulting directly from Driesch's work on the growth of damaged or partial embryos and discussed in light of Driesch's concept of "postgeneration"—that is, the way in which the embryo can modify itself in response to injury. The repair process in damaged embryos raised a natural parallel to the way in which the adult responds to injuries and changes inflicted on it from the outside. Morgan's study of regeneration in a number of organisms—planaria, earthworms, crustaceans—represented one, but certainly not the only, direction of research based on the new *Entwicklungsmechanik* school. Morgan performed a number of experiments on the factors affecting the cleavage planes of embryos and the factors affecting the cleavage of eggs. But his work on regeneration was perhaps Morgan's most unique contribution to experimental embryology. By 1896 or so, Morgan had abandoned morphological work and was a strong practitioner, as well as vocal exponent, of the experimental approach to biology as exemplified in the *Entwicklungsmechanik* school.

Interestingly, the problem of regeneration brought Morgan into direct contact with Darwinism in a form that he began to write about and to take seriously. Through his association with Brooks, Morgan had become thoroughly familiar with Darwinian theory, at least Brooks's view of it. I don't know when Morgan first read Darwin, or if he actually ever did so. But the study of regeneration brought the issues of selection and adaptation directly into Morgan's sphere of interest. He wrote: "One of the general questions that I have always kept before me in my study of regenerative phenomena is how such a useful acquirement as the power to replace lost parts has arisen and whether the Darwinian hypothesis is adequate to explain the results" (1903, p. ix). He concluded that "the theory is entirely inadequate to account for the *origin* of the power to regenerate; and it seemed to me therefore desirable to re-examine the whole question of adaptation for might it not prove true here, also, that the theory of natural selection was inapplicable? This was my starting point" (1903, p. ix-x; see also 1899). How did Morgan come to this conclusion? Why should he oppose the theory of natural selection so strongly? What do Morgan's objections reflect about the general attitude of biologists around the turn of the century toward Darwin's model for evolution?

To understand Morgan's objections to the Darwinian theory of natural selection, we must first understand his concept of species. The view that a biologist holds about the nature of species will determine much of what he or she believes about the mechanisms of evolution—the means by which species arise. Like many other biologists at the turn of the century, Morgan was a "nominalist": he held that species were only arbitrary units created for convenience by taxonomists. The only real unit in nature, he contended, was the individual: "We should always keep in mind the fact that the individual is the only reality with which we have to deal, and that the arrangement of these into species, genera, families, *etc.* is only a scheme invented by man for purposes of classification. Thus, there is no such thing in nature as a species, except as a concept of a group of forms more or less alike" (1903, p. 33). "Adaptation" was a key word in Morgan's view of species: "A group of forms more or less alike" because they shared a number of common adaptations—not because they were alike in trivial details such as number of bristles or petals. When taxonomists spoke of species, Morgan held, they spoke of groups differentiated on the basis of nonadaptive characters, and their concept of evolution thus had no relation to the development of real adaptations in nature.

> Curiously enough, we do, I think, when speaking of adaptation, at-tach one meaning to the word species and another meaning when speaking of evolution. In the latter case we often fall back upon the definitions of the systematist. When we speak of the evolution of adaptations through natural selection, however, we are thinking of organisms as groups that are structurally and functionally adapted in different ways to the environment in which they live, and differ from all other groups in these relations to the environment. These adaptive characters do not, however, in most cases lend themselves to sharp definition for purposes of identification and are shunned, therefore, by the systematist. If I am right on this point, the charac-ters of systematic zoology are, at most, only parts of adaptive struc-tures and are generally only by-products of the process of evolution; characters that belong for the most part to the dump-heap of evolu-tionary advance: and whilst they, like all characters, call for ex-planation, the student of adaptation of the living world (regarding adaptation as the fundamental problem of evolution) will pass them over as of trivial importance for his ends (1910, pp. 203-204).

As a result, he argued, it was futile to try to devise a theory accounting for the origin of groups that have no objective (only subjective) reality: "If, then, the systematist's definition of species is what we mean when we speak of species, and this definition does not concern adaptive characters (or only incidentally), clearly it is futile to attempt to explain the *origin of species* by the theory of natural selection" (1910, p. 203).

Like many of his contemporaries, Morgan tended to see species in two contradictory ways. On the one hand, his experience with organisms in nature (particularly marine invertebrates) suggested that all forms exist in an infinitely graded series, with no sharp lines between discrete groups. On the other hand, like de Vries in particular, he also tended to see groups (such as starfish, or earthworms, or maple trees) as types, bounded by a limit in their range of variability (1903, p. 291). Morgan's opinion that species are not real prevented him from understanding the basic population level on which natural selection inevitably works. Morgan's view was what Ernst Mayr has called "typological" in that he tended to think of species as bounded by a limited, circumscribed degree of variability—a view very much opposed by modern systematists and even by some systematists in Morgan's own day (Mayr, 1959, 1969), including Merriam, D. S. Jordan, and K. Jordan. Morgan's views about species persisted throughout his life. Although other aspects of his views on evolution and natural selection changed in varying degrees, his essentially nominalistic and typological views about species remained largely unchanged.

Morgan's Opposition to the Theory of Natural Selection

Most biologists around the turn of the century were familiar with the version of Darwinian theory set forth in the final edition of *On the Origin of Species*. Because more offspring were always born than the environment could support, Darwin maintained, some would have to perish. The slight inherited differences among organisms gave some individuals a slightly greater survival value than others. Those with favorable variations survived longer than those with unfavorable variations. However, the key to evolution by natural selection was not simply longevity, but rather what Darwin referred to as "differential fertility," the ability to leave a greater number of offspring in the next generation. Evolution was marked by the gradual change in frequency of certain traits in a population of organisms over time.

The one persistent problem that Darwin had left unsolved was the mechanism by which new characters could arise and be passed on in a population. Lacking a workable explanation of the origin of variations, Darwin and his followers frequently fell back on a revised version of Lamarckian theory: the appearance of a specific adaptive variation as the result of environmental influences directly on the germ tissue of the organism, or as the result of use and disuse of parts. In earlier editions of the *Origin* (notably the first), Darwin had consciously avoided reliance upon Lamarckism. But criticism from a variety of directions had forced him to adopt more of this older idea about how variations could arise.

Morgan's objections to the theory of natural selection reveal, especially in the period before 1915, his typological view of species and the nature of variation within and between species (1903). (Morgan's many criticisms of Darwinian theory have been discussed at great length in Allen, 1968, 1978.)

Selection and Continuous Variation
Among the most general objections raised by Morgan was that Darwin was confused about the types of variation on which natural selection acted. Morgan jumped into the fray between proponents of continuous and discontinuous variation, distinctly siding with the proponents of discontinuous variation.[1] Morgan felt that there were two reasons why continuous (fluctuating or individual) variations could not be considered sources of evolutionary change: first, there was no proof that continuous variations could be inherited (1903, p. 269); second, continuous variations could not occur in a large enough single step to prevent the new form from being lost through outbreeding in the next and subsequent generations. To Morgan continuous variations represented quantitative rather than qualitative differences among organisms—that is, a normal distribution about a mode. Selection of one extreme or another could not ultimately produce a new species; at most it could keep the type at the place reached by large-scale mutational changes. He thought of a species as having a fixed mode, a form like a rubber ball that could be temporarily altered by an outside agent (in this case selection) but that always returned to its original form once the outside force was relaxed. Only a distinct, qualitative break with the type could produce a new species. It is particularly in this light that Morgan became a strong supporter of de Vries's mutation theory.

Insufficiency of Ideas on Discontinuous Variation
Although Morgan could not accept evolution by small, continuous variations, he was also skeptical about the idea of evolution by small, discontinuous variations. He claimed that new variations would be quickly lost through the "swamping effect," based on a blending theory of inheritance. That is, the germinal material from each parent was mixed together like two cans of paint, so that the offspring represented a dilution or intermediate between the parental forms. Thus, in each succeeding genera-

1. Morgan (1903, p. 261) reserved the terms continuous and discontinuous for linear or vertical inheritance as distinguishing between two patterns of inheritance. Fluctuating variations were horizontal; they were phenotypic but not necessarily genotypic variations. He pointed out a few pages later that there may be some interconnection between horizontal and vertical that makes the distinction between continuous and discontinuous variations less than it is supposed to be (p. 277).

tion a new variation would become less and less visible. Because offspring can intercross with parents, swamping is unavoidable in most species (1903, pp. 286-287; also p. 195).

Natural Selection as a Negative Factor

To Morgan natural selection was only a negative factor in the origin of adaptive characters; it could select out the unfit but could not generate the new variations from which new adaptations could be derived (1903, p. 462). Selection could not explain the origin of the fit, but only the failure of the unfit to propagate. As a result, Morgan found the theory of natural selection incomplete on the origin of species, the very point that Darwin had hoped it would explain. The problem of how distinct and heritable variations arose and were inherited was to Morgan still the critical unsolved problem in understanding evolution.

Morgan argued vehemently against those Darwinian supporters who claimed that selection in and of itself could be a creative factor in evolution. To Morgan selection weeded out the unfit, but was powerless to create the fit. The theory of natural selection was silent on the most crucial element of evolution: how new adaptive variations arise.

The Incipient Stages of Highly Adaptive Organs

Morgan's fourth objection to natural selection was essentially the same that was voiced by Mivart. If selection acted on small individual differences that occur by chance, fine levels of adaptation (the vertebrate eye was always cited as the ultimate adaptation) could never have become established because their incipient stages would not have provided enough advantage for favorable selection (Mivart, 1871, chap. 2). Darwin had successfully refuted this argument in his sixth edition of *On the Origin of Species* by pointing out that no matter how little advantage an incipient structure provided, if it were of any advantage at all, it would be favorably selected (1872, chap. 7). Morgan could not see how, by random variations acted on by natural selection, something as complex and wonderfully adapted as the vertebrate eye could develop from an initial bit of light-sensitive tissue (1903, pp. 131-132). He compared the situation to the problem that had originally brought him to the question of evolution, asking how partial regeneration of a limb could provide any selective advantage. To be adaptive, regeneration must be complete (1903, pp. 380-381).

Opposition to Lamarckian Principles

Morgan also objected to natural selection because Darwin himself had leaned too heavily on Lamarckian principles, relying in many places on the concept of use and disuse to explain hereditary novelties. As Morgan wrote of one example: "By falling back on the theory of inheritance of acquired characters, Darwin tacitly admits the incompetence of natural selection to account for the evolution (of the flatfish)."[2] As a result, Morgan felt that the whole idea of selection acting on minute variation was insufficient even in Darwin's eyes. Morgan was disconcerted by the school of neo-Lamarckism that arose in the late 1800s, feeling that no evidence suggested that the effects of use or disuse could be passed on to the next generation. In 1903 he discussed the trend toward Lamarckism in the thinking of several American naturalists, including Cope, Hyatt, and Ryder (1903, pp. 230-231, 260). Neo-Lamarckians spent a good portion of their time devising elaborate hypotheses and trying to explain away objections to their theory.[3] Morgan thought that these efforts were irrelevant because neo-Lamarckians failed at the most critical part of their scientific work: they did not put theory to the experimental test. The theory of inheritance of acquired characters, although possible, remained unproved (Morgan, 1903, p. 260).

Opposition to Darwinian Methodology

Perhaps the most general criticism Morgan leveled at the Darwinian theory concerned scientific methodology. Morgan was a rigorous thinker and an experimentalist, concerned that hypotheses should be testable and should agree with all (or nearly all) the facts. He felt that many of the strict Darwinians, neo-Darwinians, and particularly, the neo-Lamarckians had engaged in flights of fancy and speculation that had no basis in fact (1903, pp. 126-163, 165ff., 171-172, 180; 1901; 1905). To Morgan the greatest offender was the German biologist Weismann whose popularity and influence were widespread around the turn of the century. Weismann had tried to unite five important areas of biology—evolution, heredity, cytology, physiology, and development—in one comprehensive conceptual scheme that included a series of hypothetical hereditary particles. Weismann's and Morgan's conceptions of the role of theory in biology were direct opposites. Weismann, working in the tradition dominated by Darwin, thought a theory served primarily to gather the facts

2. Although Morgan (1903, p. 138) had not followed the various editions of *On the Origin of Species*, he was aware that Darwin had moved more closely to the Lamarckian views between the first edition (1859) and the sixth (1872).

3. For a discussion of the neo-Lamarckian movement in America see Pfeifer (1965), where the works of Hyatt, Cope, Dall, LeConte, and King are treated at some length. For European neo-Lamarckism, one of the best sources is Herbert Spencer's *Principles of Biology* (1866, I, pp. 184-200; 402-431). See also Fothergill (1953) and Delage and Goldsmith (1912).

together in an umbrella fashion (1902, vol. 2, p. 4). On the other hand, Morgan thought a theory served primarily to give direction to further research, so that for a theory to be of any value in science it had to be framed in such a way as to be subject to verification. Morgan was a thoroughgoing experimentalist and empiricist. Although he recognized the interconnections between heredity, evolution, and development to the same degree that Weismann did, Morgan felt that those relationships would emerge most clearly only if investigators stuck close to the facts—the observations and experimental data—of biology. He had not patience with fanciful speculation unless it led directly to new observations or experiments. As Manier and I have pointed out, Morgan's experimentalism in particular, sat him squarely against the tradition of such nineteenth-century giants as Weismann, Darwin, and Nägeli.

Chance and Purpose in Evolutionary Development
Philosophically, Morgan found that one of the most difficult problems in evolutionary theory was the question of chance versus purpose. Always opposed to teleological explanations in science, he maintained firmly that variations in organisms did not arise because they were needed, but occurred solely by chance (1903, pp. 391-394).

Morgan pointed out that the term "chance" has two different meanings in relation to evolution and that confusion of these meanings had led many writers to adopt some sort of teleological explanation to account for the adaptation (Morgan, 1910). The odds simply appeared too great against the origin of complex adaptations solely by the accumulation of chance variations. The first meaning was that of the occurrence of one out of many possible events: for example, of the many possible variations that could occur, a particular one actually did take place, and whether it was variation *A*, *B*, or *C* was merely a matter of chance. The second level was that of the occurrence of a chance event at a particular time and place (unconnected with other events at that time or place), such as in the statement, "I chanced to be there." In evolutionary terms, the one chance event, a single variation, will be favorably selected only if it occurs in a particular environment at a particular time that happens to make it favorable. The occurrence of the variation and the set of conditions composing the environment are two independent sets of chance events, both of which must be considered in discussing the role of chance in evolution.

Morgan made this distinction to make a more basic point: that the concept of chance did not imply anything mysterious, outside of normal cause and effect (1910). In opposing those who held that chance events could not be studied scientifically, he pointed out that, in fact, it was the opposing idea—that of purposefulness—which was obscure and not subject to scientific study (1909). After all, Darwin himself had used the term "chance variation" as synonymous with "fluctuating variations" (Morgan, 1910). Although Darwin did not know the causes of these variations, his use of the term "chance" did not imply that they had no cause but only that they did not originate for the purpose of fulfilling a specific need.

Despite his many objections to Darwinian theory, Morgan believed that Darwin had served the admirable function, historically, of putting the question of evolution and particularly that of adaptation, on a sound scientific basis, collecting many observations and arguing his theory directly from them.

Morgan and the "Mutation Theory" of de Vries

In observing members of the genus *Oenothera* (the evening primrose), which he had collected near Amsterdam, de Vries found such extreme variations on occasion between parent and offspring that he claimed the variants must be wholly new species. From this example, de Vries generalized to all organisms, and developed his "mutation theory" (1901-1903).

De Vries made a point of distinguishing mutations from individual, fluctuating variations. The latter could be regarded merely as temporary adaptations to the environment, having nothing to do with the origin of species. Selection could not make permanent any changes that depended upon this type of variation (1901-1903, vol. 1, pp. 5-6). Mutations, on the other hand, occurred less frequently than individual variations and could not be regarded merely as extremes of some spread around the mean established for the entire species. Mutant forms had a newly established mean of their own that represented a distinct break from the mean of the parent species. A consequence of the large-scale difference between the mutant and the original forms was that the offspring would not be able to interbreed with the parents.

De Vries distinguished between several types of mutations (Allen, 1969). However, the term "mutation" as used by de Vries was quite different from the use of the term today, or even in 1915 or 1920. Biologists do not believe that new species can arise in a single large-scale change between parent and offspring. De Vries's mutations might more appropriately be called "macromutations," "monstrosities," or "sports." At any rate, the large-scale variations among the *Oenothera* he cultivated led de Vries to a general theory of evolution by large-scale jumps. Like Morgan and many others, de Vries used the term "species" in two different ways.

He felt that the species groups designated by systematists were largely arbitrary, based for the most part on differences in trivial and non-adaptive characters. However, "real" species did, in fact, exist and were distinguished from one another on the basis of marked differences in character. These species were true or elementary; these units underwent evolution by mutation.

Selection played a role in de Vries's theory, but was more minor than in Darwin's. New mutations might be favorable or unfavorable in a given environment and thus would be subject to the action of selection. Not every mutation survived, according to de Vries—only those which conferred favorable advantages on their bearer. Thus, de Vries preserved the one element of Darwin's theory that most biologists at the time could agree on: selection existed and weeded out unfit variation from the fit.

Morgan became a strong proponent of de Vries's mutation theory when it was published between 1901 and 1903, because it appeared to resolve his objections to Darwinian theory. De Vries's theory put forward a direct alternative to Darwinian theory and achieved an amazing popularity for over a decade. It was intended to supersede both the theory of natural selection in accounting for the origin of species and the Mendelian theory (which de Vries had helped to rediscover) in accounting for the inheritance of specific traits. Morgan's acceptance of de Vries's theory over Darwinian and Mendelian theories between 1900 and 1910 indicates much about his conception of the two theories as well as his philosophy of science.

The major features of de Vries's theory that Morgan found attractive as alternatives to Darwinism indicate the relationship that Morgan perceived, especially in the first decade of the century, between heredity and evolution. As a theory of discontinuous variation, de Vries's hypothesis relegated individual and minute variations to a negligible evolutionary role. The large-scale mutations that de Vries observed in *Oenothera* represented qualitatively new variations that were wholly distinct from the parental form. They were new forms, arisen fully blown, ready to take their chances in the world.

As a result of de Vries's emphasis on discontinuity, his theory avoided the problem of swamping that plagued Darwin and his followers. Not only was a mutant form infertile with its parents, thereby preventing its being lost through back-crossing, but also, de Vries hypothesized, there were "periods of mutation" in which the same mutation would very likely occur in a number of organisms simultaneously. Thus, a mutation could be preserved by cross-fertilizing with another like itself in the population.

Morgan also maintained that de Vries's theory met Mivart's objections to Darwinism—that is, that incipient stages of highly adaptive organs would not in themselves be adaptive. Because a new structure could appear more or less fully developed, in one step, the problem of utility of incipient stages became negligible.

Morgan felt that the mutation theory answered the objections of paleontologists concerned with gaps and apparent sudden jumps in fossil types between strata of rocks. What appeared to be sudden, qualitative differences between species in adjacent strata were in fact the natural result of mutations.

Because de Vries's mutations produced much more rapid changes in structure than possible (in the same time) by the Darwinian model, the evolutionary time scale could be considerably telescoped. De Vries's theory could thus solve one of the most persistent problems that Darwin's critics (especially Lord Kelvin) had raised—that is, that there had not been sufficient geological time for evolution to have occurred by selection of small-scale individual variations (de Beer, 1963, chap. 8; Burchfield, 1975, chap. 3).

Morgan noted that de Vries's theory appeared to receive considerable corroboration in the pure-line experiments of the Danish botanist Johannsen (1903) to examine the persistent evolutionary question about the types of variation (large or small) on which selection acted. Johannsen showed that selection of fluctuating, individual variations only separated out the pure lines already existing in a heterogeneous population of organisms. He emphasized that there was thus a limit to which selection could produce change in a population of organisms. The longer selection was continued, the less progress could be made in any one line. Johannsen also noticed that if selection were relaxed—that is, if the pure lines were allowed to hybridize again—the differences between them would soon disappear. Selection seemed to produce changes in a population that persisted only as long as the selection process was rigorously maintained. It could not cause the species to transcend that threshold level of variation between one species and another.

Johannsen's conclusion opposed the Darwinian theory on two accounts. It showed that the selection of fluctuating variations was not able to produce a new species and that the effects of selection by itself were purely negative. The pure-line experiments provided important experimental evidence confirming the anti-Darwinian prejudices of many investigators and contributed significantly to the acceptance of de Vries's mutation theory. A number of biologists, including Morgan, saw de Vries's and Johannsen's arguments as the best evidence to date against the Darwinian model of selection acting on small individual differences

(Morgan, 1903, p. 298; also Casey, 1905; Conklin, 1912; Davenport, 1905; and MacDougal, 1905).

The idea that selection could not produce new species was strengthened by the assertion that there was a fault in Darwin's equating of artificial and natural selection. Several writers in the late nineteenth century claimed that the most rigorous artificial selection (as in animal and plant breeding) had never produced a single case of a wholly new species (Plate, 1908, pp. 36-56; Morgan, 1903, pp. 19-20). De Vries used this claim in maintaining that the forms produced by artificial selection were not as permanent as those produced by natural selection. Acting on slight individual variations, artificial selection produced only temporary types that, if allowed to hybridize, would quickly revert to their original form (a la Johannsen). To de Vries, the creation of the fit was the result only of new mutations. He argued that the species differences created by mutation were more in agreement with those differences found among species in nature than with those forms produced through artificial selection.

Morgan and many of his contemporaries found de Vries's theory attractive because it accounted quite well, at least on the surface, for many of the substantive objections to the Darwinian theory they had encountered. De Vries's theory was especially attractive to typologists. New species arose virtually as a full-fledged "type," with no knotty problem of intermediaries that were bothersome deviations from the norm. Most important, de Vries had set his theory up in a testable way that was amenable to experimental analysis. This apparent alternative to the more speculative methods of many of the neo-Darwinians encouraged many of the younger generation of experimental biologists, especially Morgan.

Morgan, Genetics, and Natural Selection

Morgan's conversion to Mendelian genetics after 1910 proved to be an important step in his eventual acceptance of the Darwinian theory of natural selection. Treading this new road, like that of Mendelism itself, was no easy or overnight process for a skeptic like Morgan. For at least several years after becoming convinced of the chromosomal basis of Mendelism, Morgan remained unwilling to accept a number of points about Darwinian theory. More specifically, he did not see immediately how Mendelism could be applied to the study of evolution in any general way.

Between 1912 and 1915 Morgan seems to have made a substantial (but by no means complete) change in his attitude toward natural selection. In 1912 Morgan still held to de Vries's mutation theory, although he was becoming aware of the serious difficulties that it was encountering.[4] Morgan still believed, however, that discontinuous variations—that is, mutations of some sort, either large or small—were the raw material for evolution. That same year young Julian Huxley visited Morgan's lab at Columbia. Huxley was fresh from studying the evolution of protective coloration and mimicry, which he used as a special case to amplify some of Darwin's basic principles. Huxley was most impressed with the *Drosophila* mutants in Morgan's laboratory and viewed them as throwing considerable light on the origin of variability. He argued openly with Morgan (to the delight of Morgan's students) on the issue of natural selection, chastising the older man for his persistent skepticism. Muller (1962) recalls that a few days after Huxley's departure, Morgan mumbled something about his English guest as "that young whippersnapper." But when he published his Vauxeum lectures (1916), Morgan saw much more clearly that Mendelian phenomena were compatible with and helped to explain certain aspects of Darwinian theory. Several factors were responsible for Morgan's change of mind.

In 1909, just as Morgan began breeding his *Drosophila* stocks, he began to doubt the evolutionary role of discontinuous variations in general, and de Vries's mutations in particular:

> It has often been urged, and I think with much justification, that the selection of individual, or fluctuating variations could never produce anything new, since they never transgress the limits of their species, even the most rigorous selection—at least the best evidence we have at present *seems* to point in this direction. But a new situation has arisen. There are variations within the limits of species that are definite and inherited, and there is more than a suspicion that by their presence the possibility is assured of further definite variations in the same direction which may further and further transcend the limits of the first steps. If this point can be established beyond dispute, we shall have met one of the most serious criticisms of the theory of natural selection.

The occurrence of small but definite variations that seem to be inherited in a predictable manner gave Morgan a new insight into natural selection —an insight he had previously lacked. Contrary to his earlier belief, Morgan now recognized that these small variations were in fact heritable, and could thus be acted upon by selection.[5]

4. See Morgan (1912). Between 1910 and 1912 Davis's cytological studies began to suggest that the supposed "mutations" that de Vries observed in *Oenothera* could be explained by complex chromosomal arrangements (Davis, 1910, 1911, and 1912).

5. Morgan (1925) is a revised form of *A Critique of the Theory of Evolution* (1916). Differences between the books are not substantial in terms of Morgan's understanding of natural selection, except in a few particular cases.

The new type of mutation that Morgan observed in *Drosophila* was smaller than de Vries's species-level jumps, but larger than the individual fluctuating variation of the old-style Darwinians. Morgan's observations of *Drosophila* indicated that heredity could still be discontinuous and yet the source of variability on which selection acted. Neither species-level variations (de Vries) nor infinitely graded variations (the orthodox Darwinians) were the only sources of variation. In fact, neither were probably sources of variation in nature at all—the small-scale "micromutation" such as scute bristle, vermillion eye, or dumpy wing provided the raw material of evolution.[6]

By 1912 the mutation theory had drawn serious criticism, principally through the work of Davis (1910, 1911, 1912). Morgan wrote that the English edition of de Vries's *Mutation Theory* was already behind the times, largely because Davis had shown that the large-scale mutations observed in *Oenothera* appeared to be the result of complex hybrid rearrangements. Muller even showed that the concept of balanced lethals (that lethal genes are maintained in the population because their lethal nature is only apparent as homozygous recessives), which he had developed to explain certain *Drosophila* results, could also be applied successfully to many cases in *Oenothera*. De Vries's variations seemed to be in complete agreement with Mendelian principles; they certainly did not represent species-level differences. The decline in validity of the mutation theory was a strong factor encouraging Morgan to seek a convergence of Darwinian and Mendelian explanations.

By 1914 the idea of genic interaction had begun to gain considerable support within the Morgan group. Muller and Sturtevant had used the idea of modifying factors to explain Castle's results with hooded rats (Sturtevant, 1965), and the concept of position effect was noteworthy in demanding that genes influence each other's functioning. Muller and Sturtevant seemed to have understood almost immediately that the presence of modifying genes (and the whole concept of gene interaction) could account for the apparent occurrence in any species of a continuous range of variations that were, nonetheless, inherited discretely. Darwin's assertion that selection acted on continuous, but still inherited variation, appeared to be not so erroneous after all.

---

6. It is perhaps unfortunate that Morgan chose to retain the word "mutation" in referring to those small Mendelian differences he observed, for it totally confused the issue of variation by failing to distinguish between large-scale variations that de Vries considered new species and small but definite variations that occurred within the limits of an existing species. Within the term "mutation," Morgan included a whole range of variations that bred true. But many of his readers, familiar with the term only as de Vries had used it, were understandably confused about the type of variations Morgan thought were acted on by natural selection.

In *A Critique of the Theory of Evolution*, Morgan (1916) showed how modifier genes could account for the production of many fine gradations between the definite unit characters of the classical Mendelians. The strength of expression of any character depended on the modifying factors present in the germ cell. Hence, selection could produce almost any intermediate forms by reducing or increasing the number of modifier genes. Selection could stabilize the number of modifiers in a population, it could increase them to a maximum, or it could eliminate them (Morgan, 1916, pp. 165-169). In this discussion Morgan gave Muller credit for having worked out the modifier gene conception, and went on to show how many cases of supposed continuous variation could be interpreted in terms of modifier genes (Morgan cited races of Indian corn, pigeon tail size, and size in *Paramecium*). The modifier gene concept helped to eliminate the longstanding artificial distinction between the proponents of continuous and discontinuous variation. It was no longer an either/or choice.

The concept of modifying factors also allowed Morgan to meet the objection to natural selection arising from Johannsen's pure-line experiments of 1903. Limits to the effects of selection could be attributed to the fact that modifying factors had been accumulated to their maximum (for that particular population) or reduced to zero. Selection had then to await the appearance of more variation (Morgan, 1905).

The work on gene linkage and the chromosome theory by Morgan's students led him to another idea that may have been important in applying Mendelian principles to natural selection. Violation of the expected linkage ratios (as a result of crossing-over and recombination) had an important bearing on the evolutionary process. Recombination would produce new linkage groups in which combinations of characters, perhaps never before existing together in the same organism, could be obtained. The idea of crossing-over and recombination showed Morgan and his students that an almost infinite number of possible combinations of characters could eventually occur in any population. For any given set of characters, crossing-over would give all the possible intermediate combinations between two parental forms. By 1916 Morgan understood the importance of this factor clearly: "By crossing different wild species or by crossing wild with races already domesticated, new combinations had been made. Parts of one individual had been combined with parts of others, creating new combinations" (p. 162). The new combinations of old characters, as well as the appearance of additional mutants and their successive recombination, provided the raw material on which natural selection could act.

Another important factor in Morgan's changed views on natural selection was his slow but increasing profound awakening to the difference between the genotype and phenotype of an organism. As Churchill has suggested, genotype refers to vertical inheritance (transmission), whereas phenotype refers to horizontal inheritance (translation). The clear distinction between genotype and phenotype was made by Johannsen (1911). In the first decade of the twentieth century many workers in both heredity and evolution confused genotype and phenotype. This confusion manifested itself in a failure to understand the difference between the unit factor (gene) passed on from parent to offspring, and the adult character whose development that gene may guide. Thus Morgan, for example, opposed Mendelism as "preformationist" because he did not make enough of a distinction between the inherited factor and the actual adult character to which it gave rise.

In a similar way, failure to distinguish between genotype and phenotype confused Morgan's attempts to understand natural selection. Johannsen had argued that selection acts directly on the phenotype, producing changes in the genotype after a number of generations. The phenotype is the result of the combined effect of many genes (genic interaction) plus environmental influences (such as amount of food). Ultimately, modifications of the genotype are the result of selection acting on the phenotype. Changes in the phenotype, however, merely reflect environmental factors and thus are not necessarily permanent. Changes in genotype are passed on faithfully, and thus are permanent. The genotype-phenotype distinction thus helped clear away the old problem, for evolutionists, of how to determine which variations are inherited and which are not. Breeding tests helped make those distinctions clear. Yet the point is they could be made clear, which Morgan had not seen when he wrote about evolution in 1903 or 1905.[7]

Morgan probably first encountered Johannsen's phenotype-genotype distinction in 1911, after Johannsen had delivered his 1911 paper to the American Society of Naturalists at its Christmas meeting at Ithaca in 1910. Morgan attended that meeting, and even gave a paper as part of a symposium on genotype and phenotype. More important, Johannsen spent part of the following winter term—1912—at Columbia.[8] Because Morgan found Johannsen's ideas and personality attractive (as he wrote

7. For a more detailed discussion of the role of the genotype-phenotype distinction in early twentieth-century evolutionary and genetical studies, see Allen (1979) and Churchill (1974).

8. Johannsen had originally been invited to the United States by George H. Shull (Shull to Conklin, summer 1910; Conklin Papers, Princeton University). For the winter term 1911 Johannsen held a lectureship at Columbia for visiting scholars (Rosenvenge and Winge, 1928).

to Driesch), it is highly unlikely that the two men did not discuss the issues of evolution, selection, heredity, variation, and Morgan's new work on *Drosophila*. Though it is clear they did not agree on some points (Johannsen was skeptical of the chromosome theory), Morgan may have used the opportunity to explore further Johannsen's genotype-phenotype concept. By 1915 or so, Morgan had made the distinction clearly in his own mind. As he wrote: "Failure to realize the importance of these two points, namely, that a single factor may have several effects, and that a single character may depend on many factors, has led to much confusion between factors and characters, and at times to the abuse of the term 'unit character.' It cannot, therefore, be too strongly insisted upon that the real unit in heredity is the factor, while the character is the product of a number of genetic factors and environmental conditions" (Morgan and others, 1915, p. 210; see also Morgan, 1917).

Morgan's students, particularly Muller and Sturtevant, influenced Morgan to revise his ideas about the mechanism of natural selection. As undergraduates, both Muller and Sturtevant had studied Lock's *Variation, Heredity, and Evolution*, then thought to be the most up-to-date work relating evolution and heredity.[9] Lock tried to show how Mendelian genetics and Mendelian natural selection together formed a complete mechanism for evolution. From the outset, Muller and Sturtevant were exposed to the idea that there was no fundamental conflict between the theory of particulate inheritance (Mendelism) and the theory of natural selection.

Muller's reports of these discussions with Morgan are vivid. According to Muller, Morgan was stubborn; he would not accept the idea that natural selection could create new species, and continually regarded it, as he had done in *Evolution and Adaptation*, as a purely negative force: "All of us [himself, Sturtevant, and to some extent Bridges] argued with Morgan on that . . . Morgan would come back and back . . . it seemed to us as if he somehow couldn't understand natural selection. He had a mental block which was so common in those days" (1965, typescript p. 8). He was also confused on some very basic issues of Mendelian genetics and natural selection, for example, about the effect of dominance on gene frequency in a population. He appears to have believed that a dominant gene would take over a population after a while, using this erroneous assumption to try to prove that evolution could occur without natural selection (Muller, 1965, typescript p. 10). Morgan also apparently had

9. Both Muller and Sturtevant have attested to the importance of Lock's book (1907) in helping them see the relationship between Mendel's laws and evolution by natural selection.

some difficulty understanding and accepting the modifier-gene hypothesis, even as it originally related to genetics. Altenburg pointed out that as late as 1910 Morgan attempted to explain multiple factor cases by assuming a lack of segregation, and considered them as evidence against Mendelian genetics.[10] Coming to grips with the multiple factor and modifier hypothesis was crucial to Morgan's understanding of natural selection. According to Muller, "One of the things which must have helped convince Morgan of the validity of natural selection of a Darwinian kind, was the fact that in the selection of the stock for truncate or beaded wing (both of which are due to modifying factors) we gradually got more and more truncate, or more and more beaded wing. They bred truer and truer. And when we analyzed it we found that there were a whole lot of little genes in there besides the major gene. That rubbed it in . . . That made the small mutations important" (1965, typescript p. 18). Muller, Sturtevant, and Weinstein in these early years convinced Morgan of the importance of these results.[11]

After 1916, Morgan continued to develop his ideas and understanding of the mechanisms of evolution. He wrote two more books: *Evolution and Genetics* (1925) and *The Scientific Basis of Evolution* (1932). His three works, written roughly ten years apart, helped to trace the intellectual refinement of Morgan's views of natural selection. As genetic and other types of data increased, Morgan was able to bring together more of the threads of a comprehensive theory. In his later work, Morgan was increasingly aware of the importance of dealing with evolution on the population rather than the individual level (1909, p. 376).

In *A Critique of the Theory of Evolution*, Morgan did not mention anything directly about populations of organisms and their role in evolution. Although in 1909 he had remarked that "each of us is the descendant of a large population . . . a species moves along as a horde rather than as the offspring of a few individuals in each generation," this idea did not play a role in Morgan's concept of natural selection at the time. Morgan spoke of animal groups, of the past as being the ancestors of modern groups—but he used the term "groups" in a way somewhat different from our present-day use of "population." Morgan seemed to think of a group as several species that existed as one part of a complex and continuing wave of organic life. It was a dynamic concept—species were really just groups that were constantly changing—but as such it was

10. Altenburg to Muller (March 24, 1946), Muller papers, Lilly Library, Indiana University, Bloomington, Indiana; quoted in Carlson (1974, p. 14).

11. Sturtevant to Sonneborn (May 5, 1967, pp. 1-2), Sturtevant Papers, Cal Tech Archives.

not a population approach to the problem of evolution Morgan had enunciated in 1916.

By 1925, however, Morgan had become aware of Haldane's work. Haldane was then trying to develop a mathematical procedure for dealing with evolution on the population level (Morgan, 1925, p. 142). Though Haldane's analytical methods were in their infant stage, Morgan saw the importance of this kind of thinking for clarifying many evolutionary problems. For example, he saw that such an analysis would help to show conclusively that a dominant gene does not necessarily have any better chance of surviving in a population than a recessive. He pointed out that this question could only be answered by an analysis of large amounts of numerical data. Morgan had even begun to recognize that what we today would call "selection value" had to be assigned to different genotypes (1925, pp. 141-142).

Although Morgan was aware of Haldane's work, he never seriously attempted to understand the substance of population genetics as applied to evolution. Morgan was poorly versed in mathematics, and was, in the last analysis, not really interested in it (Muller, 1965, typescript p. 10). He could see the importance of a quantitative approach to evolution, and even the necessity of setting the theory of natural selection on a mathematical base, but he could not follow the complex mathematical arguments of Haldane, Wright, or Fisher. Mathematical and statistical analysis was totally foreign to his type of biological training and inherent interest.

If Morgan's thinking on evolution and natural selection was refined and expanded along some lines, he never fully understood certain areas, such as the problem of speciation. Morgan believed that species were arbitrary units created by taxonomists for convenience. He was a "nominalist" in Mayr's terminology: that is, someone who believes species are not real units in nature. Morgan never really abandoned this idea. Given this bias, it was logical that he continued to minimize the origin of species as a real or significant problem. In 1923, Morgan stated clearly his nominalist position:

How far these new types [Mendelian variations] furnish the variations that make new species may depend on what we call "species." If, as some systematists frankly state, species are arbitrary collections of individuals assembled for the purpose of classification; or if, as other systematists admit, there are all kinds of species both in nature and in books, it would be absurd for us to pretend to be able to say how such arbitrary groups have *arisen*. It is possible that some of them may not have arisen at all—that they may have only been brought together by taxonomists (pp. 237-238).

In his usual manner, Morgan went on to throw down the gauntlet by claiming that the systematists might produce one classification system for organisms and the geneticists a completely different one—but both would be valid. The classification system depended, after all, on the criteria chosen to draw the boundaries between groups. Because these criteria depended on one's viewpoint, they were, in the last analysis, arbitrary. So certain of the correctness of his viewpoint was Morgan, that he refrained from using the word "species" in the remainder of the article.

Adaptation, not speciation, was the important outcome of evolution by natural selection to Morgan. He failed to see that an interplay existed between the two, making them components of a complex system. The biological advantage of speciation in the long run is adaptive; it reduces competition. The specialization that occurs with speciation allows separate populations to adapt specifically to noncompetitive niches. Speciation is not only an outcome of adaptation, it is also an adaptation itself. Morgan persisted, however, in drawing a distinction between speciation and adaptation that ignored the biological reality—the existence of distinct biological groups that, judged collectively on a number of criteria, can be distinguished from each other biologically (that is, are ecologically, genetically, and to a large extent reproductively distinct entities). By retaining the nominalism of his youth, Morgan persistently downgraded the importance of species as biological units or speciation as an integral part of the evolutionary process.

Morgan also never fully understood the role of isolation in species formation. Although Jordan had made this point clearly as early as 1906, Morgan never seems to have taken the idea seriously. In 1923, Morgan still argued that the observed infertility between distinct species could be the result of one or two mutations. He totally failed to understand the role of geographic barriers and physical separation of populations over long periods of time in producing divergence. As Jordan noted after reading Morgan's 1923 paper: "I do not see why geneticists in general ignore the most potent fact in Geographical Distribution, that 'new species' are formed across barriers. Very seldom do closely related forms exist together but *near together* forming 'geminate species.' Of this we have thousands of illustrations. The supposition exists, that perhaps some primal difference or mutant may have existed in the original migrants."[12]

12. Jordan to Morgan (March 28, 1923), David Starr Jordan Papers, Stanford University. Geographic isolation was necessary, Jordan went on to say, to preserve the new mutant types that appeared in a particular population. Jordan apparently made little impression on Morgan. Nothing about geographic distribution or the effects of isolation appears in Morgan's next book, *The Scientific Basis of Evolution* (1932). Morgan was simply unable to grasp some points unless he was argued with persistently and forcefully.

An even more fundamental issue with Darwinian theory was never fully settled in Morgan's mind. Although he did express the view in 1916 that the principle of selection, acting on Mendelian mutations, could produce gradual changes in a line of organisms, he was never at ease with the idea of selection itself. The concept, perhaps the very term "selection," bothered him; it sounded purposeful, and with his strong dislike of teleological thinking, Morgan reacted against purposefulness in evolutionary theory. Sturtevant recognized Morgan's strong feelings on this matter: "There was one respect in which Muller and I had to 'educate' Morgan. He was never fully happy about natural selection, since it seemed to him to open the door to explanations in terms of purpose. We convinced him that there was nothing teleological or contrary to Mendelism in natural selection—but he remained unhappy with it, and arguments had to be repeated again and again—an experience that I think was good for both Muller and me in that it made us very careful about how we stated the case for selection."[13]

Obviously, the constant discussion of this issue in the lab helped Morgan to come around to the idea that selection can work in natural populations without a guiding purpose behind it. In 1916 Morgan got the role of selection correct. He carefully steered away from the term "selection," or any reference to mystical or creative forces (pp. 187-195). He defined selection as the increase in the number of individuals that results after a beneficial mutation has occurred (owing to the ability of living matter to propagate) and believed that this preponderance of certain kinds of individuals in a population makes some further results more probable than others (1916, p. 194). Yet sixteen years later, Morgan regressed to his earlier, more ambivalent, position on selection. He wrote: "Under the circumstances, it is a debatable question whether still to make use of the term natural selection as a part of the mutation theory (i.e., Mendelian theory), or to drop it because it does not have today the same meaning that Darwin's followers attached to his theory" (1932, p. 150). Elsewhere in the same book, in his endless quest to make certain that no one viewed natural selection as creating any new variations, he wrote: "Natural selection may then be invoked to explain the absence of a vast array of forms that have appeared, but this is saying no more than that most of

13. Sturtevant to Sonneborn (May 5, 1967, pp. 1-2), Sturtevant Papers, Box 4, "Morgan" folder. The same point has also been made by Alexander Weinstein in a personal conversation, May 8, 1979. Weinstein discussed many aspects of evolutionary theory with Morgan in preparation of *A Critique of the Theory of Evolution* (1916).

them have not had a survival value. The argument shows that natural selection does not play the role of a creative principle in evolution" (pp. 130-131).

By 1932 no Darwinian claimed that selection itself was a creative principle (though a few had made this claim in the early years of the century); that is, it somehow actively created new variations in a particular direction. Morgan repeatedly came back to this point and used it to downplay the role of selection itself in favor of mutation pressure as the main force in evolutionary change, which reveals his deep-seated misgivings about the selection principle. Mutations were concrete, familiar events that he had encountered every day for over twenty years. Selection was still a vague, invisible, and slightly teleological-sounding process that he could not fully accept.

The change between *A Critique of the Theory of Evolution* (and its revised version of 1925) and the much later *Scientific Basis of Evolution* may indicate as much about Morgan's pattern of working as about a change in his evolutionary thinking. Morgan had always found the idea of selection incompatible with his antiteleological notions of nature. In preparing his Vauxeum lectures of 1915 for publication, Morgan had passed the manuscript around to all the members of the fly group for criticism. Weinstein and other members of the group corrected, discussed, and even rewrote sections of the text—all with Morgan's knowledge and approbation. They particularly criticized his views on selection. In response, Morgan apparently revised his thinking on selection, for he let the corrections stand (personal conversation, May 8, 1979). But it is not clear that he fundamentally changed his mind. In 1932, Morgan had only Sturtevant (Bridges had never taken as much active part in such theoretical questions as the others) among the old group still working with him. It is not known whether he sought Sturtevant's advice on the manuscript of the *Scientific Basis*, as he had sought the group's advice on *A Critique* fifteen years earlier. But if he did, it appears he no longer took advice so readily. Whatever the exact explanation, we know that Morgan's view of selection was always skeptical. His embrace of it in the two books of 1916 and 1925 represent acceptance of the logic behind the selection principle, as put forward by Muller, Sturtevant, Weinstein, and others. His 1932 work appears to reflect Morgan's truer feeling about selection, written without the help or initial advice from the opposing viewpoints.

Philosophically, Morgan was not only an empiricist and a staunch experimentalist. He was also fundamentally a skeptic and a materialist. His skepticism took the particular turn, as it did for many biologists of his generation, of strong antireligious feelings. Morgan was judicious (or diplomatic) enough never to write openly about his hostility to religion, but it emerged clearly in his personal conversation, and sometimes subtly in his writings. Religion was to Morgan, like Driesch's entelechy or other vitalistic notions, a form of mystery. The role of science, of biology in particular, was to combat mystery, and that meant, particularly in the context of late nineteenth- and early twentieth-century America, to combat religion. Morgan was militantly antireligious. As Dobzhansky, who worked with Morgan both at Columbia and Cal Tech, wrote: "Now, about his militant atheism—his idea was religion feeds on mystery. The way to combat religion is to combat mystery, hence to show that the biological phenomena are not mysterious, but they are scientifically explicable. That was the reason why Morgan worked in biology, particularly in genetics and evolution" (1962, p. 255). Morgan's strong antireligious feelings may well be the source of much of his philosophy of science—of his materialism, his experimentalism, his emphasis on physicochemical explanations. The bearing of this attitude on Morgan's evolutionary views may very well be that to combat religious views on creation, biology must be careful not to fall into the same trap of inventing obscure or abstract "principles," "factors," or other nonentities to account for the diversity and adaptation of life. Scientists must remember that they are nothing more than priests of a new religion if they think that by naming something they have explained its existence. Even a term like selection could represent the error of explaining by naming if there were no empirical data to support its existence. Morgan's philosophy of science, as discussed earlier in this chapter, stood on an empirical foundation of its own, but that view may have been formed initially, and was certainly fed continuously, by his strong opposition to religion and mystical explanations of any kind.

Morgan's early views on evolution illustrate some of the problems inherent in Darwinian theory as it was understood (or misunderstood!) in the period between 1870 and 1920. Morgan's objections also suggest some of the common misconceptions that biologists, especially of a more experimentalist, laboratory persuasion, brought to Darwinian theory. Morgan's change of attitude on Darwinism between 1910 and 1925 illustrates some of the important contributions that genetical thinking—of the classical school—made toward the eventual casting of evolutionary theory into a genetical mold. Morgan's role in the synthesis was not directly positive; in many ways his actual evolutionary thinking was confused. But his realization that genetics and evolution could, and must, be synthesized, and his own work in genetics that helped to make that synthesis possible, must be recognized.

References

Allen, G. E. 1968. Thomas Hunt Morgan and the problem of natural selection. *Journal of the History of Biology* 1:113-139.

———— 1969. Hugo de Vries and the reception of the 'mutation theory.' *Journal of the History of Biology* 2:55-87.

———— 1978. *Thomas Hunt Morgan: the man and his science.* Princeton: Princeton University Press.

———— 1979. Naturalists and experimentalists: the genotype and the phenotype. *Studies in the History of Biology* 3:179-209.

Burchfield, J. D. 1975. *Lord Kelvin and the age of the earth.* New York: Science History Publications.

Carlson, E. A. 1974. The *Drosophila* group: the transition from the Mendelian unit to the individual gene. *Journal of the History of Biology* 7:31-48.

Casey, T. L. 1905. The mutation theory. *Science* 22:307-309.

Churchill, F. B. 1974. Wilhelm Johannsen and the genotype concept. *Journal of the History of Biology* 7:5-30.

Conklin, E. G. 1912. Problems of evolution and present methods of attacking them. *American Naturalist* 46:121-128.

Darwin, C. 1872. *On the Origin of Species,* 6th ed., corrected. London: John Murray.

Davenport, C. B. 1905. Species and varieties, their origin by mutation. By Hugo de Vries. A review. *Science* 22:369-372.

Davis, B. M. 1910. Genetical studies of *Oenothera.* I. Notes on the behavior of certain hybrids of *Oenothera* in the first generation. *American Naturalist* 44:108.

———— 1911. Genetical studies of *Oenothera.* II. Some hybrids of *Oenothera biennis* and *O. Grandiflora* that resemble *O. Lamarckiana. American Natturalist* 45:193-223.

———— 1912. Genetical studies of *Oenothera.* III. Further hybrids of *Oenothera biennis* and *O. grandiflora* that resemble *O. Lamarckiana. American Naturalist* 46:377-427.

de Beer, G. R. 1963. *Charles Darwin.* London: Thomas Nelson.

Delage, Y., and M. Goldsmith. 1912. *The theories of evolution,* trans. A. Tridon. New York: Huebsch.

de Vries, H. 1901-1903. *Die Mutationstheorie.* Leipzig: Von Veit. (1910 English translation, *The mutation theory,* trans. J. C. Farmer and A. D. Darbishire. Chicago: Open Court.)

Dobzhansky, Th. 1962. *Reminiscences of Theodosius Dobzhansky.* Oral History Research Office, Columbia University.

Fothergill, P. G. 1953. *Historical aspects of organic evolution.* London: Hollis and Carter.

Johannsen, W. 1903. *Über Erblichkeit in Populationen und in reinen Linien.* Jena: Fischer.

———— 1911. The genotype conception of heredity. *American Naturalist* 45:129-159.

Lock, R. H. 1907. *Recent progress in the study of variation, heredity, and evolution*. New York: Dutton.

MacDougal, D. T. 1905. Discontinuous variation in the origin of species. *Science* 21:540-543.

Mayr, E. 1959. Agassiz, Darwin, and evolution. *Harvard Library Bulletin* 13:165-194.

———— 1969. Footnotes on the philosophy of biology. *Philosophy of Science* 36:197-202.

Mivart, St. G. 1871. *The genesis of species*. New York: Appleton.

Morgan, T. H. 1899. Some problems of regeneration. *Biological Lectures Delivered at the Marine Biological Laboratory of Woods Hole in 1897 and 1898*. Boston: Ginn, pp. 193-207.

———— 1901. Regeneration and liability of injury. *Science* 14:235-248.

———— 1903. *Evolution and adaptation*. New York: Macmillan.

———— 1905. The origin of species through selection contrasted with their origin through the appearance of definite variations. *Popular Science Monthly* 67:54-65.

———— 1909. For Darwin. *Popular Science Monthly* 74:367-380.

———— 1910. Chance or purpose in the origin and evolution of adaptations. *Science* 31:201-210.

———— 1912. Some books on evolution. *Nation* 95:542-544.

———— 1916. *A critique of the theory of evolution*. Princeton: Princeton University Press.

———— 1917. The theory of the gene. *American Naturalist* 51:513-544.

———— 1923. The bearing of Mendelism on the origin of species. *Scientific Monthly* 16:237-246.

———— 1925. *Evolution and genetics*. Princeton: Princeton University Press.

———— 1932. *The scientific basis of evolution*. New York: Norton.

————, H. J. Muller, A. H. Sturtevant, and C. B. Bridges. 1915. *The mechanism of Mendelian heredity*. New York: Holt.

Muller, H. J. 1962. A biographical appreciation of Sir Julian Huxley. *The Humanist* 2 and 3:51-55.

———— 1965. Interview with G. E. Allen (typescript).

Pfeifer, E. J. 1965. The genesis of American neo-Lamarckism. *Isis* 56:156-167.

Plate, L. 1908. *Selectionsprinzip und Probleme des Artbildung*, 3d ed. Leipzig: W. Engelmann, pp. 36-56.

Rosenwinge, L., and O. Winge. 1928. Wilhelm Ludwig Johannsen. *Taler i Videnskabernes Selshabs Mode den 3 Februar 1928*, pp. 9-10.

Spencer, H. 1866. *The principles of biology*. New York: Appleton.

Sturtevant, A. H. 1965. *A history of genetics*. New York: Harper & Row.

Weismann, A. 1902. *Vorträge über Descendenztheorie*. Jena: Fischer.

Hypotheses That Blur and Grow*

Hampton L. Carson

I was a beginning graduate student in zoology at the University of Pennsylvania in 1936. For twenty-five years Clarence E. McClung had been director of the laboratories. His career and attitudes had been shaped at the University of Kansas, where he was influenced by prominent but evolutionarily rather conservative paleontological and entomological systematists. His early discovery of the sex chromosomes (McClung, 1899) led him into a life's work devoted to chromosome behavior in meiosis. He was among the first to articulate a formal statement of the chromosome theory of heredity. Nevertheless, his conservative attitudes toward systematics, cytology, and evolution were retained throughout his scientific life.

McClung, a strong Lamarckian, channeled a considerable amount of his research efforts in later years into exploration of these ideas. Specifically, McClung sought to obtain evidence that somatic cells could influence germ cells. The idea was that in this manner the "experience" of the individual could be "brought into effective relation with the racial organization represented by the chromosomes of the germ cells" (1938).

As an apprentice cytologist, I set to work making a comparative study of the apical cell of the insect testis, which appeared to be elaborating substances that were being taken up by the neighboring spermatogonia. My resulting paper (1945) took refuge in a lengthy descriptive account; the Lamarckian interpretation was gently rejected.

McClung's interest in cytotaxonomy was reflected in a long-term project at Pennsylvania devoted to collecting and maintaining a very large "library" of fixed testes of many species of the family *Acrididae*, a cytologically favorable group of grasshoppers. This material was unfortunately not studied by very many people. A graduate student, P. B. A. Powers, who died rather suddenly in 1938, made significant use of the material for measuring the lengths of the karyotypes of different species.

The intellectual milieu surrounding the McClung group reflected a virtually puritanical devotion to the recording of exact cytological details. Hard rules were laid down; observations had to be made with scrupulous care and detached from any sort of theoretical influence. This approach contrasted sharply with the speculative and imaginative approach pro-

*The title of this selection is the motto beneath a mural on the stairway at the Zoological Laboratories, University of Pennsylvania.

moted by Darlington, whose book was greeted at Pennsylvania with intensely adverse criticism.

Most of the graduate students in the department learned their genetics from P. W. Whiting, the charming pioneer of the genetics of cats and wasps. Although his lectures were somewhat disorganized and abstruse, Whiting gave considerable attention to basic population genetics. C. W. Metz, who replaced the retiring McClung as director in 1940, also took over supervision of the work of several graduate students, including myself. I considered this circumstance fortunate because Metz not only had a background in *Drosophila* genetics, sorely lacking at Pennsylvania, but also was a far more liberal thinker on evolutionary matters than McClung. Metz, however, was very much preoccupied with the study of cytological mechanisms, chromosome structure, and chromosome behavior. In particular, he emphasized research dealing specifically with the genus *Sciara*; for example, selective segregation and the problems of the interpretation of a bizarre mechanism of sex determination. These matters concerned him much more than the way in which cytogenetics and population studies could contribute to an understanding of evolution. Yet I managed to persuade him to let me do a cytogeographical analysis of the populations of a single species.

Whiting and Metz set the stage and arranged the props for the arrival of the neo-Darwinian synthesis at Pennsylvania. Nevertheless, Dobzhansky actually raised the curtain for the graduate student audience in a dramatic seminar given during his memorable 1941 visit to the Pennsylvania laboratory arranged by Metz. Suddenly the neo-Darwinian idea seemed to come into sharp focus as we graduate students ran off to read Dobzhansky's book and try to understand Fisher and Wright, strange new names at the time. More than that, Dobzhansky's grand mix of genetics, chromosomes, geography, populations, and selection enabled me to bring my somewhat peripheral but long-term interest in birdwatching, plant collecting, and a general excitement over natural history into a new and meaningful professional context. Neither my studies of ecology nor those of chromosomes were the same from that time forward.

References

Carson, H. L. 1945. A comparative study of the apical cell of the insect testis. *Journal of Morphology* 77:141-161.

McClung, C. E. 1899. A peculiar nuclear element in the male reproductive cells of insects. *Zoological Bulletin* 2:187-197.

———— 1938. The apical cell of the insect testis—a possible function. *Travaux de la Station Zoologique de Wimereux* 13:437-444.

PART THREE

Final Considerations

13 Interpretive Issues
in the Evolutionary Synthesis

In the prologue Ernst Mayr raised many basic interpretive issues related to the evolutionary synthesis, including such questions as: Was the evolutionary synthesis a scientific revolution? In what sense was the evolutionary synthesis really a synthesis? Was Marxist ideology important in the synthesis? Could the synthesis be understood as primarily an internal development within biology, or were external influences too important to be neglected? Such questions arose constantly during the conference, and some have already been incorporated into previous sections. Anticipating the rise of general philosophical and interpretive questions, Mayr invited, in addition to scientists, established historians like William Coleman and John Greene, and philosophers Dudley Shapere and David Hull to participate in extended presentations on general interpretive issues.

I tentatively raised the question of whether "evolutionary synthesis" adequately conveyed an accurate sense of what occurred in evolutionary theory in the 1930s and 1940s. I argued that "synthesis" meant quite different things when applied to the coming together of Mendelian genetics and Darwinian natural selection on the one hand, and Mendelian genetics and paleontology on the other. The regularities of Mendelian inheritance enabled Fisher, Haldane, and Wright among others to say what selection of various intensities might do to gene frequencies in a population. Making certain assumptions about natural populations, these thinkers could even make predictions about what naturalists might find in the field.

No such entailment held in the synthesis of Mendelian heredity and paleontology, however. All that Simpson (1944) could argue was that Mendelian heredity was consistent with what was known of the fossil record. Thus the synthesis between Mendelism and Darwinian selection appeared more basic than that between Mendelism and paleontology.

I also suggested that the evolutionary synthesis appeared to be all of a piece because in the 1930s and 1940s the Darwinian theory of natural selection operating on small heritable differences was believed to explain data from all fields of evolutionary biology. Yet the evidence to detail this belief was unavailable at that time. The synthesis between Mendelism and Darwinian selection had been extended to cover all of evolutionary biology. What was involved here was the sociology of knowledge—how did biologists come to believe in an unproved general hypothesis? Shapere discusses the meaning of the evolutionary synthesis and other issues in this chapter.   W.B.P.

The Meaning of the Evolutionary Synthesis
Dudley Shapere

Possibly the problems we face in defining the evolutionary synthesis result from an oversimplified way of looking at scientific innovation. Questions seemed to be posed in terms of two mutually exclusive, opposing alternatives: on the one hand, some kind of positive unification of two different areas, and on the other hand, some kind of negative removal of barriers to such a positive unification. Perhaps a positive unification implies the achievement of a "complete" synthesis in the following sense: that all the (correct) data of each of the scientific areas in question must be logically deducible from a relatively few basic ideas of a synthesized Darwinian-Mendelian theory. Such a view of "completeness" is, of course, stringent to the point of being absurd. In effect, it ensures that no scientific theory has ever been complete. (After all, when are "all" the data of a scientific field in and certified as correct? Must every one of them be deduced, in the strictest sense, without any intervening simplifications or approximations, in every minute detail?) That definition of a "complete" synthesis (a synthetic theory) can never be satisfied. For the issue is decided in advance by the criterion, and only seems to be a debatable issue because the criterion has not been made explicit.

Other criteria of completeness of a theory might be formulated, more relaxed criteria that would allow us to say that certain theories are complete or that would define realizable circumstances in which they might achieve that status. For example, we might demand only that present data, or some clearly circumscribed set of data for which the theory is held responsible, be deducible, though simplifications, approximations, and other such devices would be allowable. Still other criteria or variations on these might be suggested. Although we need not select, in any final way, some one of these criteria as the correct one, we must be clear about what criterion is employed in any particular discussion, and about the points brought out by employing that criterion, and, finally and no less important, about the points obscured by employing that criterion.

On the other hand, one might have something entirely different from completeness in mind in talking of the attainment ("achievement" in that sense) of a theory, synthetic or otherwise. For example, a theory, or a synthetic theory, might be thought of as achieved for a certain body of data if we have obtained an understanding or a way of understanding those data. This achieved understanding might consist in the possibility of raising questions about those data that were not raised or taken seriously before, of thinking about those data in new terms, of developing a new vocabulary for those data that will give insight into distinctions, relations, and prior misunderstandings. In this sense the achievement of the new theory might not at all be complete: there might still be much left to be done, but a new program of research, a new way of looking at a certain area of science, would have been arrived at.

Looking at the episode in this way brings out the possibility of that accomplishment having positive aspects even though it is not complete: such an achievement, while not the development of a completed deductive synthesis (and perhaps not complete in any other useful sense), is nevertheless more than a mere negative removal of barriers to the ultimate accomplishment of such a deductive synthesis. It is, on the contrary, a positive reorientation that affords fruitful insights which may ultimately lead to the achievement of a theory that is more deductively unified, for example. And in the case of the synthetic theory of evolution, perhaps what we ought to be asking is not whether paleontology (or any other discipline) was brought into some tight deductive unification with Darwinian-Mendelian theory but rather to what extent the removal of barriers to seeing paleontological data in Darwinian-Mendelian terms, and the consequent application of the latter terms to those data, reoriented the kinds of questions asked, the kinds of research engaged in, the expectations about the sorts of things one should expect to find, and so forth, in paleontology and other fields.

We have debated not only historical points but also more general conceptual ideas, such as the meaning of "synthesis"; the role of mathematics in science; whether a certain achievement should be called a theory; the extent to which a theory can be refuted by observation, and the extent to which it is reasonable to "bend" the theory or the observations to permit the former to account for the latter; the notion of "experiment" and the clarity of the distinction between "experimentalist" and "naturalist"—that is, the utility of those ideas as tools for interpreting certain historical developments; the exact interpretations of certain concepts central to the synthetic theory—for example, "population" and "evolution"; the short-comings of "inductivism" in science; "functional" versus "nonfunctional" formulations of biological propositions; and the concept of an individual.

All of this brings out a rather well-known point: namely, that an attempt to gain understanding does not consist merely in gathering together "the facts," in this case, historical facts; understanding also requires clear and unambiguous conceptual tools for describing and analyzing those facts; understanding consists in a combination, and even more an interaction, between facts and concepts. Despite the lip service that is commonly paid to this point, many people seem to think that, as far as concepts go, either they are already clear and unambiguous, or else all one has to do about them is to lean back in one's armchair, look up at heaven, and think. Things are not that easy: for each topic in my list, there is a vast philosophical literature discussing alternative views and interpretations. It is not necessary to completely master that literature before entering into discussions, but the development of a sensitivity to those issues and alternatives can be ignored only at peril. Otherwise, our understanding of the facts before us is very liable to be left in obscurity. In this respect, philosophy of science can play an essential role in the attempt to understand such developments as the evolutionary synthesis. (By "philosophy of science," of course, I mean a kind of analysis that is really relevant to the above issues; and that excludes a great deal of the literature that goes under the title of philosophy of science.)

The word "synthetic" is the operative term in the attempt to bring out the achievement of the synthetic theory: that "theory" is said to have brought about a synthesis or unification of ideas or information of some sort from a large number of different domains such as genetics, zoology, botany, paleontology, and cytology.

There are syntheses and syntheses; what is it, exactly, that is brought out by calling this particular episode in the history of science a synthesis? Provine suggested that the word "synthesis" here is at least misleading and perhaps even inappropriate. He suggested that what was accomplished was not a positive or logical unification of these different areas, but rather only a negative achievement, the removal of barriers to the ultimate achievability of such a synthesis. It was, he claimed, a demonstration that (for example) the data of paleontology are not inconsistent with an evolutionary theory based on natural selection. Perhaps Provine's dichotomy is misleading—although the synthetic theory was a "negative" achievement in that sense, it was also more than that. At the very least, even in removing barriers to "logical unification," it provided a positive program, a way of reinterpreting data, of raising questions and constructing hypotheses and lines of experimental research.

But even describing the "synthetic" achievement of the synthetic theory in this way, as the development of a research program, or—to use Conant's words—of a "policy" (rather than the adoption of a "creed"), can hinder and obscure a full understanding and appreciation of that achievement. Let me bring this out by considering a few of the many different types of cases in the history of science that are described as syntheses or unifications.

There are a number of different sorts of cases in which bodies of *data* (or what are considered to be data) are related or unified, though there is still the need for an explanatory *theory* of those bodies of data; and the "unification" or "synthesis" consists in providing grounds for supposing that such a unified explanation can be found, and should be sought, for those data. For example, Oersted's experiment relating electricity and magnetism provided at least a strong reason for thinking that a unified explanation of those two domains could be expected. The reason was not conclusive, of course: there still remained objections to the possibility of a unified theory. Again, Sutton provided point-by-point correlations between Mendelian genes (taken not as "theoretically explanatory" of breeding data, but as themselves data to be explained) and cytological data (regarding chromosomes) that constituted very strong grounds for expecting a unified account of those bodies of data. There are differences between these two cases as regards the sorts of reasoning leading to the postulated unifications. Oersted's reasoning was "causal" (an electric current can produce a magnetic effect) in a sense that John Stuart Mill came close to capturing; Sutton's reasoning may be called "correlational" or "structural" (similarities between characteristics and behavior of chromosomes and allelomorphs). In both cases a "unification of domains" is achieved: reasons have been provided for supposing that there is a deeper (explanatory, theoretical) relationship to be found between two bodies of information hitherto supposed to have been independent (or whose suspected relationship has hitherto been inadequately supported). Such unifications may be called "domain syntheses." They can, of course, be more or less "complete" in the sense that the reasons for unification are more or less convincing. (There are also other senses in which domain syntheses can be said to be more or less "complete.")

Other theoretical syntheses or unifications in science have various sorts and degrees of strength. Some examples are Newton's synthesis of planetary astronomy and terrestrial mechanics, Maxwell's synthesis of electricity, magnetism, and optics, and Bohr's synthesis of atomic theory and spectroscopy. These may be called "deductive" or "logical" syntheses or unifications—though these words, too, have their dangers (the most important of which is that they may lead us to ignore the occurrence in such unifications of approximations, simplifications, and other conceptual devices). In some cases, the laws found to hold in two or more different areas are shown to follow from a small set of assumptions; in other cases, the vocabulary of one area is reinterpreted in terms of, and its laws shown to follow from, the fundamental propositions of a theory of another area. In still other cases, an entirely new theory of the interrelationships of two hitherto distinct areas is proposed.

It is fair to say that the synthetic theory is not precisely comparable, in respect to the achieved synthesis, to any of these sorts of cases. A central aspect of the theory was the unification of Mendelian genetics and Darwinian evolutionary theory, but there was no logical deduction (even in the looser sense described above) of one from the other. Nor was it a case of deduction of both those theories from deeper assumptions, nor was it achieved by anything that could be clearly called the postulation of a new theory regarding the relationship of evolution and genetics. Yet the unity achieved was far stronger than a mere "domain synthesis"; for example, the unification was tighter than that achieved by Sutton, which only strongly indicated the possibility of a unification. One central aspect of the case is the way that, once the barriers to reconciliation had been removed, Mendelian genetics and Darwinian evolutionary theory complemented each other. Darwinian evolutionary theory required supplementation to account for transmission of characteristics and the existence and preservation of variation, whereas Mendelian genetics said nothing about the long-term effects of variation. Thus each did something important for the other, by taking care of a certain "theoretical incompleteness" in the other. (Previous attempts had been made—for example, to supplement this weakness in Darwinian theory—but they had not been successful.) Such a mutual "plugging-in" of entire theories from different domains is rather unusual, in at least two respects. First, taking care of a theoretical incompleteness is usually a less dramatic enterprise, involving the supplementation of a theory by a single proposition or small number of them rather than by a whole theory of a different domain. Second, when two theories become unified, it is not common for both theories to do so much for each other—one usually does more explanatory work than the other. Of course, none of this means that all the work that needed to be done was accomplished by the synthesis.

As to the other fields concerned in the synthetic theory's unification—such as zoology, botany, paleontology—the synthesis seems to have been of a different character from that achieved with genetics and evolutionary theory. (We need not, after all, expect the same sort of synthesis to have been achieved in the case of all the fields concerned.) It is with regard to these, perhaps, that the case seems strongest for saying that the

Mendelian-Darwinian reconciliation provided a research program. So let us look at this way of characterizing the situation: granting that the synthetic theory accomplished more than a negative removal of barriers, can this historical achievement be characterized adequately as the introduction of a new research program in science? A paradigm case of such a research program was the adoption of Newtonian mechanics as a program or model for other areas of science, including biology, in the eighteenth and nineteenth centuries. It was as though people in various areas of science acted on the following suggestion: "Let's try to consider our subject [such as life processes] as ultimately matter in motion and exerting forces." And this can be called a "research program" or a "policy," in the sense that it is a recommendation on how certain areas of investigation should be approached—kinds of questions to ask, kinds of answers to look for, kinds of research to engage in—as opposed to a substantive proposition or creed. Grammatically, the statement is more like an imperative statement than it is like a factual claim (proposition, assertive statement); and referring to it as the statement of a "program" or "policy" emphasizes this characteristic.

But, I gather, a program of this sort is very different from the kind of thing that was brought about by the synthetic theory. In the latter case, attempts to move in the direction of a synthesis were faced with a number of specific arguments, within each subject area, that seemed to militate against the possibility of synthesis of that area with the others. These arguments included the distinction between two types of variation; the supposition that there are major saltations that play a fundamental role in evolution; the hypothesis of the inheritance of acquired characters; the absence of continuous gradations between species as revealed by the fossil record; and the absence, or at least extreme rarity, of laboratory mutations that could be unambiguously interpreted as adaptive rather than deleterious. In rejecting these general propositions, the founders of the synthetic theory had to undermine the specific evidential bases of those claims before the areas could be brought together. When those objections had been removed, moreover, there remained clear and specific reasons within the fields that *called for* a synthesis. Ernst Mayr, among others, has directed attention to considerations in the various areas brought together which called for populational rather than typological thinking.

All these events are very unlike the employment of Newtonian mechanics as a research program in the eighteenth and nineteenth centuries. It is of course true that opponents of that program raised objections, of various levels of specificity, against the applicability of Newtonian theory to their field; and it is also true that the proponents tried to advance arguments in favor of its applicability. Nevertheless, the overwhelming reason people advocated the Newtonian program in those other fields was the success of Newtonian mechanics in the domains of astronomy and mechanics, and not any call of evidence from the domains of biology or electricity or magnetism and so forth. In this sense the synthetic theory was more than a mere program. It was not simply imposed on a number of fields from without, as Newtonianism was imposed on biology from without in the eighteenth century. To the extent that it was a program, that program was adopted, and even, to some extent, shaped, in the light of considerations intrinsic to the various fields concerned (or at least some of them), considerations that made adoption of the program for that field seem reasonable. To call the synthetic theory, as applied to those areas of investigation, a "program" or a "policy" or a "proposal" or a "paradigm" threatens to call attention away from the substantive reasons why it was a promising program, and even a program that was called for. (Of course, "sociological" factors were also involved in bringing about the synthetic theory of evolution, as they were in the enthusiasm about the universal applicability of Newtonianism. And such factors, in the case of the synthetic theory, have been cited repeatedly. With all this interest in institutional, social, psychological, and educational factors, we must not forget that the synthetic theory was primarily a scientific achievement, and it is the scientific factors, the scientific reasons involved, that must be our primary concern if we are to understand that achievement.)

Certainly the synthetic theory consisted partly of negative removal of barriers and misunderstandings; it also functioned as a research program in various ways, in various fields. But it was more than an imposed attitude too: there were good reasons for its adoption as a belief-governing policy. (Grammatically, I am here emphasizing the assertive, and not merely the imperative, aspects of the program.) All this is to say that the usual ways of thinking about scientific change and innovation do not quite capture the entirety of the achievement of the synthetic theory of evolution, which in turn suggests that there is something special to be learned about the nature of scientific change and the scientific enterprise from this case that cannot be learned from most other cases.

The synthetic theory is a very interesting and important case for understanding the processes of scientific development, partly because so many fields were involved. Probably the nearest thing to it, in terms of number of areas concerned, was the unification of electricity, magnetism, light, and certain aspects of chemistry that came about through the ideas of Faraday and Maxwell. A comparison of that case with the evolutionary synthesis would reveal many similarities and some important differences. A more complete understanding of the processes of scientific unification might come out of such an analysis.

I should like to comment on three ideas that have come up in our discussions: the notion of synthesis or unification, the relation between syntheses and scientific revolutions, and the concept of scientific explanation. "Synthesis" is one of those words that involve what philosophers call a "process-product ambiguity": that is, in using the word "synthesis," one is sometimes talking about the process of synthesizing and at other times about the product that results from that process. Although the synthetic theory usually is spoken of as the product of a certain scientific development, questions about both the product and the process have been raised. These questions need to be distinguished explicitly.

First, regarding the process through which the synthetic theory was arrived at, questions have been raised about the contributions of particular fields (such as paleontology) to the emergence of that theory. Did paleontology, for example, contribute to the synthesis (that is, the synthesizing, the thinking that led to the synthetic theory), and if so, exactly how? In what way or ways did it contribute evidence that seemed to call for such a theory? What paleontological problems existed that suggested or even demanded that ideas of a certain sort (perhaps ideas already prevailing in another field) be brought into play? Were there any general ideas that paleontology contributed to the theory? In general, such process questions should be raised about all the areas concerned: questions as to whether there were problems in a particular field calling for something like the synthetic theory, whether there was in that field evidence suggesting the synthetic theory or some aspect of it, whether there were general ideas in that field that contributed to the formulation of the synthetic theory, and whether there were ideas current in other fields that people in (or out of) the field thought might be applied in their field.

Of course, other questions can be asked. Gould, for example, raised an entirely different point about the ways in which various fields might have contributed to the development of the synthetic theory. He suggested that, although paleontology did not contribute in any of the above ways to the development of the theory, at least people thought that the data of paleontology should be taken into account in the development of that theory. That is, people thought that an evolutionary theory was responsible to the data of paleontology; but precisely because certain of its data (lack of continuity in the paleontological record) challenged the standard evolutionary theory, paleontology contributed to the feeling that a better theory had to be sought. Such a consideration, of course, tells us nothing about how to construct such a theory: those sorts of indications have to have come from elsewhere.

The most fascinating discussions have been those concerning the product arrived at—that is, what was achieved or accomplished. The case of the synthetic theory brings out clearly the woeful inadequacy of our common stock of ideas for talking about scientific development—ideas that philosophers have become fixated on, as we all have, so that they blind us to what actually goes on in scientific thought. First, philosophers (and scientists as well, though not always so explicitly) have tended to think of scientific syntheses or unifications as relations between theories. The term "theory" is distressingly vague; talking about scientific achievements as theories often seems like using a screwdriver with a round point, frustrating rather than assisting us in our work. Was Mendelian genetics a theory? In what sense? Someone might think it was only a brief summary of certain facts, and not a theory at all. Apart from this vagueness, much more is involved in major scientific syntheses than mere theories, in any reasonable intuitive sense of that word. In some areas of concern to us, it is hard to point to some general theory that held sway in a particular field that was brought into the scope of the synthetic theory. It was more like bringing the "facts" of a certain field under the purview of a more general theory—not at all a bringing together of two different theories. There are relations to be considered between techniques or methods, between previously separate bodies of evidence, problems, explanatory concepts, and so forth; and the relating of these is not necessarily identical with establishing relationships between theories.

Second, philosophers (and sometimes scientists) have limited themselves by thinking that the only kinds of relations of unification between theories are relations of reduction, where an old theory is absorbed into (its concepts defined in terms of, and its postulates deduced from) a newer or more general one. In the evolutionary synthesis, however, there was a mutual modification or supplementation of different theories (and, more generally, of the concepts, techniques, problems, and so forth, of different areas).

Once the scope of the problem is broadened beyond relations between theories and beyond relations of reduction, there still remain ambiguities. What do we think of when we talk about achieving a synthesis in science? One possibility is that the outstanding problems of two different areas are resolved by one synthetic overview. Sometimes, in science, this is not what happens; yet we still call the achievement a synthesis. For example, the suggestion has been made that, for paleontology, the achievement of the synthetic theory lay in its drastic and fundamental modification of the kinds of problems that paleontologists dealt with. This achievement is very different from actually *solving* problems that were already existing in the field. Perhaps that can be said to have been done for Darwinian evolutionary theory by Mendelian genetics, which provided an answer to the standing problem, in the former theory, of the transmission and preservation of variations; the achievement there was,

in a sense, a completion of what was incomplete before. But if new problems, new ways of looking at the data, new ways of interpreting the data,
new approaches to research are introduced, as is alleged to have happened in paleontology, then it is more like a new beginning than it is like
a completion. (We should be also careful about these words "beginning"
and "completion," however.) One is tempted to think here in terms of a
revolution, a new order, in paleontological thinking. It was nonetheless a
synthesis, a bringing together, of ideas of different fields. But we must
not forget that it was a different sort of bringing together than happened
between evolutionary theory and genetics. In the latter case, the bonding,
in important part at least, consisted in a mutual completion; but the relations between that synthesis and paleontology seem to have been more
like an "application" of that synthesized outlook to this other area. Or
perhaps it is better described as providing plausibility arguments to show
that the ideas of the synthetic theory of genetics and evolution might well
apply here also, and should be used as the bases of a reformulation of the
problems, methods, and expectations of the field. (In these connections,
however, we must remember that the synthesis achieved between genetics and evolutionary theory was much more than a mere conjunction of
two complementary theories; it also injected into that conjunction a
number of new ideas and approaches, most notably, perhaps, populational thinking. In such recasting of modes of thinking in those areas, the
synthetic theory was no less revolutionary in genetics and evolutionary
theory than it was in paleontology.)

We must also ask whether the synthesis was of still a different character in still other fields. Furthermore, we must ask whether the synthetic
theory was conceived in the same way by different people, particularly
those working in different areas. As Mayr asked, did such people as
Rensch and Simpson interpret the synthetic theory in the same basic
ways?

We tend to think of unification and explanation together: a synthesis
or unification in science is supposed to be automatically a better or more
complete explanation than we had before; our problems, or at least many
of them, tend to disappear when a scientific synthesis has been achieved.
But is this necessarily true? It has been pointed out that embryologists
gave comparatively little attention to the synthetic theory. They were
willing enough to admit that embryological processes could be understood or accounted for in some general sense in evolutionary terms.
Nevertheless, they thought (presumably on good grounds) that specific
questions of crucial interest to them would not be answered merely by
bringing embryology within the scope of the synthetic theory; that
theory was simply irrelevant to their most pressing concerns. This think-

ing suggests that the intuitive connection between synthesis and explanation must be looked at more critically; for the case of embryology seems to provide evidence that, if a field is brought under, or is conceived as plausibly explainable in terms of, some other theory, it does not necessarily mean that all the questions, or even the most important questions, of that field will be answered by bringing it under this larger viewpoint. What is accomplished by a synthesis or unification may not automatically remove all or even the crucial questions already existing in a field.

In general, then, there is much to be learned from an analysis of the case of the synthetic theory of evolution; and we must not only learn about the facts, but also, in the very act of studying them, forge the conceptual tools that will help us to deal with them. Because the tools passed on to us by our philosophical heritage seem crude and inadequate, the examination of the case must go hand in hand with a more general analysis of the nature of scientific reasoning and its achievements.

There seem to me to be at least three essential components of the evolutionary synthesis, corresponding fairly closely to the historical sequence of events: the removal of objections to the compatibility of Darwinian evolutionary theory and Mendelian genetics, and the demonstration that those two sets of ideas are mutually supportive (the bases of this aspect were already laid by Bateson, Fisher, Haldane, and Wright); the extensive and detailed casting of those sets of ideas in terms of new modes of thought, particularly populational thinking (though again the roots of this aspect had been laid earlier, it seems to have been primarily Dobzhansky's 1937 *Genetics and the Origin of Species* that was recognized as having accomplished this); and the demonstration that various other fields—including paleontology, zoology, systematics, botany, among others—are not incompatible with the ideas brought together in the first and second aspects, but on the contrary could be understood in terms of, and in some cases even called for, those ideas (Simpson, Rensch, Mayr, Stebbins, among many others). The concepts of "understanding" and "calling for," of course, as well as other central concepts in the interpretation of the case, require further analysis along the lines I have indicated in my earlier remarks—perhaps somewhat different analyses for each of the different fields concerned. However, this picture of the synthesis is almost certainly incomplete and oversimplified and subject to revision.

14 Epilogue

William B. Provine

One certain conclusion emerged from the conference. All participants, whether scientists or historians, young or old, agreed that a consensus concerning the mechanism of evolution appeared among biologists during the 1920-1950 period. Darwin, despite all his influence, was unable in his lifetime to produce a corresponding consensus, and none coalesced until the second quarter of the twentieth century. Whatever it is called—evolution, the modern synthesis; the evolutionary synthesis; or twentieth-century Darwinism—every participant agreed that a comprehensive and compelling view of the mechanism of evolution appeared during this time. For the sake of simplicity, I will hereafter refer to this development as the evolutionary synthesis, despite the flexibility inherent in the term "synthesis."

The evolutionary synthesis is unquestionably an event of first-rank importance in the history of biology. Although Darwin's theory of evolution by natural selection had been widely known since 1859, the consequences of the theory had generally not been incorporated deeply into most areas of biological thought. With the evolutionary synthesis, and the acceptance of natural selection operating on small differences as the primary mechanism of evolution, evolutionary theory began to permeate almost all of biology with new meaning.

When compared to other obviously important developments in recent biology, such as the revolution in molecular biology or sociobiology, the evolutionary synthesis is more basic to the rise of modern biology. Darwin never doubted that a molecular substructure underlay his theory of evolution by natural selection, or that psychology and behavior were strongly influenced by the same mechanisms that guided the evolution of physical characteristics, particularly in heredity and development. Molecular biology is important precisely because it reveals mechanisms that govern processes known to be important for evolution. Just as Darwin's theory of evolution by natural selection was the most important development in nineteenth-century biology, I suspect future historians will point to the evolutionary synthesis as the primary development in twentieth-century biology.

The influence of the evolutionary synthesis is so pervasive and fundamental to all of modern biology that the origins of the synthesis are shrouded in obscurity. Twentieth-century Darwinism resembles Darwin's views in 1859 sufficiently closely that the 1859-1929 period appears dominated by ignorance and inability to understand what Darwin had said. In modern textbooks the gap between early Darwinism and twentieth-century Darwinism is filled only by the rediscovery of Mendelian heredity, which solved the problem of heredity. The evolutionary synthesis somehow seems to have been a part of biology for a long time, almost since Darwin.

Molecular biology, on the other hand, seems new. And molecular biologists and historians have produced a sizable literature on the origins of molecular biology. Ranging from Watson's *Double Helix* (1968) and Stent's *Coming of the Golden Age: A View of the End of Progress* (1969) to Olby's *Path to the Double Helix* (1974), the literature examines the revolution in molecular biology from many angles. One of the very first publications on the historical analysis of molecular biology contained the recollections of almost every major participant (see Stent and others, 1966). Yet, until now, the evolutionary synthesis has received almost no attention from the scientists who participated in it.

Past Scholarship concerning the Evolutionary Synthesis

One might hope to find in the major works of the synthesis an explicit analysis of the historical origins of the synthesis—that is, by such authors as (listing chronologically): Fisher, *Genetical Theory of Natural Selection* (1930); Wright, "Evolution in Mendelian Populations" (1931); Ford, *Mendelism and Evolution* (1931); Haldane, *Causes of Evolution* (1932); Dobzhansky, *Genetics and the Origin of Species* (1937); Darlington, *Evolution of Genetic Systems* (1939); Huxley, ed., *The New Systematics* (1940); Mayr, *Systematics and the Origin of Species* (1942); Huxley, *Evolution: The Modern Synthesis* (1942); Simpson, *Tempo and Mode in Evolution* (1944); White, *Animal Cytology and Evolution* (1945); Rensch, *Neuere Probleme der Abstammungslehre* (1947); Jepsen, Mayr, and Simpson, eds., *Genetics, Paleontology, and Evolution* (1949); Stebbins, *Variation and Evolution in Plants* (1950); and Huxley, Hardy, and Ford, eds., *Evolution as a Process* (1954).

Even a rapid glance at these works reinforces the well-known assertion by Kuhn (1962) that the actual historical origins of a scientific field generally will not be found in the major books that embody the fundamental beliefs of the field. None of these books provides even a preliminary historical analysis of the origin of the synthesis, except insofar as earlier works are cited to show how they led logically to the modern view. Thus the history in these works is chronological-logical, and makes no pretense to being a historically accurate view of the origin of the synthesis. These works are fundamental to any historical analysis of the evolutionary synthesis, but they do not explicitly indicate how the synthesis originated.

As early as 1950 some participants in the synthesis began to look back and suggest how this important development occurred. The Golden Jubilee of Genetics meeting in 1950 produced many papers on the history of genetics (published in Dunn, 1951). The essay of greatest interest is Dobzhansky's "Mendelian Populations and Their Evolution." This paper is not really a historical essay, but it contains a sketchy outline of how the synthesis had proceeded. Dobzhansky basically argued that Darwin had discovered the true explanation for the origin of species in his mechanism of natural selection of relatively small differences. But Darwin's theory had a major drawback: it depended on the theory of blending inheritance, which guaranteed that any available supply of heritable variations in a population would be blended away, eliminating half of the variance in each generation on an average. The rediscovery of Mendelian inheritance supplied just the right key to resolve this problem about the supply of heritable variations. A direct mathematical consequence of Mendelian heredity was that segregating genes were maintained at constant frequencies in large, random breeding populations with no selection (Hardy-Weinberg equilibrium). The work of the Morgan school elucidated the processes of mutation and their relation to cytology. Then came the synthetic work of Chetverikov, Fisher, Haldane, and Wright, who showed theoretically the relations between genetics and evolutionary processes. Dobzhansky concluded that these thinkers "may be considered founders of the modern analysis of evolutionary phenomena" (Dunn, 1951, p. 575).

In the 1955 Cold Spring Harbor symposium on population genetics Dobzhansky was more explicit:

> The foundations of population genetics were laid chiefly by mathematical deduction from basic premises contained in the works of Mendel and Morgan and their followers. Haldane, Wright, and Fisher are the pioneers of population genetics whose main research equipment was paper and ink rather than microscopes, experimental fields, *Drosophila* bottles, or mouse cages. This is theoretical biology at its best, and it has provided a guiding light for rigorous quantitative experiment and observation (1955, p. 14).

Dobzhansky's view fit perfectly with the writings of Fisher, Haldane, and Wright. In 1930, Fisher outlined a view very similar to that presented by Dobzhansky in the 1950s. I suspect the difference was that Fisher believed that he was presenting no more than a logical analysis, whereas Dobzhansky believed he was presenting a historically accurate view. One conclusion is certain: most geneticists in the 1950s accepted a view similar to Dobzhansky's. They believed that the evolutionary synthesis was a function or product of advances within the field of genetics and that these advances were then applied to other fields like systematics, paleontology, embryology, cytology, and morphology. Almost all textbooks cited Fisher, Haldane, and Wright as the cofounders of modern evolutionary theory.

Surprisingly, only Waddington and Mayr from fields other than genetics challenged this dominant view. Waddington (1953) suggested that attributing modern evolutionary theory to Fisher, Haldane, and Wright was a distortion of the historical facts. The only response to Waddington's challenge was Mayr's extension of it in his introductory address at the 1959 Cold Spring Harbor symposium on genetics and twentieth-century Darwinism. With geneticists all around, Mayr explicitly challenged the view that geneticists, especially Fisher, Haldane, and Wright, were entirely responsible for the evolutionary synthesis (1959). Wright and Haldane responded, but neither furnished much historical evidence to support the view that Mayr had challenged.

In the mid-1960s a number of works detailing the history of genetics appeared. Those by geneticists included Dunn's *Short History of Genetics* (1965), Sturtevant's *History of Genetics* (1965), and Wright's "Foundations of Population Genetics" (1967). All three works, to the extent that they analyzed the connection of genetics to the evolutionary synthesis, adopted the view that advances within genetics, particularly the theoretical work of Fisher, Haldane, Wright, and sometimes Chetverikov, contained the key discoveries that led to the synthesis. Dunn argued that Dobzhansky's *Genetics and the Origin of Species* "showed how the gap between theory and observation could be closed and brought population genetics the recognition not only of geneticists but especially of biologists of all kinds. Thereafter, the basic conception of the gene pool, composed of all the alleles, many of them differentiated into wild-type and mutant, in a Mendelian population, became the model on which evolutionary studies were based. The extent to which such ideas and methods have transformed evolutionary biology can be seen in Ernst Mayr's *Animal Species and Evolution* (1963)" (1965, p. 206). Dunn had read Mayr's book carefully and even reviewed it for *Isis*. The same general view may be seen in both Sturtevant's and Wright's accounts: the real advances took place in genetics and were exported to other fields of evolutionary biology, thus creating the evolutionary synthesis. Any person searching the literature about the evolutionary synthesis would come away with the impression that the issue was settled. Advances within genetics had led directly to the evolutionary synthesis.

Yet in none of these comments about the evolutionary synthesis was any detailed evidence cited in favor of the general interpretations. Dunn, Sturtevant, and Dobzhansky might be historically correct in their views, but they did not give sufficient supporting evidence. On the other hand, Mayr and Waddington might be correct in their views, but they too had presented insufficient evidence. In scientific circles there the issue stood until Mayr organized this conference.

The basic textbooks of the history of biology, Nordenskiøld's *History of Biology* (1927) and Singer's *Story of Living Things* (1931), were both written too early to give an account of the evolutionary synthesis. Indeed, both books exemplify typical attitudes toward Darwinism before the synthesis. Singer exhibits an intense dislike of the idea of natural selection and clearly does not understand it; Nordenskiøld devotes as much space to neo-Lamarckism as neo-Darwinism, and argues that modern genetical investigations are largely divorced from larger evolutionary questions. Both books are helpful for understanding presynthesis attitudes, but they do not help in understanding the synthesis itself.

Three works from the 1950s dealing with the history of evolutionary theory included the synthesis: Zimmermann's *Evolution: Die Geschichte, ihrer Probleme und Erkenntnisse* (1953), Carter's *A Hundred Years of Evolution* (1957), and Fothergill's *Historical Aspects of Organic Evolution* (1952). All three contain sections on evolution in the 1930s and 1940s. Carter and Zimmermann are too general and superficial to be of much use. Fothergill's account is more comprehensive and contains a useful bibliography, but his book suffers from an overzealous belief in the inheritance of acquired characters. His book is most useful for a sympathetic account of neo-Lamarckians. All three accounts were too early to use the materials from the 1959 Darwin centennial, and none utilizes the growing body of unpublished documents now available.

Newer work by historians of science drawing on rich unpublished source materials has slowly begun to appear. Adams has written on genetics and evolution in Russia; Allen on Morgan and Goldschmidt; Carlson on Muller; Olby on the rise of molecular biology; and Provine on the origins of theoretical population genetics. These specialized studies are important and useful, but none contains an analysis of the evolutionary synthesis. They point out clearly the great need for a broader framework in which to view the complex relationships involved in the evolutionary synthesis.

This need was explicitly noted by Mayr in "Recent Historiography of Genetics" (1973). He pointed out that the evolutionary synthesis was far more complex and many-stranded than literature on the history of genetics indicated. He made this point specifically in reference to my *Origins of Theoretical Population Genetics* (Provine, 1971). I distinctly recall agreeing with Mayr on this point, but assuming that I probably would not be working in that area for many years, if at all.

Contribution of the Conference on the Evolutionary Synthesis

The near vacuum of detailed historical analysis combined with the widely held interpretation that Fisher, Haldane, and Wright had accomplished the evolutionary synthesis (a misleading view, in his opinion) led Mayr to organize this conference.

Much more occurred during the conference than is recorded in this volume. Mayr sent a particularly revealing questionnaire to each prospective participant who had been part of the synthesis. The answers to these questionnaires are very helpful to anyone wishing to study this period in depth, especially since some people, like Rensch, Huxley, Astaurov, Simpson, Dunn, and Stern, fully or partially completed the questionnaire but could not attend the conference. Mayr (and I, to a much lesser extent) corresponded with the participants, creating a very valuable communications file. And the audiotapes of the conference provided much material not included in this volume. All of these documents —the questionnaires, correspondence, and tapes and transcripts—have been deposited at the Library of the American Philosophical Society in Philadelphia.

During the conference participants talked incessantly during breaks and meals. I learned a tremendous amount in conversations with Mayr, Dobzhansky, Ford, Darlington, Stebbins, Boesiger, Lerner, Weinstein, and others. These conversations were one important reason why many of the younger historians of science attending the conference have since turned their major research interests to topics connected with the evolutionary synthesis. For example, during one break I was speaking with Dobzhansky about his relationship with his friend Dunn, who had died only a few weeks before. After a very emotional story about Dunn's generosity to his colleagues and students, Dobzhansky suggested that I contact Dunn's widow, Louise, about Dunn's reprint collection, which he feared might be thrown away or otherwise disposed of. Through Dorothea Bennett, Dunn's collaborator, I reached Mrs. Dunn. Her late husband's office at Columbia's Nevis Biological Research Station was being dismantled, and Mrs. Dunn kindly turned the reprint collection over to me. The collection has been enormously useful to my research and very helpful to many of my students and colleagues.

General Contributions

The evolutionary synthesis was a very complex process; its historical development cannot be encompassed accurately by any simple thesis. The synthesis occurred on many levels. Some investigators were mostly concerned with phenomena at the chromosomal level in individual organisms, whereas others examined entire populations in nature and yet others looked at macroevolutionary phenomena. The evolutionary synthesis was more than a simple application of new concepts in genetics to other facets of evolutionary biology, as earlier accounts have suggested. Furthermore, the synthesis proceeded at very different rates in different fields, or in the same fields but different countries, or even in the same fields in different divisions of the same university. Accurate descriptions of such phenomena must take account of the specificity of detail involved. No simple historical thesis, however brilliant, can describe all essential elements of the evolutionary synthesis.

Contrary to the view found in many textbooks that the modern synthesis flowed easily from Darwin's views once Mendelism was added to them, the conference proved conclusively that Mendelism and Darwinism were synthesized only with considerable difficulty. Many evolutionists did not perceive the very successful science of experimental genetics as the essential key to a sound evolutionary theory. Many of the participants in the synthesis revealed during the conference that their teachers, colleagues, and even they themselves held views quite incompatible with the evolutionary synthesis in the late 1920s and well into the 1930s and even 1940s. Such views were widespread among influential biologists. The other side of this conclusion is, of course, that the evolutionary synthesis occurred as a specifiable development. I think every participant in the conference agreed with this conclusion, even if the evolutionary synthesis meant different things to each one. To begin with, then, the conference demonstrated that the evolutionary synthesis occurred, and that it was a very complex development.

Mayr's Contribution

Mayr opened the conference by presenting the first substantive, comprehensive interpretation of the evolutionary synthesis. Any subsequent attempt to trace the evolutionary synthesis must begin with his keynote address (see the prologue to this volume), which encompasses evolutionary theory from the late nineteenth century up through the 1940s. Mayr's address unified the whole conference because every participant heard it or read it, and many responded to Mayr's views in their own presentations. Thus Mayr played a crucial role during the conference, and his opening statements will continue to be the cornerstone of future historical research on the synthesis. Of course, not all participants agreed with Mayr's views. Many disagreements, mostly of interpretation, appeared during and after the conference, but they simply indicate the centrality of the keynote speech in the results of the conference.

Contributions of the Various Biological Disciplines

Considered in the light of scholarship about the evolutionary synthesis before this conference, the novel substantive contributions of the selections concerning the roles played by various fields in the synthesis are obvious to even a cursory reader.

The field of genetics was unquestionably important, even if the old view that geneticists alone created the synthesis must be abandoned. Lewontin's presentation (chapter 1) suggests new ways to understand the role theoretical quantitative genetical theory played, or might have played, in the synthesis. The systematics section (chapter 4) shows conclusively that the field of systematics must be recognized as a major independent contributing factor to the evolutionary synthesis. Dunn's reading (1965) of Mayr's Animal Species and Evolution indicated to him how influential genetics was in systematics, which transformed and stimulated genetic studies of natural populations.

Stebbins' presentation in the botany section (chapter 5) indicates that botanists had developed most of the prerequisite key concepts of the evolutionary synthesis, but these concepts were not put together into a large synthetic view until his Variation and Evolution in Plants (1950). An unstated corollary I have drawn from his selection is that the evidence was available by 1940, and probably by 1937, for a botany book corresponding to Dobzhansky's Genetics and the Origin of Species.

Cytology played a multifaceted role in the synthesis, as indicated by the accounts of Darlington and Carson (chapter 2). Developments in cytology were crucial in the refutation of de Vries's theory of discontinuous species formation. Chromosomal rearrangements were essential in many studies of the genetics of natural populations of both animals and plants. Darlington's influential ideas about the evolution of genetic systems were based on his cytological researches. Discoveries in cytology lent credence to the whole field of genetics (genes did enjoy a material existence), thus raising the confidence of nongeneticist biologists about the developments within genetics. Cytological work shed much light on the problem of species formation, especially regarding isolating mechanisms at the chromosomal level.

At first glance, paleontology appears to be the best example illustrating the view that the evolutionary synthesis was merely the application of genetics to other fields. Chapter 6 indicates clearly that Simpson's *Tempo and Mode in Evolution* (1944) was enormously influential among paleontologists. By the mid-1940s Dobzhansky's *Genetics and the Origin of Species* had been through two widely read editions, which prepared geneticists for the extension of the synthetic view to paleontology. Paleontologists, on the other hand, were far less familiar with researches in theoretical and applied population genetics and typically considered modern genetical research to have little direct bearing on macroevolution. Simpson's book removed the barriers between the new synthetic view of microevolution and the study of macroevolution. The consequences may easily be observed as early as the 1947 Princeton conference on genetics, paleontology, and evolution.

The Situation in Various Countries

Because the evolutionary synthesis was genuinely different in different countries, the history of the evolutionary synthesis is difficult to analyze. Just compare Ford's experience as a young biologist in England, knowing or working with Lankester, Goodrich, Huxley, and Poulton, with the situation Boesiger described in France. The contrast is stark, and reveals a central problem for understanding the synthesis. The evolutionary synthesis did not occur, according to the presentations in Part Two, at the same rate or in the same way in the United States, England, France, Germany, Russia, or the Scandinavian countries.

At first glance, ideas about natural selection in France in the twentieth century appear monolithic; most evolutionists there have conveniently been classified as neo-Lamarckians. But Boesiger and Limoges (chapter 10) demonstrate that important distinctions existed between schools of evolutionary thought in France. Any attempt to portray this diversity as monolithic obscures the actual historical development of evolutionary thought in France. A similar situation existed in France in the nineteenth century, as Conry (1974) indicates. Important differences in schools of thought can be found even within one university. Stebbins wrote about the opposed views of his biology teachers in Harvard College and the biologists in the Bussey Institution of Harvard University (chapter 12).

This diversity of views within and between countries raises special difficulties for understanding the evolutionary synthesis. An enterprising researcher can produce counterexamples from at least one country or school of thought to almost any general assertion about the synthesis. Thus a historically accurate view of the synthesis must incorporate specific details of this diverse development.

General Interpretive Questions
What precisely do we mean by the term "synthesis"? Was the evolutionary synthesis a scientific revolution? Shapere shows that the term "synthesis" was used during the conference with many different meanings, from signifying logical entailment to simple removal of barriers between disciplines. Discrimination between the various meanings of the word "synthesis" is a great help in understanding more precisely what happened in evolutionary theory in the 1930s and 1940s. One note of unanimity at the conference may perhaps need to be revised. Although all participants seemed to agree that an evolutionary synthesis had occurred, they may have had different syntheses in mind. The evolutionary synthesis may therefore have appeared more cohesive during the conference than it actually was.

Conflicting Interpretations and Questions for Further Research
Like most first steps into the historical analysis of an important development, the conference raised more questions than it answered. In line with the suspicions expressed by Mayr in his opening words, many conflicting interpretations and differences of emphasis remained at the end of the conference. Further meticulous research will be required to resolve these issues.

For example, the conference illuminated but did not settle the relative roles played by the fields of genetics and systematics. Did systematists supply key concepts missing from the conceptual framework of geneticists? What role was played by the theoretical models of the mathematical population geneticists? How influential was the work of Sumner and Goldschmidt in bridging the gulf between systematists and geneticists? Were the relative roles played by systematics and genetics reversed in some countries or schools of thought?

Another unresolved issue was the assessment of the relative contributions to the synthesis of botany and zoology. Not until 1950 did a major book by a botanist appear portraying the full scope of the evolutionary synthesis already apparent in the zoological literature. Yet botanists unquestionably played a significant part in preparing the way for the synthesis.

The role of cytologists is especially difficult to assess: they seem to have played no major or crucial role in the synthesis. Darlington's work in the late 1920s and 1930s was widely known, but also widely mistrusted. How influential was White's *Animal Cytology and Evolution* (1945) compared with Darlington's *Recent Advances in Cytology* (1932, 1937) and *Evolution of Genetic Systems* (1939)? Was the work pioneered by Sturtevant and Dobzhansky using cytology in the genetic analysis of natural populations the most important contribution of cytology to the synthesis?

Simpson has written that he thought the conference did not provide an adequate analysis of his role in the evolutionary synthesis. Perhaps he is correct. Only additional research will reveal the impact of his *Tempo and Mode in Evolution* (1944), as compared to the writings of other paleontologists and the chapters on macroevolution in Mayr's *Systematics and the Origin of Species* (1942) and Rensch's *Neuere Probleme der Abstammungslehre* (1947). Another remaining problem is analysis of the extent to which the new systematics and genetics were really synthesized with paleontological data: is more than a consistency argument found here?

Clearly, a systematic appraisal of differences in the synthesis in various countries is needed. Why in England, where Darwinism still had traditional strength in the 1920s, were neo-Lamarckians so vociferous? How important were Darwin's heritage and mystique in the development of the evolutionary synthesis in England? Was it important that biologists like Huxley, Fisher, and Ford were Darwinians from a very early age, whereas Mayr and Rensch in Germany were initially to some extent neo-Lamarckians? One crucial research project will be assessment of the comparative influence of Russian biologists and the Morgan school, especially Sturtevant, in the formation of the views Dobzhansky expressed in his highly influential *Genetics and the Origin of Species* (1937).

Historical analysis of the reciprocal influences of social and philosophical issues upon the synthesis has scarcely begun. Philosophers will certainly play an important role in the determination of what is even meant by the term "evolutionary synthesis."

Ernst Mayr said in the prologue that the evolutionary synthesis illustrates graphically many important processes in the development of scientific thought, and that he hoped the conference would make a major contribution to the understanding of these processes. I have no hesitancy in asserting that this conference has made an enormous original contribution to the history of science.

References

Carter, G. S. 1957. *A hundred years of evolution*. New York: Macmillan.
Conry, Y. 1974. *L'Introduction du Darwinisme en France au XIXe Siècle*. Paris: Vrin.

Darlington, C. D. 1932. *Recent advances in cytology*. Philadelphia: Blakiston.
——— 1939. *The evolution of genetic systems*. Cambridge: Cambridge University Press.
Dobzhansky, Th. 1937. *Genetics and the origin of species*. New York: Columbia University Press.
——— 1955. A review of some fundamental concepts and problems of population genetics. *Cold Spring Harbor Symposia on Quantitative Biology* 20: 1-15.
Dunn, L. C., ed. 1951. *Genetics in the twentieth century*. New York: Macmillan.
——— 1965. *A short history of genetics*. New York: McGraw-Hill.
Fisher, R. A. 1930. *The genetical theory of natural selection*. Oxford: Oxford University Press.
Ford, E. B. 1931. *Mendelism and evolution*. London: Methuen.
Fothergill, P. G. 1952. *Historical aspects of organic evolution*. London: Hollis and Carter.
Haldane, J. B. S. 1932. *Causes of evolution*. London: Longmans Green.
Huxley, J. S., ed. 1940. *The new systematics*. Oxford: Oxford University Press.
——— 1942. *Evolution: the modern synthesis*. London: Allen and Unwin.
———, A. C. Hardy, and E. B. Ford, eds. 1954. *Evolution as a process*. London: Allen and Unwin.
Jepsen, G. L., E. Mayr, and G. G. Simpson, eds. 1949. *Genetics, paleontology, and evolution*. Princeton: Princeton University Press.
Kuhn, T. S. 1962. *The structure of scientific revolutions*. Chicago: University of Chicago Press.
Mayr, E. 1942. *Systematics and the origin of species*. New York: Columbia University Press.
——— 1959. Where are we? *Cold Spring Harbor Symposia on Quantitative Biology* 24:1-14.
——— 1973. The recent historiography of genetics. *Journal of the History of Biology* 6:125-154.
Nordenskiöld, E. 1927. *History of biology*. New York: Tudor.
Olby, R. C. 1974. *The path to the double helix*. Seattle: University of Washington Press.
Provine, W. B. 1971. *Origins of theoretical population genetics*. Chicago: University of Chicago Press.
Rensch, B. 1947. *Neuere Probleme der Abstammungslehre*. Stuttgart: Enke.
Simpson, G. G. 1944. *Tempo and mode in evolution*. New York: Columbia University Press.
Singer, C. 1931. *The story of living things*. New York: Doubleday.
Stebbins, G. L. 1950. *Variation and evolution in plants*. New York: Columbia University Press.
Stent, G. 1969. *The coming of the golden age: a view of the end of progress*. Garden City, N.Y.: Natural History Press.
——— and others, eds. 1966. *Phage and the origins of molecular biology*. Cold Spring Harbor, N.Y.: Biological Laboratory.
Sturtevant, A. H. 1965. *A history of genetics*. New York: Harper & Row.

Waddington, C. H. 1953. Epigenetics and evolution. *Symposia for the Society of Experimental Biology 7*. New York: Academic Press.

Watson, J. D. 1968. *The double helix*. New York: Athenaeum.

White, M. J. D. 1945. *Animal cytology and evolution*. Cambridge: Cambridge University Press.

Wright, S. 1931. Evolution in Mendelian populations. *Genetics* 16:97-159.

———— 1967. Foundations of population genetics. In *Heritage from Mendel*, ed. R. A. Brink. Madison: University of Wisconsin Press, pp. 245-263.

Zimmermann, W. 1953. *Evolution. Die Geschichte ihrer Probleme und Erkenntnisse*. Munich: Alber.

How I Became a Darwinian

Ernst Mayr

I have no recollection of when I first learned about evolution. My parents were not churchgoing types, but my father—even though he was a judge by profession—was an enthusiastic naturalist. On Sundays we would regularly go on excursions to collect ammonites in a limestone quarry, or to visit the location of some rare flowers, or to watch a heron colony. Before I was ten years old I knew most of the local birds by sight, and by the time I was sixteen or seventeen not a call note or song was unknown to me. In high school I read Haeckel's *Welträtsel* naively and avidly, not as a guide to evolutionary studies but to have ammunition in arguments about the Bible and religion!

When I entered the university in 1923, I took the medical curriculum (to satisfy a family tradition). I chose Greifswald at the Baltic for my studies for no other reason than that, of all German universities, it was situated in the ornithologically most interesting area. Even though I was inscribed as a medical student, I was first and foremost an ornithologist. Anything that had to do with birds interested me, and if this included evolutionary phenomena, it was only coincidental to my ornithological interests. I do not recall that during my studies in Greifswald (1923-1925) I had conversations or controversies with anybody about evolutionary problems. (See chapter 9 for my recollections about my teachers.)

After I went to Berlin in February 1925, and switched to biology, I had exactly 16 months to complete my Ph.D. thesis and the obligatory course work. This tight schedule allowed for a minimum of thinking about such "extraneous" matters as the mechanisms of evolution. One of my co-students, Kattinger, and I occasionally argued about this subject but we were quite in agreement that mutations could not be the answer to speciation and adaptation. A yellow *Drosophila melanogaster* with crumbled wings and white eyes was still a perfectly good *D. melanogaster*, we maintained. *D. simulans*, visually indistinguishable from *melanogaster*, however, was clearly a different species. We were quite convinced at that time that species characters were in an entirely different category from the mutations with which the mutationists "played" in the genetics laboratories. Like Darwin we believed in a categorical difference between continuous and discontinuous variation.

None of the professors in the Zoological Institute of the University of Berlin, so far as I know, was interested either in evolutionary mechanisms or in the process of speciation, including C. Zimmer, director of the museum, who as a busy administrator had fallen badly behind in reading the current literature. I took the last course given by the magnificent zoologist Karl Heider on the classification of the *Vermes* and learned all about the phylogeny of *Protostomia* and *Deuterostomia* but nothing about evolutionary mechanisms. The interests in the Zoological Institute were either strictly morphological (Ernst Marcus) or purely physiological (K. Herter).

Biology in Berlin was taught at that time at two distant locations. One was at the Invalidenstrasse in the north of Berlin, where most of the medical institutes, the Agricultural University, and the big Natural History Museum were located. The nearby Zoological Institute was where I studied and did my thesis. The other place was Dahlem, at the far south of Berlin, where the Botanical Institute was located as well as the Kaiser Wilhelm Institut für Biologie. The two places were two different worlds. The only course I took in Dahlem was one by Kniep on the lower fungi. Goldschmidt in 1925-26 was in Japan and Curt Stern at Columbia University with Morgan. Evolutionary mechanisms were not mentioned in Kniep's course and I had no evolutionary conversations with anyone in Dahlem. The only person who excited our interest (it was either at that time or in 1930 when I was back in Berlin for a short period) was Jollos. His heat-induced *Dauermodifikationen* seemed to form a bridge between phenotypic adaptation and genetic change. The single time I heard him lecture was an exciting occasion for us, and the large lecture hall was filled to capacity. We were rather disappointed when later attempts to confirm his results were unsuccessful (Plough and Ives, 1935). I never heard of Muller's X-ray mutations.

There was considerable interest in species and speciation among the curators of the Berlin Museum, but straight taxonomy was for most of them in the foreground of their interests. However, this was not true of Erwin Stresemann, curator of birds, under whom I did my degree work. In 1919 he had reviewed the species concept and had concluded that degree of morphological difference as a species criterion had been quite rightly abandoned in the late 1890s by the leading ornithologists and replaced by that of reproductive isolation. Rephrasing a similar statement by Hartert in 1910, he concluded that "forms of the rank of species have physiologically diverged from each other to such an extent, that they can come together again [after the removal of the geographical isolation], without mixing with each other" (Stresemann, 1919, p. 64). He repeated: "Morphological divergence is thus independent of physiological divergence" (p. 66). He reconfirmed in a series of papers from 1920 to 1936 that reproductive isolation is the decisive species criterion.

At this period Stresemann was perhaps the most active of all ornithologists in combining geographically various nominal species into *Rassenkreise*—that is, polytypic species. Geographic speciation was so completely taken for granted by Stresemann that he mentioned it (1936) only when he encountered divergent ideas, such as the claim by Lowe (1936) that Darwin's finches had originated as a hybrid flock (a la Lotsy). Stresemann concluded that hybridization, the result of a premature removal of geographic barriers between incipient species, invariably leads to a secondary intergradation of these populations, and not to speciation.

As progressive as Stresemann was in practicing population systematics and in his concepts of species and speciation, he was rather backward in his understanding of the mechanisms of evolution. He probably would have called himself an orthodox Darwinian, but he felt quite strongly that there were severe limits to the power of natural selection. Even though he repeatedly pointed out how often geographic variation obeys Bergmann's and other climatic rules, he left it open by what mechanism this adaptation is achieved.

Stresemann expressed quite frequently his skepticism of the unlimited power of selection. He vehemently rejected the claim of mimicry between an oriole (*Oriolus*) and a honey-eater (*Philemon*) in the Moluccas, for "this [interpretation] belongs to the realm of unlimited imagination [*Phantasie*], which has always characterized the extreme selectionists" (1914, p. 399).

Like many of his contemporaries he believed that mutation pressure could occasionally overcome selection pressure. A species of hawks (*Accipiter novaehollandiae*) that is polymorphic in Australia, consisting of gray and white individuals, is all white in Tasmania: "Selection has not interfered in this development: the snow-white hawk has achieved a complete victory in Tasmania over the pigmented, that is protectively colored, form" (1926). He ascribes this to "inner causes which finally led to a change in the body chemistry of all individuals." Stresemann did not appreciate the principle of differential reproductive success, but rather assumed that the same mutation would eventually take place in all individuals of the population. He used the word "mutation" in a strictly de Vriesian sense, assuming that it was a relatively rare but drastic phenomenon, affecting some species but not others.

Stresemann was right in assuming that conspicuous differences between geographic races and closely related species did not necessarily have to be built up by the gradual accumulation of very small differences, as postulated by some of the extreme Darwinians, but could be "monogenic" as we would now say. He was, of course, mistaken in assuming that these saltations (that is, de Vriesian mutations) were in contrast to Darwinian gradual variation. He was also wrong in ascribing these changes to "internal" mutation pressures: "The disposition toward the acquisition of a new characteristic of coloration is hidden in all individuals of the population. It is realized in some of them earlier than in the others, but the uniformity of the population is ultimately restored on the new basis" (1926, p. 165).

Stresemann's interpretations document the nature of the misconceptions that reigned in the 1920s even among the most enlightened systematists. Very similar ideas on polymorphism and mutation pressure were expressed at approximately the same time by Chapman (1923, 1928) in New York.

I became an assistant at the University Museum in Berlin on July 1, 1926, but left for New Guinea and the Solomon Islands in February 1928. I did not return until the end of April 1930.

Bernhard Rensch, also an assistant at the museum, was away most of 1927 on an expedition to the Lesser Sunda Islands. As a result, we did not meet often, although he probably had more influence on my thinking than anyone else. I greatly admired his 1929 book, which I read in 1930 when I returned from the Solomon Islands. His 1933 and 1934 publications had an equally great influence on me. Rensch more than anyone else revived the idea of a correlation between the geographic variation of various characters and the climatic conditions of the respective areas. The then prevailing genetic interpretation (de Vriesian mutationism) was not at all compatible with these findings and Rensch was thus forced to accept a Lamarckian interpretation. This I found completely logical at the time. When Rensch around 1935-36 became aware of pleiotropy and the existence of very small mutations, he had little trouble adopting the Darwinian interpretation and maintained a consistent Darwinian viewpoint from then on. This conversion was reinforced by some experimental work he did at that time in Timoféeff-Ressovsky's laboratory (see chapter 9).

In a paper on the origin of bird migration (written in joint authorship with W. Meise mostly in 1927 but not published until 1930), I still maintained that some phenomena, like regular annual transoceanic migrations, could not be explained by natural selection. I think this is the last antiselectionist statement found in my publications.

Timoféeff-Ressovsky was at Berlin-Buch during that same period but our contact was exceedingly slight. I no longer recall having made his acquaintance, but when I saw him in Moscow in 1972 he reminded me that we had met. I think he had attended a lecture of mine in which I described the range expansion of the serin finch (*Serinus canaria*), the topic of my Ph.D. thesis (1926). Range expansions were at that time a subject of great interest to Timoféeff-Ressovsky.

I was always a voracious reader of the current literature and I rather suspect that (like Simpson) I received more stimulation from reading than from talking with any of my associates (including Rensch). I stayed in Berlin just long enough (to the end of 1930) to work out part of my New Guinea collections and then left for the United States.

New York

The position I was offered by the American Museum was to prepare taxonomic revisions of the birds collected by the Whitney South Sea Expedition, particularly to describe the new species and subspecies. When the Rothschild Collection (Tring, England) was bought by the American Museum in 1932, I was given the additional responsibility of supervising the cataloging of this collection (280,000 bird skins!) and consolidating the Old World portion (about five-sixths of the total) with the existing Old World collections. For the first ten years after my arrival in New York (January 20, 1931) I was almost exclusively engaged in purely taxonomic research, but dealing primarily with an area (New Guinea and the Solomon Islands) that may be the most favorable in the world for the study of speciation. At least this was the considered opinion of Ernst Hartert, director of the Rothschild Museum, and at that time probably the world's outstanding ornithologist (Stresemann, 1975, chap. 14).

Several colleagues at the American Museum might have influenced my thinking appreciably. Chapman was sixty-seven when I arrived, but I do not recall that I ever discussed evolutionary problems with him. In several papers in the 1920s Chapman had dealt with speciation in birds in terms of two kinds of variation. He was convinced that general coloration and size were the result of the "direct action of the environment." This variation "has convinced me of the profound influence exerted by observable environmental factors (chiefly climatic) on the species" (1923). He clearly was a neo-Lamarckian. (For Chapman's views on discontinuous variation, mutation, and natural selection, see chapter 4.) His statements, I think, reflect the thinking of the older generation of naturalists rather accurately.

Robert Cushman Murphy (born in 1887), another colleague of mine, also had not kept up with the latest developments in Morgan's laboratories, even though he was twenty-three years younger than Chapman. I remember a lecture in which he proposed that evolutionary changes happened when an organism adopted a new function to accommodate a new mutation (see also Murphy, 1936, p. 336). He was rather shocked when I defended the opposite viewpoint—that is, that the selective value of its genes might change when an animal adopts new habits. Cuénot's idea that new mutations preadapt a species for an adaptive shift was still widely accepted among naturalists in the 1930s.

Among all my colleagues in the bird department the one who clearly had the best biological education was James P. Chapin (born in 1889), who had received his Ph.D. in the zoology department at Columbia University. He was a Darwinian without any reservations and presented his views eloquently in the introductory section of "Birds of the Belgian Congo" (1932). He ascribed adaptive geographic variation to natural selection for "no other theory has half the logical force" (p. 300). "The slight differences that characterize many recognizable subspecies are doubtless grounded in the germ plasm, where they may have been produced by mutations of small degrees. Numbers of slight Mendelian characters will simulate blended inheritance. Environment selects rather than directs the variations" (p. 298). One can read between the lines what arguments he was trying to refute. I do not recall having any lengthy evolutionary discussions with Chapin. Most of our discussions concerned details of avian taxonomy.

Several other members of the staff were interested in evolutionary problems, for example, G. K. Noble in herpetology. He was a person of extremely strong opinions and could not tolerate disagreement. When I was invited in 1939 to give a lecture in 1940 on speciation phenomena in birds, he came to my office and lectured me on what I should present. I was appalled at the time because it was a strictly de Vriesian interpretation of speciation. New species were made by mutations!

One of my most interesting colleagues at the American Museum was the entomologist Frank Lutz, who had formerly worked at Cold Spring Harbor in Davenport's laboratories. He always claimed (Davenport, 1941) that he had given Morgan the *Drosophila* stocks that later became famous. Lutz had worked on *Drosophila* and had published a lengthy report on his researches in the Carnegie Institution publications (1911). Lutz was an imaginative person, but also a person of strong and often heterodox opinions. He flatly rejected the idea that there was such a thing as mimicry. I no longer remember what he said about speciation, but I rather suspect that it was something highly heterodox.

Around 1939-40 I joined a sandwich luncheon group at the museum that included Frank Beach, Mont Cazier, Charles Bogert, sometimes Ned Colbert, Jim Oliver, and quite frequently Herman Spieth. Only Spieth and I had intensive evolutionary discussions; in 1946 I persuaded him to give up mayflies (then his field of research) for the study of *Drosophila* behavior.

Everyone asks me what my contacts were with Simpson and finds it difficult to believe that there were virtually none during the first ten years of our association at the American Museum. Although most members of the staff lunched together and Simpson and other paleontologists (such as Granger, Brown, and Colbert) sat at the same long staff table, I do not recall any scientific conversations with him. As I remember it, my first scientific contact with Simpson was in 1941 when I sent him the last manuscript chapter of my *Systematics and the Origin of Species* to criticize. According to my recollections, Simpson very courteously criticized various formulations of mine and returned the chapter to me with a set of helpful comments. My more direct contact with Simpson did not start until after Walter Bucher had organized the National Research Council Committee on Evolution. By that time my book had been published, or at least it was in press.

Contacts with Columbia University

My recollections are rather vague, but I suspect that my contacts with the department of zoology at Columbia University were of considerable importance for the later development of my thinking. After I had been in New York for quite a while (probably after 1936), I began to attend the genetics seminars organized by L. C. Dunn. At first I went only occasionally but later with great regularity. Although the emphasis in these seminars was on developmental genetics, owing to the special interests of Dunn and his students, I absorbed by osmosis a great deal of what was then the most modern genetics. However, I do not recall a particular person, lecture, or publication emanating from the Columbia group that had a major impact on me.

The most exciting publication I encountered at that period was a paper by Dobzhansky (1933) on geographic variation in lady-beetles. I exclaimed, "Here is finally a geneticist who understands us taxonomists!" When Dobzhansky gave the Jesup lectures at Columbia University in 1936, it was an intellectual honeymoon for me. He came down to the museum and I was able to demonstrate to him the magnificent geographic variation of South Sea Island birds, particularly in the genus *Pachycephala*. I was delighted with the book that came out of his lectures, *Genetics and the Origin of Species* (1937), and found that Dobzhansky's interpretation agreed on the whole extremely well with the ideas that I had formed independently, from the literature, from the study of my own material, or from my contacts in New York.

A few years later I presented my ideas in "Speciation Phenomena in Birds" in the *American Naturalist* (1940) and gave in 1941, together with Edgar Anderson, the Jesup lectures at Columbia. The bibliography of the resulting volume (1942), as well as the more detailed discussions in it, clearly reveal the influences that had shaped my thinking in the preceding ten years. Without Dunn's encouragement to expand the lectures into a book, I do not think I would have written *Systematics and the Origin of Species* (1942).

Richard Goldschmidt visited New York repeatedly and spent one entire Saturday morning with me at the museum in 1936 or early 1937. I finally took him home with me, and he shared with us a simple meal of lentil soup with frankfurters. In 1939 he gave the Silliman lectures at Yale University, published as *The Material Basis of Evolution* (1940), in which he defended his thesis of systemic mutations and minimized the importance of geographic speciation. I had tried to argue this point with him at several informal meetings before and after the lectures (for which Spieth and I traveled to New Haven), but never got to first base. His attitude struck me as very peculiar because in his earlier publications, Goldschmidt seemed to have been a thorough adherent of the concept of geographic speciation, not surprising after his many years of association with the systematists at the Munich Museum. However, in 1933, seemingly rather suddenly, he reversed himself and developed his new theory.

Until Goldschmidt's Silliman lectures I had simply taken it for granted that anybody who seriously thought about the subject would have to adopt the concept of geographic speciation. It was the official philosophy in the bird department of the Berlin Museum where I had grown up, and it did not matter whether or not an inheritance of acquired characters played a role in this process. In the late 1930s Goldschmidt denied it; several other contemporary publications, for example, by Robson and Richards (1936), minimized it; and the paleontologists (Osborn, Beurlen, and Schindewolf) ignored it altogether. I suddenly realized that it was important to present massive documentation in favor of geographic speciation so that at least this particular uncertainty could be eliminated from the panorama of evolutionary controversies. Even though personally I got along very well with Goldschmidt, I was thoroughly furious at his book, and much of my first draft of *Systematics and the Origin of Species* was written in angry reaction to Goldschmidt's total neglect of such overwhelming and convincing evidence.

History tends to repeat itself. David Starr Jordan had once before (1905) complained that the massive findings of the taxonomists about speciation are so totally ignored by the laboratory zoologists: "It is now nearly forty years since Moritz Wagner (1868) first made it clear that geographical isolation was a factor or condition in the formation of every species . . . we know on the face of the earth. This conclusion is accepted as almost self-evident by every competent student of species or of the geographical distribution of species . . . The principles set forth by Wagner . . . have never been confuted, scarcely even attacked . . . but in the literature of evolution of the present day they have been almost universally ignored" (1905). This neglect was still true in the 1930s. Rensch's work, for instance, perhaps because it was in German, was almost totally ignored and so was my own work prior to 1942. Also largely overlooked were the fine studies of Schmidt on *Zoarces*, of Goldschmidt on the geographic variation of *Lymantria*, and of Sumner on *Peromyscus*. Sumner's work was continued by Dice and his students (including Frank Blair), all of whom demonstrated the importance of geographic variation, geographical barriers, and adaptation to local environments. Unfortunately, Dice waited until he was well over seventy before he began to bring all these researches together in book form, but by that time his findings were no longer news and his manuscript, so far as I know, was never published.

People have often asked me what impact Fisher, Haldane, and Wright had on my thinking. My answer is quite embarrassing. I knew nothing of Fisher until I read Dobzhansky (1937), and even then I did not read Fisher's book. In 1931 or 1932 I lectured at Yale University about geographic speciation in South Sea Island birds and I still remember my puzzlement when Grace Pickford asked me what effect population size had on the process of speciation. I was later told that I gave a very stupid answer, because at the time I had not yet heard of Wright and genetic drift. I did not become aware of Haldane's work until about 1947. Mathematical population genetics affected me only indirectly through Dobzhansky's book. It helped to convince me that even very small selective advantages are of evolutionary consequence. This contribution may seem rather small in terms of the total theory but it actually was crucial. It, plus the recognition that many, if not most, mutations are very small (which I also learned from Dobzhansky and from Chapin), were for me major bridges between the thinking of the naturalists and that of the laboratory geneticists.

Dobzhansky's thinking was so acceptable to me because, of course, the Russian school of population geneticists had grown out of taxonomy and talked in terms of species and natural populations and their variation. In fact, the Russians were interested in exactly the same phenomena as the evolutionary taxonomists; they were geneticists who spoke the language of the taxonomists and had adopted population thinking. Dobzhansky's *Genetics and the Origin of Species* was so influential because it built a bridge between genetics and systematics. Fisher's genetic theory of natural selection, with its emphasis on the fitness of individual genes, simply did not speak to the taxonomist, which perhaps is the reason why—with the spectacular exception of Ford—Fisher was relatively unsuccessful in making an impact on British systematics.

I have no idea how reliable these recollections of mine are. I speak of events and publications that happened forty or forty-five years ago. It will require a critical study of the literature and the available correspondence to determine the real truth. Furthermore, it will be difficult to define what the "real truth" is. The new synthesis undoubtedly looked very different to Rensch, to Dobzhansky, to Simpson, to Ford, and to me. Different people, different personal interests, different aversions, and different research materials all affect our thinking and ultimate conclusions. I am sure that even now, thirty-five years later, many aspects of the synthesis continue to look different to the various participants.

June 1974

References

Baur, E. 1930. *Einführung in die Vererbungslehre*. Berlin: Borntraeger, pp. 387-402.

Chapin, J. P. 1932. Birds of the Belgian Congo. *Bulletin of the American Museum of Natural History* 65:265-300.

Chapman, F. M. 1923. Mutation among birds in the genus *Buarremon*. *Bulletin of the American Museum of Natural History* 48:243-278.

——— 1928. Mutation in *Capito auratus*. *American Museum Novitates* 335.

Davenport, C. B. 1941. The early history of research with *Drosophila*. *Science* 93:305-306.

Dobzhansky, Th. 1933. Geographical variation in lady-beetles. *American Naturalist* 67:97-126.

——— 1937. *Genetics and the origin of species*, 1st ed. New York: Columbia University Press.

Goldschmidt, R. 1933. Some aspects of evolution. *Science* 78:539-547.

——— 1940. *The material basis of evolution*. New Haven: Yale University Press.

Jollos, V. 1934. Inherited changes produced by heat-treatment in *Drosophila melanogaster*. *Genetica* 16:476-494.

Jordan, D. S. 1905. The origin of species through isolation. *Science* 22: 545-562.

Lowe, P. R. 1936. The finches of the Galapagos in relation to Darwin's conception of species. *Isis* 78:310-321.

Lutz, F. E. 1911. Experiments with *Drosophila ampelophila* concerning evolution. *Carnegie Institution of Washington Publication* 143:1-40.

Mayr, E. 1926. Die Ausbreitung des Girlitz. *Journal für Ornithologie* 74:571-671.

———— 1940. Speciation phenomena in birds. *American Naturalist* 74:249-278.

———— 1942. *Systematics and the origin of species*. New York: Columbia University Press.

———— and W. Meise. 1930. Theoretisches zur Geschichte des Vogelzuges. *Vogelzug* 1:149-172.

Murphy, R. C. 1936. *Oceanic birds of Southern America*. New York: American Museum of Natural History, vol. 1, p. 336.

Plough, H. H., and P. T. Ives. 1935. Induction of mutations by high temperature in *Drosophila*. *Genetics* 20:42-69.

Robson, G. C., and O. W. Richards. 1936. *The variation of animals in nature*. London: Longmans, Green.

Stresemann, E. 1914. Beiträge zur Kenntnis der Avifauna von Buru. *Novitates Zoologicae* 21:358-400.

———— 1919. Über die europäischen Baumläufer. *Verhandlungen der Ornithologischen Gesellschaft in Bayern* 14:39-74.

———— 1926. Übersicht über die Mutationsstudien I-XXIV und ihre wichtigsten Ergebnisse. *Journal für Ornithologie* 74:377-385.

———— 1936. Zur Frage der Artbildung in der Gattung Geospiza. *Orgaan der Club van Nederlansche Vogelkundigen* 9:13-21.

———— 1975. *Ornithology. From Aristotle to the present*. Cambridge, Massachusetts: Harvard University Press.

Wagner, M. 1841. *Reisen in der Regentschaft Algier*. Leipzig.

———— 1868. *Die Darwinsche Theorie und das Migrationsgesetz der Organismen*. Leipzig: Duncker and Humblot.

Curt Stern*

Ernst Mayr

In the 1920s and 1930s, there were two centers of biology in Berlin—one in the north around the Museum für Naturkunde, the Zoological Institute, the College of Agriculture, and the Medical Institute; the other in the south around the Kaiser Wilhelm Institut für Biologie and the botanical institute at Dahlem. The naturalists were concentrated in the north; the experimentalists, physiologists, and others in the south. Curt Stern's experiences provide an eyewitness report of someone who worked in the south, in contrast to Mayr and Rensch. Although Stern was not an evolutionist in the narrow sense of the word, he always took a considerable interest in evolutionary questions.

Curt Stern was born August 30, 1902, in Hamburg. He received his Ph.D. in Berlin (1923) with a thesis under Max Hartmann on the morphology and physiology of certain protozoans. From 1923 to 1933 he was a scientific investigator at the Kaiser Wilhelm Institute in Berlin-Dahlem, loosely attached to Richard Goldschmidt's department but doing independent research. Originally Stern's thinking was heavily influenced by Hartmann, one of the most outspoken opponents of Kammerer.

Stern's career was decisively influenced by a two-year fellowship in Morgan's laboratory at Columbia University (November 1924-November 1926). He also stayed with Morgan in Pasadena (December 1932-March 1933). In his words:

> During my two years in the fly room with Morgan at Columbia University, Bridges and Sturtevant discussed various aspects and, particularly in the case of Sturtevant, evolutionary aspects of the classical genetics that was conducted at Columbia University in the mid-1920s. Sturtevant definitely was interested in evolutionary questions and so was Morgan. Sturtevant in fact was actively interested in speciation problems with ants, not because he was interested in ants themselves, but because he wanted to have another species, very different from *Drosophila*, to be used in considerations of the species concept and various similar things. Bridges, on the other hand, was not interested in evolution. The discussions in general were not of evolutionary problems but of chromosomal properties.

*Aware that two entirely different intellectual traditions prevailed at the two centers of biology in Berlin, I asked Curt Stern, of the University of California (Berkeley), to record his impressions of the attitude toward evolution at Berlin-Dahlem. Although he was unable to attend the conference, his very kind response to my questionnaire was given in rich detail, which I have converted into this account.

E.M.

Much of Stern's genetic research has evolutionary significance—not only his classical studies on crossing over, but also his work on interspecific sterility (1936), and his research on isoalleles (a term he coined in 1943). Stern commented about factors that delayed the coming of the synthesis:

> After 1910 the interest in evolution decreased. The reason was not a disenchantment so much with the problems of evolution, but with having reached a deadlock concerning the contributions of different factors to the possibilities of evolution. Bateson in 1914 gave an address in which he explained evolution in terms of loss of genes, so that in every evolutionary step there would be more and more genes lost. This explanation did not seem likely, but the interpretation of Johannsen's work with pure lines seemed to represent a difficulty that was not easily overcome. The best minds became preoccupied with different areas of research such as chromosomal genetics, *Entwicklungsmechanik*, and other branches of biology flourishing in the 1920s.

How drastically the interest in evolution had declined among the experimentalists is well illustrated by Max Hartmann's work on general biology (1927). In 756 pages, he devotes only five to evolution.

Stern continued:

> The synthesis was delayed greatly by the different opinions of paleontologists and other biologists regarding the question of inheritance of acquired characteristics. There was an attempt in 1929 to get various opposing groups together; the German genetic society and the German paleontological society met jointly in Tübingen for several days to discuss particularly the question of inheritance of acquired characteristics. Weidenreich spoke in favor, Federley against. Federley was an excellent cytogeneticist and Weidenreich was an excellent paleontologist, but these people simply could not get together to make this meeting a conciliatory synthesis. Added to this divergence, the geneticists were accused by the paleontologists of dealing only with superficial traits that had no relation to the essential organic nature of organisms and of evolution. Thus the theory of natural selection was attacked both from the narrow point of view of paleontologists and from the broader point of view of biologists in general who thought that evolution was too complex a process to be easily accounted for by mechanistic interpretations (see Federley, 1930).
>
> While Weidenreich had recounted the [point] that the real genetic contribution showed that there was an inheritance of acquired characteristics, he shared the opinion of many other older biologists of the period—that is, that genetics played only a very inferior role in evolution. He was convinced that genetics explained only the most

superficial aspects of evolution and that actual evolutionary phenomena had to be based on much more fundamental considerations. It was, in other words, not just a lack of communication between evolutionists and other types of biologists.

The lack of communication was probably not too serious. The geneticists, the paleontologists, and the systematists were basically familiar with the genetic facts that had a bearing on evolution but, despite this familiarity, they rejected the application to actual cases.

Stern thinks that the eventual progress in understanding "was made primarily through a clarification of concepts, such as mutation, somatic, and germinal," rather than by new discoveries.

The opponents of Darwinism found it particularly difficult to accept an explanation of adaptation. According to Stern:

The complexity of adaptations impressed biologists and they found it difficult to give natural selection a role in the origin of the most complicated and unexpected types of adaptations. While teaching, I discussed at length the parable of the monkey at the typewriter who on a completely random basis composed the whole library of books. By itself, this method would appear to be a very improbable way of getting a library, but the additional thing that had to be considered was not only the random typing by the monkey but also the presence next to him of a censor who with each letter typed looked at it and decided whether it made sense—whether the letter, for instance, completed a word that was not yet a unit, and eliminated all those types at every step which did not fit into the general picture of adaptations.

Concerning his contact with population genetics, Stern wrote:

I learned first about the Russian population genetics of the Chetverikov school during the Berlin Congress in 1927. Chetverikov gave a paper about finding mutants in heterozygous recessive condition in the Caucasus, thus proving that mutations were not just products of laboratory environments but occurred in nature just as well (Chetverikov, 1928).

The full significance of these findings took some time to be realized. Timoféeff-Ressovsky, who at that time worked in Germany, followed up Chetverikov's work by some studies of his own in population genetics (see Timoféeff-Ressovsky, 1927), but by and large his evolutionary period began only after he had been for many years a radiation biologist interested in the effects of X rays on chromosomes but not so much with the aspects of wild populations of *Drosophila* as he studied later.

Stern also wrote about mutations and their evolutionary significance:

I do not remember when I first came across the concept of mutation, but probably during my first semester (1920) listening to lectures of Max Hartmann. I read about de Vries first in Richard Goldschmidt's textbook of genetics (1920), and in the textbook by Baur
(1919). From the very beginning of the *Drosophila* work, mutations
did not pose any problems for me. They were not necessarily large
deviations from a standard type, but could be very fine in nature.
Muller had worked on mutations produced by temperature effects
since 1920 and mutation was not anything that was not compatible
with evolutionary expectations.

I am not sure that I ever fully gave up the idea that a mutation was
nearly always a deleterious disturbance of the genotype. But I did
recognize that there are alleles that are very similar in effect and can
be shown only by special methods to be really different. I called
them isoalleles (Stern, 1943).

The concept of the small, frequently invisible mutation was
brought home to me particularly clearly in a study by Baur on the
snapdragon, in which he investigated wild populations in Spain for
the occurrence of variations of a very fine quantitative nature (1924,
1932). It was a pity that Baur discontinued his experimental work
when he organized the new Müncheberg Institute for Plant Breeding.

The frequency of mutations was measured for the first time to
some degree by Muller who showed not only that they were rare
events, but more important that they were events that repeated
themselves in appearing, so in spite of their rareness they increased
up to certain limits in populations. Muller's experiments (1927) with
X rays producing mutations were regarded as models of what really
was going on in the germ plasm.

I asked Stern about his early views on variation. He replied:

I never thought that there were two kinds of variation, cytoplasmic and chromosomal; nor, as was sometimes proposed, that
chromosomal mutations were of a more fundamental evolutionary
nature than so-called cytoplasmic mutations that represented continuous variation. I first learned about the work of Schmidt, Sumner, and Goldschmidt when Goldschmidt's monographs on geographic variation appeared (summarized in Goldschmidt, 1934). I
remember that I was much relieved by reading Sturtevant's paper
(1918) on selection of modifiers. He explained orthogenesis in terms
of Darwinian selection and thereby removed them from the somewhat mystical aspects to their reasonably better understood ones.
Payne's work at the same time (1918) with *Drosophila* also contributed to the demonstration that different kinds of variations all
obeyed the chromosomal genetic aspects and were not outside of
them.

Stern responded to my question about contemporary views on the role of the environment in evolution:

I do not know when I first realized that the environment acts solely as the agent of natural selection. I think I grew up in an atmosphere that supported this idea; it did not cause any revolutionary upset in my mind. Jollos published an inaugural lecture at the University of Berlin in 1922 titled *Selektionslehre und Artbildung*. This article was not in itself very influential, but it showed the way some people thought along modern lines.

I think I was a total selectionist from my student days on, but I was aware of the disagreement of others who could not conceive selection as being able to create some of the morphological adaptations, for instance, the vertebrate eye. In 1929-30 I had a literary feud with Weidenreich (in *Natur und Museum*), who regarded Muller's X-ray mutations as evidence for the inheritance of acquired characters. I "certified" that Weidenreich had not been able to understand genetics when he referred to the *Drosophila* mutants as *kinkerlitzchen*. Fisher's book (1930) had no influence in Dahlem.

Little interaction existed between Berlin-Buch and Dahlem, I think, for primarily external reasons. The distance from Dahlem to Buch was too great to permit getting together more than once a year or so. Also, as I pointed out earlier, Timoféeff-Ressovsky's work was then primarily on radiation genetics and not on evolutionary aspects.

Stern wrote about the nature of evolution and the species problem:

In the 1930s I saw evolution as a strictly random process, not containing a trend toward progress and improvement or of some other sort. And yet it required some aspects that formed, so to speak, a bonus over what was accomplished by strict random processes. In other words, random processes led to a variety of genetic constitutions and only a very small minority were of such a nature as to improve the situation. To some degree this was part of Sturtevant's article on orthogenesis in *Science* (1924).

My own work on fertility and sterility in *Drosophila* with different Y chromosomal segments (1936) was stimulated by Bateson's famous sentence: "The production of an indubitably sterile hybrid from completely fertile parents which have arisen under critical observation from a single common origin is the event for which we wait" (Bateson, 1922, p. 394). Dobzhansky's *Genetics and the Origin of Species* (1937) was an important source leading to my becoming more deeply interested in the species problem.

Comparing Stern's impressions with those of Rensch and Mayr vividly highlights the existence of two worlds side by side in Germany in the 1920s. That there was urgent need for building a bridge is obvious.

References

Bateson, W. 1914. Presidential address to the British Association, Australia. Melbourne meeting. In *William Bateson, F.R.S.: naturalist*, ed. B. Bateson. Cambridge: Cambridge University Press, 1928, pp. 276-296.

———— 1922. Evolutionary faith and modern doubts. In *William Bateson, F.R.S.: Naturalist*, ed. B. Bateson. Cambridge: Cambridge University Press, 1928, p. 394.

Baur, E. 1919. *Einführung in die experimentelle Vererbungslehre*, 3d ed. Berlin: Borntraeger.

———— 1924. Untersuchungen über das Wesen, die Entstehung, und die Vererbung von Rassenunterschieden bei *Antirrhinum majus*. *Bibliotheca Genetica* 4:1-170.

———— 1932. Artumgrenzung und Artbildung in der Gattung *Antirrhinum*, Sektion Antirrhinastrum. *Zeitschrift für Induktive Abstammungs- und Vererbungslehre* 63:256-302.

Chetverikov [Tschetwerikoff], S. S. 1928. Über die Genetische Beschaffenheit wilder Populationen. *Verhandlungen des V. Internationalen Kongresses für Vererbungswissenschaft*. Leipzig: Borntraeger, vol. 2, pp. 1499-1500.

Dobzhansky, Th. 1937. *Genetics and the origin of species*. New York: Columbia University Press.

Federley, H. 1930. Weshalb lehnt die Genetik die Annahme einer Vererbung erworbener Eigenschaften ab? *Zeitschrift für Induktive Abstammungs- und Vererbungslehre* 54:20-50.

Fisher, R. A. 1930. *The genetical theory of natural selection*. Oxford: Oxford University Press.

Goldschmidt, R. 1920. *Einführung in die Vererbungswissenschaft*, 3d ed. Leipzig: Engelmann.

———— 1934. Lymantria. *Bibliographia Genetica* 11:1-186.

Hartmann, M. 1927. *Allgemeine Biologie*. Jena: Fischer. (2d ed., 1933.)

Jollos, V. 1922. *Selektionslehre und Artbildung*. Jena: Fischer.

Muller, H. J. 1927. Artificial transmutation of the gene. *Science* 66:84-87.

Payne, F. 1918. An experiment to test the nature of the variations on which selection acts. *University of Indiana Studies* 5:1-45.

Stern, C. 1936. Interspecific sterility. *American Naturalist* 70:123-142.

———— 1943. Genic action as studied by means of the effects of different doses and combinations of alleles. *Genetics* 28:441-475.

Sturtevant, A. H. 1918. *An analysis of the effects of selection*. Washington, D.C.: Carnegie Institution of Washington, publication no. 264.

———— 1924. An interpretation of orthogenesis. *Science* 59:579-580.

Timoféeff-Ressovsky, H. A. and N. W. Timoféeff-Ressovsky. 1927. Genetische Analyse einer freilebenden *Drosophila melanogaster* population. *Wilhelm Roux' Archiv für Entwicklungsmechanik der Organismen* 109:70-109.

J. B. S. Haldane, R. A. Fisher, and William Bateson

C. D. Darlington

I had a prolonged personal relationship with J. B. S. Haldane and R. A. Fisher and a short but very important contact with William Bateson. I had twelve years of constant discussion with Haldane. He was working in the same lab doing his long-division sums at an adjoining desk, and I spent my vacations with him and enjoyed his conversation. He read all my papers and books as a matter of duty—maybe it was pleasure too. The odd thing is that I usually got no ideas from him in connection with my research work. Once or twice we discussed natural selection, but it seemed rather obvious to me that, for example, the rings of *Oenothera* were built up by each one having a selective advantage over its predecessor, and when he had that idea too it seemed very obvious. We were both absolutely dogmatic about selection theory, only I was applying it in a new field (see our parallel papers: Darlington, 1932; Haldane, 1932).

I had forty years' experience with Fisher, gradually modifying my relationship with him as time went on. When Haldane joined the Communist party, taking the *Journal of Genetics* with him, Fisher and I collaborated in founding *Heredity* in 1947. Although we arrived at certain ideas independently, we never discussed them with each other throughout that time. Just once, in 1930 in a short conversation, he told me that he agreed with my views on *Oenothera*. Once or twice, twenty years later, I tried to suggest to him that the principles of selection were not in fact going to operate with the absolute rigor he expected, but I could never get him to discuss it. I didn't dare to suggest that my views were much more Wrightian than those of Wright himself, because our relationship would have been broken off even earlier than it was. How curious that these complete barriers existed between people who were nominally connected with one another. There was almost no exchange of ideas between us during that time.

People have difficulty in understanding Bateson's complexity. They try to think of him as a consistent character, but in fact he evolved rapidly, rose to a climax like a rocket that exploded in 1906 with the word "genetics," and later somewhat declined. I knew him in his last years of decline. He defined genetics as the study of heredity and variation together with their bearing on evolution. He always had that in his mind in regard to genetics.

I came to know Bateson because of his visit to America in January 1922. I have a picture of him at Ithaca, with Bridges showing him the chromosomes, Wright standing next to him, and Emerson looking on.

And there is the old lion (only sixty but looking seventy) being confronted with these terrible things and appearing nonplussed. He was, however, gravely in decay at that time and did a number of inconsistent things. Immediately when he returned to England, he appointed a cytologist to look at chromosomes. He appointed Frank Newton, who taught me about chromosomes and died four years later, at thirty-three, without completing any contribution of his own beyond the analysis of the prophase of meiosis in diploid tulips.

The very next year Bateson's son Gregory brought the neo-Lamarckian experimenter Paul Kammerer to London for a confrontation with Bateson. This story has recently been told by Koestler (1973), giving an erroneous picture of Bateson as representing the established science of the day. Of course, Bateson was, from the point of view of English biology at that time, an outcast. Opposed to Lamarck and critical of Darwin, he couldn't get a professorship in any university; he had been forced to leave Cambridge and he had failed to get into Oxford. He was working at a horticultural institution, shut off from the possibility of teaching genetics. Although he recommended Edward Murray East as head of the Bussey Institution of Harvard in 1909, that did not help him in England.

After the war, about 1918, Bateson took up the studies Baur had been developing with chimeras and with plastid inheritance. He worked busily to the last day of his life in the same lab with Newton and me, although, I'm afraid, he was muddled. He had discovered something, going a little further than Baur: cytoplasmic inheritance in flax. He was so unwilling to make this novel discovery that he did what people often do—he coined a word to avoid making the discovery. Like Goethe's principle of *"denn eben wo Begriffe fehlen,"* he coined the word "anisogeny" (Bateson, 1926). There was a terrible moment a few months after he died when one of his assistants (Chittenden) had to explain to another assistant (Miss Pellew) that anisogeny was probably a case of genic reaction with cytoplasmic inheritance. Bateson's anisogeny shows his declining aspects. Yet he did bring Newton and the study of chromosomes into the John Innes. Newton and I started discussing Janssens' crossing-over theory in connection with our chromosome work with tulips and hyacinths and our breeding work with *Prunus* and *Rubus* and *Primula*. But Bateson never realized that this work would give the clue to the problems of species.

References

Bateson, W. 1926. Segregation. *Journal of Genetics* 16:201-235.

Darlington, C. D. 1932. The control of the chromosomes by the genotype and its bearing on some evolutionary problems. *American Naturalist* 66:25-51.

Haldane, J. B. S. 1932. The time of action of genes, and its bearing on some evolutionary problems. *American Naturalist* 66:5-24.

Koestler, A. 1973. *The case of the midwife toad.* New York: Random House.

Morgan and the Theory of Natural Selection

Alexander Weinstein

The Origin of Species was not required or recommended reading in any undergraduate or graduate course in zoology or botany that I took when I was a student at Columbia University from 1910 to 1916. One of the textbooks of the time advised students to postpone reading the *Origin* till after they had read Romanes' *Darwin and after Darwin*, or some other such book. There were, I think, two reasons for this advice. One was the opinion—which goes back to Huxley—that the *Origin* is a difficult book and that therefore the students needed an easier preliminary treatment. The other reason was that the subject should be evaluated not only in the light of the information available to Darwin, but also of the evidence that had accumulated since Darwin wrote.

An attempt at such an evaluation was made by Morgan in his *Evolution and Adaptation* (1903). He said that "the line between fluctuating variations and [de Vriesian] mutations may be sharply drawn" (p. 297), and that variations other than Darwin's individual or fluctuating variations "may give us an explanation of evolution without competition, or selection, or destruction of the individuals of the same kind taking place at all" (p. 128). This gave Morgan the reputation of being an opponent of Darwin's explanation of evolution and an adherent of de Vries's mutation theory. When I took the elementary course in zoology as a freshman in 1910-11, Bashford Dean, who gave the lectures in the second half of the course (which dealt with vertebrate zoology and evolution), said that "so eminent an experimental zoologist as Professor Morgan believes that evolution has occurred by mutation"; and he added that most biologists believed that about 80 percent of evolution is the result of natural selection and about 20 percent the result of mutation.

If we examine Morgan's book carefully, we see that his reputation in this respect does not represent his ideas with complete accuracy. Referring to the fact that animals and plants are adapted to their environment, he says:

> I can see but two ways in which to account for this condition, either
> (1) teleologically, by assuming that only adaptive variations arise,

or (2) by the survival of only those mutations that are sufficiently adapted to get a foothold. Against the former view is to be urged that the evidence shows quite clearly that variations (mutations) arise that are not adaptive. On the latter view the dual nature of the problem that we have to deal with becomes evident, for we assume that, while the origin of the adaptive structures must be due to purely physical principles in the widest sense, yet whether an organism that arises in this way shall persist depends on whether it can find a suitable environment. This latter is in one sense selection, although the word has come to have a different significance, and, therefore, I prefer to use the term *survival of species* (p. 463).

This statement does not seem to me to make clear the distinction between the original significance of selection and the "different significance" it has come to have. Apparently Morgan is trying to play down the struggle of organisms against one another and to emphasize their struggle against their environment, or rather their finding an environment against which they will not have to struggle. This would be in accord with his attempt, in the passage on page 128 quoted above, to rule out "competition" and "destruction of individuals of the same kind." In any case, the struggle against environment and the finding of a more suitable environment are included in Darwin's conception of the struggle for existence and natural selection.

The fact that nonadaptive as well as adaptive variations arise is also, of course, part of Darwin's theory of natural selection, and Morgan does not deny that it is. But he says that some of Darwin's "more ardent but less critical followers . . . have contented themselves, as a rule, with pointing out that certain structures are useful, and this has seemed to them sufficient proof that the structures must have been acquired because of their value" (p. 462). This, while not entirely clear, seems to mean that adaptive structures are supposed by these "more ardent but less critical followers" to arise directly in response to the need of the organism—the notion Morgan characterizes as teleological in the passage on page 463 quoted earlier. Since he does not attribute this teleological notion to Darwin, the difference between Darwin's and Morgan's views of the origin of adaptive structures reduces to this: that Darwin supposes the structures to be the result of the selection of successive small variations, and Morgan supposes them to come into existence in one step. Both allow that there have been other variations that have been rejected; both allow that there must be selection.

Morgan considered two other theories of evolution: Lamarck's theory of the inheritance of acquired characters, and Nägeli's theory of orthogenesis. He says of Lamarck's theory:

We may fairly sum up our position in regard to the theory of the inheritance of acquired characters in the verdict of "not proven." I am not sure that we should not be justified at present in claiming that the theory is unnecessary and even improbable (p. 260).

With reference to Nägeli's theory he writes:

His hypothesis appears, therefore, entirely arbitrary and speculative to a high degree (p. 337).

Despite Nägeli's protest that his principles are purely physical and that there is nothing mystical in his point of view, it must be admitted that his conception, as a whole, is so vague and difficult in its application that it probably deserves the neglect which it generally receives (p. 338).

But Morgan was not consistent about orthogenesis. In another passage he is willing to entertain a similar notion and even says that it is necessary for the mutation theory:

Thus, while the mutation theory must assume that some new characters will go on heaping up, we lack the experimental evidence to show that this really occurs. It would be also equally important to determine whether, if after several mutations have successively appeared in the same direction, there would be an established tendency to go on in the same direction in some of the future mutations. But here again we must wait until we have more data before we attempt to build up a theory on such a basis (pp. 461-462).

Morgan also gives an account of Mendelism, basing it on the first edition of Bateson's *Mendel's Principles of Heredity* (1902). Morgan's account is brief, but correct as far as it goes, and expresses no misgivings. He says Mendel's results have been confirmed, that Mendel's theoretical interpretation is so simple that there can be little doubt he has hit on the real explanation, and that he has discovered one of the fundamental laws of heredity (pp. 284-285). Morgan mentions the comparison that had been made between the segregation of genes and the reduction of chromosomes in the formation of eggs and sperm (p. 433).

In Morgan's *Experimental Zoology*, published about four years later, the treatment of Mendelism is less correct. He no longer accepts Mendel's explanation of the results, but tries to account for them without segregation of genes and purity of germ cells. He also doubts that the chromosomes are the bearers of the genes (1907, pp. 72-80). He continued to express such doubts in papers published up to the time he obtained Mendelian results in *Drosophila* in 1910.

Although Morgan became doubtful about Mendelism, he became less doubtful about the inheritance of acquired characters. He gave an account of Brown-Séquard's experiments, in which epilepsy had been induced in guinea pigs and had apparently been transmitted to their offspring. Morgan concluded: "The experiments appear to have been carried out with such care and the results are given in such detail that it seems that they must be accepted as establishing the inheritance of acquired characters" (1907, p. 54).

Morgan's *Experimental Zoology* was published in January 1907. At almost the same time (December 1906) there appeared another treatment of evolution in the light of post-Darwinian work, Lock's *Recent Progress in the Study of Variation, Heredity, and Evolution*. Lock had been a student of Bateson's; but unlike Bateson he thought there was a relation between genes and chromosomes, and he tried to synthesize Mendelism with cytology as well as with biometry and the mutation theory. Lock's book came out at a time when a controversy was being waged between Mendelians and those who called themselves Darwinians. A. R. Wallace, in a 1908 article in *Contemporary Review* in which he defended Darwinism against the charge of being an "unsuccessful hypothesis," recommended that those interested in the subject first read Lock's work as "the only recent book giving an account of the whole subject from the point of view of the Mendelians and Mutationists," and that they then read Wallace's own book entitled *Darwinism*. A page quoting Wallace's favorable reference was prefixed to later editions of Lock's book.

The book must also have impressed E. B. Wilson, for he used it as a text in an undergraduate one-semester course on heredity and evolution. Muller took the course in the first semester of the academic year 1909-10 and conceived a high opinion of Lock. Wilson never gave the course in subsequent years, so it was never taken by any other student who became a *Drosophila* geneticist.

In the same semester that Muller was taking Wilson's course, Sturtevant and Bridges were taking the first half of the elementary course in zoology. It happened that the regular lecturer for this part of the course, the protozoologist G. N. Calkins, was on sabbatical, and for the first and only time the lectures were given by Morgan. This is how Sturtevant and Bridges became acquainted with Morgan; but Sturtevant has recorded that no genetics was included in Morgan's lectures. Of course Morgan was then only beginning his *Drosophila* researches; he did not publish anything on them until 1910, when he obtained his first mutation. Still, the complete omission of the subject from the elementary course seems strange. Morgan had treated Mendelism in his *Evolution and Adaptation* and his *Experimental Zoology*. The second edition of Bateson's *Mendel's Principles of Heredity*, giving a full account of Mendelian researches up to that time, appeared in 1909, and Lock's book must have been known to Morgan because it had gone through two editions and was being used in Wilson's course. Sturtevant has recorded that he became acquainted with Lock's book as a student, though he has not stated when or how. I don't recall that Bridges ever mentioned the book.

When I took the elementary course in zoology, in 1910-11, Calkins (back from his sabbatical) included one lecture on Mendel's work at the end of the first semester. In the second semester Dean recommended as a historical treatment J. W. Judd's *Coming of Evolution*, which was actually limited to the coming of evolution in England. Dean also recommended three other books and required that the students read one of them: for those who wished an elementary treatment, *Animal Life* by Jordan and Kellogg; for those who wished something more advanced, *Evolution and Animal Life* by the same two authors; and for those who wished something still more advanced, Lock's book. I read the second and third books, and that is how I became acquainted with Lock. I do not know whether Lock was recommended or required by Dean in the previous year, when the course was being taken by Sturtevant and Bridges.

I don't recall any undergraduate student's saying he had read the *Origin of Species*, although one said he intended to. When I was a graduate student, I found in discussions with other graduate students that only one (H. B. Goodrich) had read the *Origin*. He said he had read different parts at different times, not consecutively, but was certain he had in this way covered the entire book. Those who had not read the *Origin* did not express any intention of reading it; and one (Muller) said it was not necessary to read it, because its ideas had all been incorporated in the biology of our time, and reading the book would not give us any additional ideas. This third reason for not reading Darwin is valid, at any rate to a large extent, if our aim is only to obtain a correct theory of evolution; but not if our aim is also historical, for we would not know which ideas to ascribe to Darwin, since those who quote his ideas do not always credit them to him, and on the other hand, some attribute to him ideas he did not hold.

Of course the chief reason why graduate students did not read the *Origin* was that biology had to a great extent been transformed since Darwin's day by the extensive application of experimental methods. A large literature had grown up dealing with such new fields as cytology, experimental embryology, and genetics, and students had to familiarize themselves with this literature, since two of Wilson's three graduate courses, and all of Morgan's graduate courses, were concerned with the new fields. The new literature of course had important bearings on evolution and was read partly for this reason. But it also had an interest of its own, as it threw light on vital processes other than evolution. In fact, many biologists were interested in these other processes as much as or more than in evolution. Bateson, for instance, who coined the term "genetics," often used instead the phrase "genetic physiology."

I managed to complete almost all the required courses by the end of my first graduate year, and then decided it was time to read the *Origin*. This was in the fall of 1914. Morgan saw the book on my desk in the fly room, and said, "You're reading the *Origin*? What do you think of it?" When I had finished the book, I told him I thought Darwin had proved his theory. This seemed to disappoint Morgan, and he raised two objections. One was that the theory of natural selection was tautological: it said that the fittest survive, but what it meant by the fittest was those that survive, so it was only repeating itself. The other objection was that the theory was teleological, that purpose entered into it.

I disagreed with both objections: I said that fitness could be tested in ways that were objective and did not involve the idea of purpose. For example, if we consider the case of a wolf chasing a deer, we can determine separately the maximum speed of which each animal is capable: this would measure the fitness of each animal with respect to this trait. Then we could bring the two animals together and see whether the fitter would succeed. I pointed out what has often been said, that the term "purpose" in Darwin is only a shorthand way of saying what a trait accomplishes, that actually natural selection eliminates purpose in a teleological sense. Morgan was not convinced.

On other occasions too, Morgan expressed opposition to the theory of natural selection. Once he said that the invention of the theory had not required any great amount of intelligence on Darwin's part; even Wallace had thought of it, and Wallace was a stupid man, as shown by his belief in spiritualism. But at another time Morgan said that de Vries regarded the mutation theory not as opposing natural selection but as supporting it.

Discussions of natural selection with Morgan have been reported by Muller and Sturtevant, both of whom said that he seemed impossible to convince and would raise the same objections in each discussion. Muller said he seemed to have a mental block on the subject (interview with G. E. Allen), and Sturtevant (1959) said that natural selection "was always a point of view with which he [Morgan] was basically dissatisfied."

Interest in selection was increased in 1914 by the publication of Castle and Phillips' experiments on hooded rats. These are a strain in which the color is limited to the head and a stripe down the back. The width of the stripe varies, and the experiments showed that it can be increased or decreased by selection. Castle and Phillips concluded that a single pair of genes is involved: a recessive gene that produces the hooded pattern, and a dominant gene that produces uniform color all over the body. They attributed the variation of the hooded pattern to fluctuation of the hooded gene, and the effectiveness of selection to the picking out of fluctuations in the desired direction. They even seemed to believe that selection not only picked out the fluctuations that happened to occur in its direction, but induced the gene to fluctuate in this direction. Apparently they believed that this was Darwin's notion of how natural selection works.

Castle and Phillips' paper was reported in the Journal Club (seminar) by Muller. He pointed out that the results did not prove Castle and Phillips' interpretation; for while the hooded pattern required homozygosis for the hooded gene, the variation of the pattern could be the result of modifying genes for which the strain was heterozygous, and different combinations of these could be brought about by selection. Muller concluded that the explanation based on modifying genes was the more probable, and he showed how to decide between the two theories experimentally.

Muller's report was published in the *American Naturalist* in 1914. (Sturtevant said he had suggested to Muller that it be published.) A similar analysis was included in the *Mechanism of Mendelian Heredity* (pp. 196-203) in 1915. Castle, not accepting the multiple-gene interpretation, published a rejoinder in which he continued to maintain his original view. Not until 1919 did he admit that further experiments showed the modifying-gene explanation to be correct. During this time the case was discussed in the fly room; Morgan, like the other *Drosophila* workers, supported the modifying-gene interpretation. This was only to be expected, for Castle's theory that genes fluctuate and that selection of fluctuations causes change in its direction was exactly what Morgan had objected to in *Evolution and Adaptation*.

In 1915-16 Morgan prepared his *Critique of the Theory of Evolution*, which was delivered as lectures at Princeton in February 1916 (and somewhat later at Berkeley) and was published in 1916. When he was writing these lectures, he did what he often did with his manuscripts: he passed them around to the other workers in the fly room. Muller had by then finished his work for the doctorate and was at the Rice Institute in Houston; Bridges did not take much part in theoretical discussions; so the manuscript was read chiefly by Sturtevant and myself. In the spring of 1916 Sturtevant was gone for a time on a trip, but I was there throughout the year.

Morgan seemed to want to present the theory of natural selection fairly and accurately. Whereas he had previously said that the conception of the theory had not required any great amount of intelligence, he now announced that he was going to say, "I will not allow anyone to admire

Darwin more than I admire Darwin." The statement appears in the *Critique* in the following form:

> While I heartily agree with my fellow biologists in ascribing to Darwin himself, and to his work, the first place in biological philosophy, yet recognition of this claim should not deter us from a careful analysis of the situation in the light of work that has been done since Darwin's time (p. 145).

And in the next sentence the *Origin* is referred to as a "great book."

Nevertheless Sturtevant and I both noticed that Morgan's block against natural selection was still operating. He never worded statements on the subject in a way that seemed to us correct and Darwinian, and he seemed unable to bring himself to advocate natural selection. He did, however, ask us to correct and rewrite the passages we objected to, and he accepted our revisions. But in conversation he continued to raise the same objections he had expressed before. And in the second edition of the *Critique*, published in 1925 under the title *Evolution and Genetics*, the characterization of Darwin's *Origin* as "his great book" was altered to "his famous book" (p. 119).

In his last year at Columbia Morgan told me one day in the spring of 1928 that he had been rereading *Evolution and Adaptation*. He said, "It's not a bad book; it says things that had to be said. It's still a good book." And in his *Scientific Basis of Evolution*, published in 1932 when he was at Cal Tech, he de-emphasized natural selection and emphasized variation, much as in *Evolution and Adaptation*.

On the other hand, Morgan's conversion to Mendelism, and to the chromosome theory (which he had also opposed before his work with *Drosophila*) showed no signs of being superficial. One part of the manuscript of the *Critique* contained a statement to the effect that these two theories had done much to clarify the problem of heredity. When he showed this to me and asked my opinion, I replied that I didn't think the statement was strong enough, that I thought he would be justified in saying that this work had solved the problem of heredity. This struck him as wrong, he said, because there was the whole remaining problem of the method by which genes produce their effects. I acknowledged that this was true, but said it was a separate subject, part of embryology and development; that the traditional problem of heredity was the passing on of hereditary traits from one generation to another. He thought about this and hesitated. I said, "Well, if you don't want to say that this work has solved the problem of heredity, you could say that this work has solved the traditional problem of heredity, specifying what you mean by the traditional problem." I suggested that if Weismann had been given the results of the *Drosophila* work, he would have agreed that the problem was solved. For we knew from the results that hereditary characters are determined by units—the Mendelian genes; we knew where the units are—in the chromosomes—and which units are in which chromosomes; we knew how they are arranged—linearly—and their relative positions; we knew how they are passed from one generation to another—by segregation and independent assortment, by linkage and crossing over. Morgan considered this a bit, then accepted it; and the statement that the traditional problem of heredity has been solved appears twice on page 144 of the *Critique*.

This shows that Morgan had no reservations about Mendelism and the chromosome theory. If he had had any, he could have raised them at this point as he had done in his *Experimental Zoology* and in papers written before 1910; he could have said, as many biologists were saying, that there were many characters whose inheritance had not been satisfactorily analyzed, or that seemed to be non-Mendelian. But he made no such objections. His objection was not that Mendelism and the chromosome theory did not account for all of heredity, but that they did not account for development.

Morgan's hesitation was apparently not entirely dispelled; for in a paper published in the following year, after contrasting what had been accomplished in the study of heredity with the relatively more difficult subject of embryology, he wrote:

> Do not understand me to say that I think all the problems of heredity have been solved, even with the acceptance of the chromosomal mechanism as the agent of transmission. In fact, I think that we are only at the beginning even of this study, for the important work of McClung, Wenrich, Miss Carothers, and Robertson shows that there are probably many surprises in store for us concerning *modes of distribution* of Mendelian factors. Moreover, the method by which crossing over of allelomorphic factors takes place is still in the speculative stage, so far as the cytological evidence is concerned, as are also many questions as to how the lineally arranged factors hold their order during the resting stages of the nucleus and during the condensed stages in the dividing chromosomes (1917, p. 537).

In a footnote he added, "The statement that I made in my recent book on the 'Critique of the Theory of Evolution,' that the *traditional* problem of heredity has been solved, is not in contradiction with the above statement which concerns the future problems of heredity."

Morgan's paper lists the objections that had been raised to Mendelism and the chromosome theory, and the misunderstandings that had arisen, and disproves the objections and corrects the misunderstandings by citing specific experimental evidence.

The manuscript of this paper, written in Woods Hole, was given to Sturtevant, Bridges, and me to read and to suggest a title for. In the course of our discussion I recalled Maxwell's *Theory of Heat* and some books with similar titles by other physicists. Knowing Morgan's antipathy to unsupported speculation, I said that in these titles the word "theory" was used not in the sense of speculation as contrasted with observation and experiment, but in the sense of a synthesis of what is known directly from observation and experiment and what is known by logical inference from observation and experiment. I then suggested that the paper could in the same way be titled "The Theory of the Gene." I began to add that if this was not satisfactory, I had another suggestion. Morgan stopped me at once, saying, "When you have a good title, don't look for another." He used the title for the paper, which appeared in the *American Naturalist*, and again for his Silliman lectures at Yale, which were published as a book in 1926.

On page 25 of this book Morgan summarizes what is meant by the theory of the gene. This summary, though brief, is somewhat more detailed than what is said about the mechanism of heredity on page 144 of the *Critique*; but the statement that the problem of heredity has been solved is omitted. (Both the summary and the statement had been omitted in *Evolution and Genetics* in 1925.) All this, however, does not indicate any weakening of Morgan's acceptance of Mendelism and the chromosome theory. When the *Critique* was published in 1916, these two theories were regarded with skepticism by many if not most biologists; it was therefore important to emphasize what had been proved. By 1925 and 1926 both theories were generally accepted, and it was more important to emphasize what problems remained to be solved.

It will help us to understand Morgan's attitude toward the theory of natural selection if we bear in mind that, since the publication of the *Origin of Species*, four generations of scientists have discussed, analyzed, and debated it, and have formulated and reformulated its propositions, sometimes mathematically; as a result the concepts have been more sharply defined and the issues clarified. In the nineteenth century, when Morgan was a student and a young researcher, this clarification had not proceeded as far as it has today. This is evident in some of the correspondence between Huxley and Hooker, who were Darwin's closest friends among scientists and had opportunities of discussing the theory with him (Huxley, 1900).

Huxley wrote to Hooker on March 9, 1888:

I have been trying to set out the argument of the "Origin of Species," and reading the book for the nth time for that purpose. It is one of the hardest books to understand thoroughly that I know of, and I suppose that is the reason why even people like Romanes get so hopelessly wrong.

Two weeks later Huxley wrote again:

I suppose [J. D.] Dana [the geologist] has sent you his obituary of Asa Gray [the first supporter of Darwin's theory in America].

The most curious feature I note in it is that neither of them seems to have mastered the principles of Darwin's theory . . .

I have read the life and letters [of Darwin] all through again, and the *Origin* for the sixth or seventh time, becoming confirmed in my opinion that it is one of the most difficult books to exhaust that ever was written.

I have a notion of writing out the argument of the *Origin* in systematic shape as a sort of primer of *Darwinismus*.

Hooker in his reply (March 27, 1888) did not agree with Huxley about Gray but agreed that Darwin had been misunderstood by many writers:

I did not follow Gray into his later comments on Darwinism, and I never read his *Darwiniana*. My recollection of his attitude after acceptance of the doctrine, and during the first few years of his active promulgation of it, is that he understood it clearly . . .

He certainly showed far more knowledge and appreciation of the contents of the *Origin* than any of the reviewers and than any of the commentators, yourself excepted.

Latterly he got deeper and deeper into theological and metaphysical wanderings, and finally formulated his ideas in an illogical fashion.

In Huxley's letters the words "to understand thoroughly" and "to exhaust" show that his difficulty was with the implications rather than with the explicit meanings of Darwin's statements. Morgan, in the *Critique*, mentioned the implications specifically as causing difficulty. Speaking of the idea that genes fluctuate and that evolution is the result of selection of these fluctuations, he said:

Darwin himself was extraordinarily careful, however, in the statements he made in this connection and it is rather by implication than by actual reference that one can ascribe this meaning to his views (pp. 155-156).

Morgan also referred to the diversity of interpretations of the *Origin* and implied that since the interpretations differed, some of them must be wrong:

> Darwin's Theory of Natural Selection still holds to-day first place in every discussion of evolution, and for this very reason the theory calls for careful scrutiny; for it is not difficult to show that the expression "natural selection" is to many men a metaphor that carries many meanings, and sometimes different meanings to different men (p. 145).

This multiplicity of meanings is found in the discussions of evolution in the later years of the nineteenth century which, as I have said above, were not much read by students of my generation. But an account of them was given in a book called *Darwinism To-day*, by Kellogg (1907), which was read to some extent, partly because (as Sturtevant said) it was useful for answering questions we might be asked on a doctoral examination. The theories treated by Kellogg have been described as "a bewildering array" (Wright, 1968, p. 8).

This did not bother students much, because to them these theories were mere historical curiosities that had been superseded by Mendelian work, especially that on *Drosophila*. But Morgan had read these theories, not in Kellogg's book but in the original books and papers in which they first appeared; to him they had been not historical curiosities but serious opinions put forward by contemporary authors, many of them eminent biologists. He may have become even more bewildered than students of a later generation, especially if he did not at first distinguish between the views of Darwin and those of the "more ardent but less critical" disciples who called themselves selectionists. Still, if this is how his confusion originated, it does not explain why it persisted—as it did when he stated that the argument of the *Origin of Species* was tautological and teleological.

It is interesting to inquire to what extent Morgan's ideas on evolution in 1903 and 1907 influenced the subsequent development of the subject, or foreshadowed it. His idea that genes do not change teleologically (that is, in response to the needs of the organism) is contained in Darwin's theory of natural selection; and as this continued to be the prevalent explanation of evolution, it is not easy to discover what effect Morgan's advocacy had. The idea that genes do not fluctuate is opposed to Darwin's view; but Morgan's advocacy did not prevent Castle from maintaining the fluctuation of genes in 1914 and for some years later.

The idea of stable genes became established, I think, not so much by Morgan's arguments in 1903 and 1907 as by Johannsen's experiments on beans and the work of Morgan and his associates on *Drosophila* from 1910 on. Morgan's attempt to minimize the importance of selection in evolution cannot have had much effect, as is shown by the continuing prevalence of Darwin's theory and the development of population genetics in the work of Haldane, Wright, Fisher, and others. Nevertheless, Morgan's denial of the part played by selection had an important result in turning his own attention from investigating the effects of selection to investigating the causes of the changes of genes. It may also have influenced others like MacDougal, who tried to obtain mutations in plants by exposure to radium and by the injection of chemicals.

Morgan considered variation (as distinct from selection) so important in evolution that for the Friday evening meetings at his home, at which he and some of his students read and discussed some important book every year, in 1908-09, as recorded by Payne (p. 28), he chose Darwin's *Variation of Animals and Plants under Domestication*. The notion of learning from this book how to produce mutations may make us smile today; but it shows how little information on the subject was then available. Morgan did not stop at reading the book: he tried to produce mutations experimentally, working on *Drosophila* because of its small size, short life cycle, and large number of offspring. He tried chemicals, temperature, X rays, and radium; and he got some mutations, as MacDougal also had. But in neither set of experiments was it possible to prove that the mutations had occurred as a result of the treatment.

The spontaneous variations that were found in *Drosophila* by Morgan and his students included not only mutations but also rearrangements of genes—inversions, deletions, and translocations. From these it was possible to build up strains in which mutations could be detected easily; and some strains of this kind appeared spontaneously. One of these, obtained by Muller, enabled him to prove in 1927 that X rays produce mutations in *Drosophila*. Stadler independently demonstrated in 1928 that radiations produce mutations in plants. And since then it has been proved that mutations can be produced by chemicals.

References

Bateson, W. 1902. *Mendel's principles of heredity: a defence*. Cambridge: Cambridge University Press.
———— 1909. *Mendel's principles of heredity*. Cambridge: Cambridge University Press.
Castle, W. E. 1914. Mr. Muller and the constancy of Mendelian factors. *American Naturalist* 49:37-42.
———— 1919. Piebald rats and selection. *American Naturalist* 53:370-376.
———— and J. C. Phillips. 1914. Piebald rats and selection. Washington, D.C.: Carnegie Institution of Washington, publication no. 195.
Huxley, T. H. 1900. *Life and letters*, ed. Leonard Huxley. London: Macmillan, vol. 2, pp. 192-193; New York: Appleton, vol. 2, pp. 204-206.
Jordan, D. S., and V. L. Kellogg. 1901. *Animal life*. New York: Appleton.

———— 1907. *Evolution and animal life*. New York: Appleton.

Judd, J. W. 1910. *The coming of evolution*. Cambridge: Cambridge University Press.

Kellogg, V. L. 1907. *Darwinism to-day*. New York: Holt.

Lock, R. H. 1906. *Recent progress in the study of variation, heredity, and evolution*. New York: Dutton.

Morgan, T. H. 1903. *Evolution and adaptation*. New York: Macmillan.

———— 1907. *Experimental zoology*. New York: Macmillan.

———— 1916. *Critique of the theory of evolution*. Princeton: Princeton University Press.

———— 1917. The theory of the gene. *American Naturalist* 51:513-544.

———— 1925. *Evolution and genetics*. Princeton: Princeton University Press.

———— 1926. *The theory of the gene*. New Haven: Yale University Press.

———— 1932. *The scientific basis of evolution*. New York: Norton.

————, A. H. Sturtevant, H. J. Muller, and C. B. Bridges. 1915. *The mechanism of Mendelian heredity*. New York: Holt.

Muller, H. J. 1914. The bearing of the selection experiments of Castle and Phillips on the variability of genes. *American Naturalist* 48:567-576.

Payne, F. (n.d., subsequent to 1973). *Memories and reflections*. Bloomington: Indiana University.

Sturtevant, A. H. 1959. Thomas Hunt Morgan, 1866-1945. *Biographical Memoirs of the National Academy of Sciences* 33:283-325.

Wright, S. 1968. *Evolution and the genetics of populations*. Chicago: University of Chicago Press, vol. 1.

Morgan and His School in the 1930s

Theodosius Dobzhansky

T. H. Morgan was in the autumn of his life when I joined his group late in 1927. While at Columbia University, he still carried on some experimental cultures of *Drosophila*. After moving to Pasadena, he abandoned *Drosophila* research, and during the rest of his life reverted to the work of his younger years—experimental embryology. His admirable honesty and detestation of cant made him admit perfectly frankly that *Drosophila* genetics began to escape him. However, he by no means lost interest in what was being done in the field founded by him, especially in its general scientific and philosophical implications. For several years I could mostly listen, and rarely participate, in the frequent discussions among Morgan and his collaborators, mainly Bridges, Sturtevant, and Schultz. But for about six years, between 1935 and 1940, when I had a laboratory room by myself, Morgan came in about once a week and talked freely on problems of all kinds, from science to philosophy to politics. Thanks to these, often rambling, talks, I became well acquainted with his personality and *Weltanschauung*.

Morgan was a complex and in many ways a contradictory personality. "Naturalist" was a word almost of contempt with him, the antonym of "scientist." Yet Morgan himself was an excellent naturalist, not only knowing animals and plants but aesthetically enjoying the observing of them. Philosophy was a waste of time, and also a treacherous enemy of rigorous thinking. Yet Morgan was familiar with philosophy. Morgan carried economy of laboratory expenditures to ridiculous stinginess, yet he was generous with his private funds, and during the depression years helped several students without letting them know the source of the money. Morgan was a liberal, and he hired and promoted people regardless of their origins, yet he occasionally talked like a race and class bigot partly from his desire to maintain his reputation of being slightly eccentric and a bit impish and partly because of some much deeper personal motives that may have required a psychoanalyst's couch to clarify fully.

Morgan enjoyed nothing more than to show that some generally accepted views, in science or in anything else, were invalid. More than once he admitted that he argued for or against something for the sake of intellectual gymnastics, to hear how well his interlocutor would oppose him. This trick sometimes caused disappointment and even offense to those who did not know him well. It would be too much to suppose that Morgan consciously engaged in this sort of intellectual gymnastics also in some of his scientific publications. There is however no doubt that he loved to prick what were, or seemed to him to be, overinflated but empty balloons.

There was in Morgan's Weltanschauung another part that was no joking matter to him; the direction of his scientific activity and his personality are incomprehensible without appreciating Morgan's deep-seated and uncompromising opposition to religion. It can be gleaned from some of his writings, although for obvious reasons he did not talk about it explicitly, except with a few intimates. The main goal of basic biology, in fact of natural science, was to show the invalidity of religious views of man and the universe. To do so one must dispel mysteries enveloping man and the world, because mysteries are the foundations and supports of religion. Because heredity was one of the mysteries, genetics was an important science demystifying this particular phenomenon of nature. Evolution was, needless to say, tremendously important, because it did away with the biblical story of creation of the world and of man. By the mid-1930s Morgan felt that, although genetics and evolution were by no means fully clarified, there was still untouched this mystery of mysteries —human thought, human brain, and nerve activity in general. The money given by the Rockefeller Foundation after Morgan had been awarded his Nobel Prize went to start a laboratory at Cal Tech for the study of nerve physiology, not to strengthen genetics or evolutionary studies.

The religion that Morgan fought during his scientific life was, of course, the Bible Belt fundamentalist religion. How was it possible for a man of great intellectual sophistication and many-sided education to overlook the advanced religious thought of his time? He did not overlook it, but held it to be merely an artful ploy to save the same old-fashioned fundamentalist religion. Morgan's colleagues and collaborators, mainly Sturtevant and Bridges, shared his antireligious preconceptions, but at a rather lower level of sophistication. No matter how strongly Morgan felt religion to be a harmful superstition, he held dedication to religion in students and other colleagues to be their private affair. Others took it to be evidence of a low level of intelligence.

To serve the function that Morgan assigned to it, biology had to be free of any taint or suspicion of philosophical idealism, vitalism, or teleology. It had to be strictly reductionistic. Biological phenomena had to be explained in terms of chemistry and physics. Morgan himself knew little chemistry, but the less he knew the more he was fascinated by the powers he believed chemistry to possess. There was no surer way to impress him than to talk about biological phenomena in ostensibly chemical terms, and some individuals successfully used that technique.

In his old age, the Pasadena period of his life, Morgan no longer participated actively in *Drosophila* genetics research. Problems of evolution remained, however, the center of his interests, as they were in his youth and his middle age. To doubt that a man who published at least four books on evolution was himself vitally interested in evolutionary problems is nothing short of preposterous. (And yet a symposium organized to celebrate the centennial of his birth omitted evolution entirely.) Morgan's attitude toward evolutionary studies was determined by his philosophical preconceptions. An account of evolution in strictly physico-chemical terms was in his day out of reach, but any theory that would be acceptable to him had to be at least free of "metaphysical speculations."

In the early 1900s Morgan was, incredibly, an opponent of Mendelism. I knew Morgan, of course, almost a quarter of a century later. However, I believe that I can understand the logic of this seemingly illogical turn in his ideas. The late nineteenth century was a period of speculative theories of heredity, which are remembered at present by very few biologists, other than specialists in history of biology. Some leading biologists, especially in Europe, saw fit to invent a whole collection of living particles, to which they ascribed various wonderful properties, and which they assumed to be components of imaginary mechanisms of heredity and evolution. How many contemporary biologists have ever heard about gemmules, extracellular and intracellular pangenesis, micelles, biophores, or determinants? Morgan was skeptical about these speculations, and hugely enjoyed poking fun at them. Of course, he made the mistake of placing Mendelian genes on the same heap with pangenes and biophores. For a few years he overlooked the basic difference between armchair speculations and inferences drawn from rigorously conducted experiments. He who never blundered in his life may throw at Morgan the first reproach. Anyway, Morgan cannot be accused of unwillingness to change his opinions when they were shown to be invalid. The former critic of Mendelism became its most effective supporter and developer.

Also quite logically and predictably, Morgan was inclined to assume that evolution consisted of mutations. A mutation, after all, could be observed happening, and a mutant could be submitted to experimental tests. But the mutations that Morgan observed in *Drosophila* did not at all behave like the ones de Vries found in *Oenothera*. *Drosophila* mutants required much study and an interpretation different from that given by de Vries. In this study, Morgan's collaboration with Bridges, Muller, and Sturtevant was invaluable. The greatest breakthrough was Bridges' work on nondisjunction of sex chromosomes in *Drosophila*. Bridges showed that the behavior of the chromosomes was predictable from the behavior of visible mutant genes and vice versa. The Mendelian gene was thus shown to reside in a chromosome, a body visible under the microscope.

To Morgan this discovery was as welcome and enlightening as it was unwelcome and confusing to Bateson and Johannsen, who would have genes as symbols to be used to manipulate data obtained in hybridization experiments. Morgan at last had genes that were material particles, to be sure not yet chemically analyzable and not visible under a light microscope. At any rate, these genes were no longer abstract symbols in quasi-mathemetical formulas. The contrasting attitudes of Morgan on one side and of Bateson and Johannsen on the other characterized the two schools of thought among the pioneers of genetics. It would be an exaggeration to label them materialist and idealist; it is rather a preference for the tangible and visible, contrasted with symbols that can be written on a piece of paper.

The next step was the demonstration that genes are arranged in chromosomes in a single linear file. The phenomena of linkage and recombination of linked genes were discovered not in *Drosophila* but the sweet pea *Lathyrus odoratus* by Punnett, an associate of Bateson. Quite in keeping with the tradition of the Bateson school, these phenomena were given a purely symbolic interpretation. In a very early paper (1911), Morgan gave a different interpretation of linkage in *Drosophila*. Even though Bridges' work on nondisjunction was still in the future, Morgan advanced the simple idea that the amount of recombination is a function of the distance between genes in a chromosome. Some years later this idea was given a better formulation by Sturtevant. Morgan measured the amount of recombination in ratios, Sturtevant in percentages. Why should this make any difference? It did—the percentages of recombination of three linked genes stood in the relations $AC = AB + BC$ or $AC = AB - BC$, reflecting linear orders ABC or ACB. Visible demonstration of the linear arrangement of genes in chromosomes was given around 1930, by Muller and Painter in Texas and myself in Pasadena, working with chromosomal aberrations. Rediscovery of polytene chromosomes in the larval salivary glands by Heitz and Bauer and their study by Bridges in *Drosophila* came at about the same time. These findings gave deep satisfaction to Morgan, who clearly felt that his main lifework was completed.

Morgan and his collaborators liked to say that genetics can be studied without any reference to evolution. But at least Morgan and Sturtevant (and Bridges to a lesser extent) were interested in the bearings of their work on evolutionary problems. In their views evolution was a succession of mutations, hence of changes in some presumably minute segments of the chromosomes. Of course, the problem of the role of natural selection was inescapable. But then why did Morgan, in his conversations as well as his writings, always discuss evolution in terms of mutations, and little in terms of selection? To him mutations were observable facts; he saw them in his cultures, whereas natural selection was still a theoretical inference. This emphasis led almost inevitably to an error of perspective. Most "good" mutations—that is, mutations usable in linkage studies—are cripples or at any rate reduce the viability of their carriers. They could be envisaged to be of evolutionary importance only in some extraordinary circumstances. To be sure, small mutations were also known in *Drosophila*. However, they were not good for genetic experiments, were seldom preserved, and little studied.

It seems hardly believable now that in the 1920s and even 1930s influential biologists declared "Morgan's mutations" to be products of abnormal laboratory environments; these mutations did not occur in nature at all. Chetverikov's demonstration that they did occur in nature was welcomed but produced no excitement. The reasoning went something like this: of course, mutations occur in and out of laboratories, only unreasonable people could have doubted this. Timoféeff-Ressovsky and later Dubinin extended Chetverikov's work, and still later I took up this study. Morgan was mildly interested, but Sturtevant thought the results were not worth the efforts. Most mutants found in natural populations concealed in heterozygous state were still lethals or cripples.

Morgan was not an antiselectionist, at least not in the years when I knew him. However, he conceived the action of natural selection as that of a sieve—it preserves the rare useful mutations, and lets the rest be eliminated. I remember well a conversation with him (it may have been in 1937, plus or minus a year). He challenged me to explain how it happens that mutants promoted by natural selection are useful when they arise. He was thinking in terms of what Cuénot called preadaptation, though he did not, as far as I can recall, use that word. I fear that my response to his challenge was not a satisfactory one. I should have but did not answer that natural selection can be more easily envisaged as a factor creating adaptive changes if it compounds many small genetic variants into novel adaptive genotypes. Each variant can be seen as preadapted only after the fact. There is no need to suppose that mutations originating these small variants are by chance or by plan preadapted at the origin, which poses another fundamental problem, that of the classical versus the balance model of genetic population structure.

Those in the Morgan school, and Sturtevant and Muller more emphatically than the others, were exponents of what I later called the classical model. That is, most individuals in populations of any species are homozygous and alike for most genes. Each species has its wild type, deviations from which are infrequent and either pathological or insignificant. Variations in facial features in man were more than once mentioned as examples of insignificant genetic variations. Thinking back about these old discussions, I wonder if here was a germ of the pan-neutralism, now fashionable in some circles. Be that as it may, the name "neoclassical" suggested by Lewontin for the pan-neutralist model seems to me quite inappropriate. The very essence of the classical model was the wild-type idea and the assumption of rarity and paucity of genetic variation. Modern pan-neutralists cannot and do not maintain these old theses.

Morgan was profoundly skeptical about species as biological and evolutionary realities. The species problem simply did not interest him. Sturtevant, on the other hand, was interested; he after all was the author of an early monograph of the American species of Drosophila. However, consistent with his adherence to the classical model of genetic population structure, he firmly believed that at least closely related species of Drosophila differed in very few genes. The evidence of this, which he regarded as conclusive, was the occurrence in closely related, and even not so closely related, species of Drosophila of similar mutant alleles. Indeed, mutants such as white, yellow, forked, and cinnabar types, arose in many species. If the mutants are so similar are not their wild-type alleles also similar? The findings of Beadle and other pioneers of biochemical genetics seemed to show that genetic similarities extend much beyond the confines of a genus or even a family. Enzymes with similar functions are found in quite dissimilar and phylogenetically remote organisms. Why suppose that the genes responsible for the production of similar enzymes are not identical? The evidence of their nonidentity was obtained much later, within the last two decades or so.

If species differ in few genes, then few lucky mutational changes can add up to produce new species. The chain of arguments is thus closed: we observe mutations arising within species, and we find no evidence of species differing in many gene loci. Of course, the mutations must be lucky ones to pass the scrutiny of natural selection. Species differ also in chromosomal changes, such as inversions. Yet Sturtevant found just a single inversion differentiating *Drosophila melanogaster* and *D. simulans*. Individuals within a species can differ in several inversions. Such phenomena as sterility of interspecific hybrids and lack of sexual attraction between individuals of different species remained unaccounted for. However, mutant genes that induce sterility are known, and some forms of inversion heterozygosis make the carriers semisterile.

During the 1920s and the 1930s the evolutionary views of members of the Morgan school were not identical with those of today's evolutionists. Surely, such an identity would not be expected. In a living science, an interval of thirty or forty years brings changes and novelties. No less surely, the Morgan school was then in the forefront of evolutionary studies in most though not in all respects. They missed some and underestimated the importance of other contributions that were then being made, particularly by Chetverikov, Fisher, Wright, and Haldane. Yet a majority of evolutionists of that time espoused various forms of Lamarckism, autogenesis, and finalism. To these, Morgan and his colleagues opposed a body of facts and ideas derived chiefly from their own and other studies on genetics. Genetic theory of evolution is not a synonym of synthetic theory. However, without a genetic foundation the synthetic theory could not exist. Given that foundation, a synthetic theory was probably bound to appear.

Biased and false descriptions of human relations in Morgan's fly room have appeared in talks and publications. Freedom of scientific and philosophical views prevailed in the group of scientists and students headed by Morgan. Everybody knew what problems everybody was studying and what results were being obtained. It never occurred to anybody to keep his current work secret, or for that matter his plans for work in the immediate future. Current and future work was freely and extensively discussed and commented on. Disagreements were not hidden under a cloak of false delicacy. An erroneous or reckless opinion was promptly and uncompromisingly, with some people perhaps on occasion rudely, contradicted. Nevertheless, a junior scientist was permitted to disagree with his seniors. Having been brought up in Russia, in a scientific tradition different in many ways from that of Morgan, I disagreed with Morgan's views on many occasions. And I am convinced that it was because of my willingness to state frankly views different from his, that during the late thirties Morgan liked to engage me in scientific and philosophical discussions, sometimes making deliberately provocative statements. To be a member of Morgan's school for more than a decade was, for me, formative and inspiring.

G. G. Simpson*

Ernst Mayr

George Gaylord Simpson was born in Chicago on June 16, 1902, but grew up in Denver. He attended the University of Colorado in 1918-19 and 1920-22, and received both his Ph.B. (1923) and his Ph.D. (1926) from Yale. Although he was a student of geology at both Colorado and Yale, he also did much work in the biology department at Yale. From 1927 to 1959 Simpson was on the scientific staff of the American Museum of Natural History as a vertebrate paleontologist; between 1945 and 1959 he was a professor in the department of geology at Columbia University too. From 1959 to 1970 he was Alexander Agassiz Professor, Museum of Comparative Zoology, Harvard University. Since 1970 he has been professor of geology at the University of Arizona.

Intellectual Influences

In the 1920s and 1930s, the years during which the ideas matured that he published in 1944 and in subsequent books, Simpson had contacts with

*George Gaylord Simpson was one of the key figures in the synthesis. As he was unable to attend either of the workshops, in the spring of 1974 I sent him a questionnaire, which he answered very fully. The following account is an abridged and edited version of his answers to some of the questions (many other questions and replies are here omitted). The questionnaire in its entirety will be deposited in the archives of the American Philosophical Society. For further detail on Simpson's life see his autobiography (1978).

E.M.

teachers, fellow graduate students, and colleagues. He characterized these contacts and their influence on him as quoted below. Simpson is not a "verbal" type.[1] "Reading was always more important than personal contact. I would also much rather read than hear a lecture, rather write than converse," he wrote me. During my own personal contacts with Simpson, I do not recall any scientific, and particularly evolutionary, conversations with him until about 1941, although from January 1931 on we were members of the rather small scientific staff of the American Museum and usually ate lunch at the same staff table. Perhaps I considered him a paleontologist and he me an ornithologist.

Simpson considered the question of his principal teachers who most affected the formation of his evolutionary thinking:

> At Colorado A. J. Tieje, an invertebrate paleontologist who encouraged me to transfer to Yale. At Yale my major teacher was Richard Swann Lull, vertebrate paleontologist, and author of the then leading text on organic evolution. The whole geology staff was particularly able and inspiring. Next to Lull, I was most influenced by two invertebrate paleontologists, Charles Schuchert, retired but still very much present, and Carl O. Dunbar. In biology I was most influenced by Ross Harrison, already famous then and enduringly so in history, and by L. L. Woodruff. Other older, or more advanced, persons who influenced me personally (and not only through their writings) and who were in a way teachers about evolution included W. D. Matthew, H. F. Osborn, W. B. Scott, W. K. Gregory, J. B. S. Haldane (younger in fact than Julian Huxley, but he always affected me as a senior and Julian did not).
>
> Lull was considered a great authority on evolutionary theory—and he was, in the sense of being well informed on the ideas, old and new, in this field. He never had an original idea of his own about theories of evolution, but just expounded everyone's views as if all were equal. That was very useful to me only in telling me what the established alternatives were and what to read. Harrison taught morphology in a rather nineteenth-century Teutonic way, very thorough and wonderful basic knowledge, with no emphasis on evolutionary theory. Woodruff gave a good introduction to Mendelian genetics as of the 1920s. As to evolution, he adhered to what he called "clarified Darwinism," accepting natural selection as the

1. Simpson comments on this: "Contrary to his [Mayr's] statement, I certainly am a verbal type—I have written more than 700 publications and thousands of letters, all in words . . . I can talk in several languages, in fact, and do so with pleasure in social company. Incidentally, I have given (orally) a number of courses—one in fact for several years with Ernst—and have given hundreds of lectures, in most of the states and in several languages in countries on all the continents except Asia and Antarctica."

only natural explanation of adaptation, but believing that many "variations" are neutral to selection and hinting that the total explanation may include, if not the supernatural, at least some vitalistic factor. I was already antivitalistic, but apart from that I thought Woodruff's attitude reasonable, and still think it was for that time, before the synthesis.

Matthew wrote almost nothing directly on evolutionary theory, but he talked a good deal about it and was sensibly neo-Darwinian in a somewhat turn-of-the-century manner. He influenced me a bit on that, but much more in other ways. The same could be said of Gregory, who was always more interested in morphogenesis than in other aspects of evolution but accepted neo-Darwinian and eventually synthetic theory as far as he followed them. Osborn's views on evolution always struck me as completely untenable both on theoretical and on evidential grounds, yet I respected him and he influenced me favorably, as I believe, in that he insisted that one try to learn not only facts but also their meaning—even if it finally proved (as he foresaw was possible in his own case) that one had the meaning wrong.

Scott was particularly interesting. As a young man he had been somewhat oriented toward theory and optimistic about progress toward understanding evolution. When I first knew him, he was already in his seventies (he died in 1947 at the age of eighty-nine) and was disillusioned. He thought that the course of evolution could be worked out and was largely orthogenetic, but that no explanation for it was in sight. He considered Darwin's theory the only one really to offer an explanation, but believed that it was proved wrong by the facts of the fossil record. That, indeed, added to my interest in trying to determine whether the course of evolution as far as we can reconstruct it, does in fact contradict Darwinism. (Of course I decided, after some further years of study, that in essentials it does not, but quite the contrary.)

Simpson responded to my question about influential associates:

Most of my peer associates in graduate school were studying geology, not biology. They happened to be a particularly brilliant group (within twenty years they held most of the key positions in American geology), and several have been lifelong friends, but we did not discuss evolution to any great extent, hardly beyond all agreeing that it is a fact. I saw little of biology students outside of classes.

In London, where I had a postdoctoral fellowship in 1926-27, my closest associate at the British Museum was A. T. Hopwood. He was a neo-Lamarckian and I was already a neo-Darwinian. We did have many discussions about evolution among many other things. Outside the museum I had discussions especially with Wilfrid Le Gros

Clark and Julian Huxley, both older than I but early enough in their careers to be little beyond my peer group. Both remained close friends until their deaths. Arthur Hopwood, keen in the 1920s, lost interest and we had drifted apart long before his death.

At the American Museum I do not recall any discussions as early as the 1920s or early 1930s really relevant to evolutionary theory with most of my peers (but many with some of my elders). G. K. Noble, again older than I but not too far from being a peer, did discuss evolutionary theory with me. We disagreed because he believed in evolution by saltatory mutations. After the 1930s, I had many important contacts in New York, especially with Dobzhansky and Mayr, and elsewhere with many others more or less in my age group.

I asked Simpson which workers in his area of specialization were particularly strongly opposed to the Darwinian synthesis. What were their reasons? He replied:

At the time when I began to consider this subject I believe that the majority of paleontologists were opposed to Darwinism and neo-Darwinism, and most were still opposed in the early years of the synthetic theory. I believe that I was the first to coordinate paleontology with the then synthetic theory, although since then most paleontologists have come around to this, of course now in somewhat different and more sophisticated form.

I have mentioned Arthur Hopwood, friend of my youth, and Osborn and Scott, nearer my grandfather's generation. Paleontologists, as evolutionists, had always been impressed or even overimpressed by evidence of universal and progressive adaptation in the fossil record as well as in its recent outcome. That predisposed them to neo-Lamarckism and to orthogenesis. In the United States, they were also influenced by E. D. Cope, the ablest American paleontologist of the nineteenth century and a thoroughgoing neo-Lamarckian. Although Osborn would not qualify as neo-Lamarckian, his strangely idiosyncratic theories grew from that soil, and belief in orthogenesis was one of the few beliefs left to Scott as an old man.

Then, too, one of the dominant ideas in evolutionary biology just before the synthetic theory arose and even in the early years of the theory was a belief in evolution by major mutations, unchecked by natural selection. Goldschmidt's theory of systemic mutation was a somewhat tardy manifestation that helped to keep those ideas current. To most paleontologists such a process, as general in evolution, was unthinkable, and they therefore tended to reject all genetic evidence, which nevertheless became a crucial element in the formation of the synthetic theory. However, as usual, things are not so simple, for the German paleontologist Schindewolf attempted a premature synthesis with genetics on just that basis, of major mutation without selection. Among French paleontologists other influences were at work, and still are to a surprising extent, partly as a result of a long tradition of vitalism, in turn traceable to (among other things) a turn for classical philosophy and the influence of Roman Catholic theology, oddly current even in anticlerics. The whole was and is complicated by French chauvinism, which automatically turned them against Darwinism, the product of an English gentleman, at the start, and later against the synthetic theory, developed mostly by British and American scientists. (This chauvinism, of course, is not true of all French paleontologists or biologists, but of enough of them to warrant consideration of the causes.)

Simpson also wrote about influential books:

In graduate school, besides working through Lull's comprehensive although rather nonselective text (I did not take the corresponding undergraduate course), I had of course read *On the Origin of Species* and other classics of the evolutionary literature, including Henderson's somewhat odd masterpiece, *The Fitness of the Environment*, and D'Arcy Thompson's charming but anachronistic and anti-Darwinian *On Growth and Form*. I also read much of the journal literature, usually as it appeared, but this has been so much overlain by later reading and library research that I can rarely be sure just when I read a particular paper. I do remember that I read Garstang's famous paper on recapitulation when it first appeared and was greatly impressed; I was a senior at Yale.

While still in graduate school I found Othenio Abel's books particularly interesting and useful; like Lull's books and his teaching they helped little or hardly at all in explaining the history of life but greatly in reconstructing it from the available evidence.

When I was definitely working toward a synthesis in my own mind, I read the works that became some of the sources of the synthetic theory, notably those of Fisher and Haldane. I was a bit late in getting to Sewall Wright, but did so in time. (Seven of Wright's relatively early papers are cited in *Tempo and Mode*.)

My own thinking along theoretical lines was nevertheless mostly along lines of historiography and organismal adaptation, in fossil and recent organisms, until the first edition of Dobzhansky's *Genetics and the Origin of Species* (1937). That book profoundly changed my whole outlook and started me thinking more definitely along the lines of an explanatory (causal) synthesis and less exclusively along lines more nearly traditional in paleontology. Although I have since learned a great deal from Mayr's books, they had no influence on my orientation in the synthetic theory. I had not seen Mayr's first book, *Systematics and the Origin of Species* (1942) when I wrote

Tempo and Mode in Evolution (1944). My manuscript was completed before Mayr's book appeared, and thereafter I could make no changes because I was overseas in the army until late in 1944, after my own book was published.

The publications that I have so far mentioned all seem to me to have been excellent within their scope and for their time and are of at least historical interest now. Most have been superseded, if only by later editions or similar works by the same authors. I think a few (including my own of that period) are still worth reading for their own sake and still of current interest as distinct from their historical importance (as is still true of almost all of Darwin's books). Abel's books have been practically forgotten, although at least one (*Palaeobiologie*) has real historical interest. None appeared in languages other than German, not now read easily if at all by the younger generation in America, and in Europe Abel became personally non grata after the war, not for scientific but for political reasons.

Two books that certainly influenced me because I had to consider the probability of their theses in the light of evidence available to me but that proved, in my opinion, to be seriously misleading were Schindewolf's *Paläontologie, Entwicklungslehre und Genetik* and Goldschmidt's *Material Basis of Evolution*. I admired both these men and most of their work, although I concluded that their theoretical views were untenable. The necessity of considering such opposing views was a definite factor in the foundation of the synthetic theory.

Simpson summarized his own most original and important contributions to the evolutionary synthesis:

My most original contribution was probably bringing the history of life, as seen by the fossil record and otherwise, into the growing synthesis and demonstrating that it is consistent with inferences from genetics particularly, but also from all other sources of knowledge of evolution.

My writing of greatest impact on evolutionary theory was probably *Tempo and Mode in Evolution*, later supplemented by *The Major Features of Evolution* (1953)—although *The Meaning of Evolution* has had wider circulation throughout the world and in various scientific disciplines (notably anthropology), and a textbook, *Life*, oriented around evolution, has had most influence on education in itself and in inspiring a number of other texts similar in approach.

By the 1930s, when I began to accumulate the material that became *Tempo and Mode*, I had reached a philosophical and theoretical viewpoint that seemed to me to have some originality and some importance. It therefore required expression in writing. Further study and writing followed because of continued and increasing interest both on my part and on the part of readers. I have always habitually conveyed ideas preferably in writing.

The following additional comment by Simpson was written in January 1977, in response to my request:

> My feeling is that my contributions to the theory have been published and that others can judge them with less bias or embarrassment than I. A paleontologist much younger than I has in fact expressed such a judgment, and in response to this request I paraphrase and shorten but do not modify his comments: "Simpson," he said, "has had a central role in fashioning our modern version of evolutionary theory. This provision of a paradigm for research, whether in agreement or opposition, is one of the great achievements in the history of biology. Simpson reconciled paleontology and genetics in an almost single-handed act. This completed the most important theoretical advance in paleontology since the recognition that fossils record the history of life." There are other paleontologists and biologists who think much less of my efforts.

Simpson does not think that the new synthesis was a return to the classical Darwinian interpretation (excluding certain Lamarckian tendencies):

> To consider the synthesis as a return to classical Darwinism is oversimplification to the point of falsification. There was no return but a continuous development from 1859 to now, for Darwinism, neo-Darwinism, and synthetic theories formed an intellectual sequence that always had adherents and that continuously progressed even when it had more opponents than it does now. The synthetic theory also includes essential elements that are not only non-Darwinian (quite a few) but also even anti-Darwinian (only a few).

He attributes the long delay of the synthesis to several factors:

> The fact that there were many biologists who did not follow what became eventually the mainstream of evolutionary thought (there was a "delay" only in that sense) was the result of many factors, but most essentially of two:
>
> 1. For a long time neo-Lamarckism (as distinct from true Lamarckism, by the way) was at least a conceivable alternative to neo-Darwinism and, as usual, convincing proof of a negative was hard to come by.
>
> 2. Rediscovery of Mendelian genetics, which may seem odd now that geneticists are among the principal founders and supporters of the synthetic theory, but in the crucial period from 1900 into the 1930s a majority of the authoritative geneticists were opposed to

some, at least, of the theoretical views then developing from neo-Darwinism to synthetic.

Since I think that the synthesis was not remarkably delayed but developed normally or even rather rapidly as the history of science goes, I do not think one can single out specific factors responsible for the rate of acceptance. There was certainly a period of some confusion around 1900 to 1930, but so was there from 1859 to 1900 and farther back, because of a multiplicity of hypotheses and even full-fledged theories about evolution. There is less confusion since full (even early) development of the synthetic theory, but much still exists, for example, among biochemists. Development of the synthetic theory required more and better communication, but I do not see that the communication deteriorated after the 1880s. In some respects it became more or less continually better.

In some respects, however, it has gotten worse since the development of the synthetic theory. Good communication between, say, systematists and geneticists was hardly possible between the founding of genetics (somewhat arbitrarily in 1865) and 1900 when most practitioners, notably including Mendel, had no notion about what if anything the two fields had to do with each other and when Darwin's interest in genetics only led him astray. Now the average molecular biologist and the average organismal biologist may communicate badly because neither really understands the other's field. The "delay" was not because "the best minds" went elsewhere. That evolutionary theory *did* continue to advance is because some, at least, of the best minds did *not* go elsewhere. Lack of communication may affect some good minds, but it is typical of mediocrity.

I next asked several questions: What evolutionary problems seemed to have been particularly intractable in the period just prior to the evolutionary synthesis? The role of natural selection? The origin of new species? The nature of genetic variation? The origin of new evolutionary types (or of new structures)? The role of the environment? As Simpson responded:

I cannot place these in a really fixed order, but important and then especially difficult problems included:

The origin of new "types" and structures, or taxonomically of higher taxa;

The almost systematic absence of major "missing links" in the fossil record; the reality or only apparent occurrence of saltations;

The identification of necessary and sufficient evolutionary factors (such as mutation, recombination, selection, or environmental change);

The nature of adaptation, its discontinuities, overlaps, and so forth;

Determination and significance of rates of evolution;

Development of a truly evolutionary taxonomic concept of species;

The reality or possibly spurious nature of such supposed factors as racial inertia and momentum in evolution or supposed specific and higher category "life-cycles"; the prevalence and causes of extinction (little progress has yet been made on causes);

Overall the congruence (or not) of the history of life with theories and concepts from neontology, especially in systematics and genetics.

What I personally made of these, and various other problems, is documented in my publications, especially *Tempo and Mode* and *Major Features of Evolution*.

Simpson also discussed one of the most frequent arguments of Darwin's opponents, that the selection of continuous variation could not explain the origin of entirely new structures with new functions. He wrote:

This argument did bother me, but I soon decided that it was not cogent. I did this in part by considering such prime examples as the vertebrate eye, which by comparative study can readily be shown to be plausibly derivable by selection on (essentially, or in a population sense) continuous variation. This is one of the bonuses of good training in comparative morphology.

Did the emergence of man pose a particularly difficult problem?

If other mammals originated by evolution, I never saw any reason to doubt that man did too. (I got over the effects of Presbyterian theology before I was far into my teens.)

Simpson first learned about the Russian population genetics (Chetverikov school) in the thirties:

I did not know about Russian genetics until the late 1930s when I started reading first Dobzhansky, then Timoféeff-Ressovsky. Except for Dobzhansky, I was not much influenced by them directly, at least. Doubtless through mere ignorance, I have never gotten much from Chetverikov. Russian paleontologists are strong on facts, weak on theories. (Some Russian zoologists and other biologists are now doing very interesting theoretical studies.)

He replied to my question about the concept of mutation:

Mutation was of course extensively treated in my undergraduate course in biology. I can quote the views of a teacher from a class note (1922): "Mutation is a heritable variation due to a fundamental change in the germ plasm independent of segregation and crossing-over." Of course at that time no one knew what a mutation was,

only what some of them did. For the time, that was a good definition. I met de Vries's theory at about the same time, but I do not
remember ever considering it likely.

I was early told, and I believed, that mutations are usually deleterious. I do not remember ever being told, or believing, that they are
always so. I also early encountered an opinion that they are usually
neutral (an idea curiously lost for a time and now resurrected by
some molecular biologists). I never believed that and don't now.

I don't remember when I encountered the concept of small, frequently invisible mutations, but I have entertained it for a long time.
In my university days the concept of an invisible mutation did not
come up, because if it had no visible effect there was no way to learn
of a mutation's existence. As far as the frequency of mutations is
concerned, all I remember of early statements is that mutations were
infrequent. I did not encounter quantitative statements until I began
serious reading of (somewhat) more modern genetics in the 1930s.

The idea was widespread in the nineteenth and early twentieth centuries that there were two kinds of variation, discontinuous variation
("sports," de Vries's "mutation") and continuous variation. I asked Simpson about his attitudes and those of his teachers:

> Not much was made of this point, but it was strongly implied in
> my lessons in the early 1920s that hereditary variation is essentially
> discontinuous and that apparent continuity is caused by environ-
> mental influences on a discontinuous variate with fairly numerous
> steps. With certain qualifications that is indeed correct, and I have
> had no reason to give up the essence of my original view of this mat-
> ter. Contrary to your question's implication, belief that discontinu-
> ity is equal to de Vriesian mutation does not really enter into the
> matter. (Lull did confuse the issue by calling small "graded varia-
> tions" Darwinian or continuous and large ones de Vriesian or dis-
> continuous.)
>
> The question of blending inheritance might also have been men-
> tioned, as it was connected with that of continuous variation. It used
> to be claimed that natural selection would not be effective on blend-
> ing inheritance and that this objection was removed only when it
> was discovered that all inheritance is particulate in some measure
> and at some level, a point sometimes still made in discussing the
> basis of the synthetic theory. As I was taught from the start that in-
> heritance is not blending, I was not affected, but in fact the argu-
> ment never was valid. Natural selection could act even if inheritance
> were blending.

I asked about his views, and those of his teachers, on Darwin's belief in
the inheritance of acquired characters.

Note of my teachers was an outspoken neo-Lamarckian, but

some expounded the theory as a possibility. As already noted, some of my early associates were neo-Lamarckians. I never accepted the inheritance of acquired characters.

Simpson took issue with some of my questions. For example, I asked: When did you realize that the environment acted solely as the agent of natural selection? He responded:

> The environment does not act solely as the agent of natural selection, because it also affects the development of the phenotype. Phenetic effects of the environment on an individual are, of course, not heritable, yet they can possibly affect natural selection.

And when I asked when he first realized that natural selection is the only direction-giving factor in evolution, he replied:

> This question is also poorly phrased. Natural selection is not "the only direction-giving factor in evolution." Some of the fence-sitters among my teachers and some of the neo-Lamarckians and mutationists among my later early associates believed firmly that natural selection, if effective at all, had a merely negative effect by eliminating the unfit and no role in originating the fit. This bothered me, but I figured out that it is not in fact true. I clearly stated in *Tempo and Mode* (1944) that natural selection is a creative factor, but I had reached that conclusion at least ten years earlier.

Simpson started to think about the relative importance of genes versus genotype versus individuals as the target of selection when he was a graduate student:

> My teachers in 1922-1926 already knew that selection necessarily acts on both genes and genotypes. What they didn't quite grasp, or at least didn't quite get across to me, was what selection really is, that is, genetically differential reproduction over a significantly long sequence of generations. I had to learn that for myself later.
>
> My main theoretical interest has always been on macroevolutionary problems. However, as a working taxonomist I have had to deal with species and infraspecific populations as well, and from graduate school on I have always been concerned with the numerous problems of species, especially those involving resemblances and differences between paleontological and neontological species (and practices concerning them), successive and contemporaneous species, and so on.
>
> I do not think the species problem has been solved during the synthesis, if by solution is meant a definition and criteria generally acceptable to biologists and generally applicable without equivocation to all kinds of organisms. In that sense there simply is no possible

solution. The so-called biological definition has no time dimension and is indeed nonevolutionary or preevolutionary, hence not really either an outcome of the synthetic theory of evolution or integral to it. A truly evolutionary concept of species, still open to various difficulties, has been reached by me and others and is integral with the synthetic theory.

CONTRIBUTORS

CONFERENCE PARTICIPANTS

INDEX

Contributors

Mark B. Adams
University of Pennsylvania
Philadelphia

Garland E. Allen
Washington University
St. Louis, Missouri

Ernest Boesiger*
University of Languedoc
Montpellier, France

Richard W. Burkhardt, Jr.
University of Illinois
Urbana

Hampton L. Carson
University of Hawaii
Honolulu

Frederick B. Churchill
Indiana University
Bloomington

William Coleman
University of Wisconsin
Madison

C. D. Darlington
Oxford University
Oxford, England

Theodosius Dobzhansky*
University of California
Davis

E. B. Ford
Oxford University
Oxford, England

Michael T. Ghiselin
Salt Lake City, Utah

Stephen Jay Gould
Harvard University
Cambridge, Massachusetts

Viktor Hamburger
Washington University
St. Louis, Missouri

Richard C. Lewontin
Harvard University
Cambridge, Massachusetts

Camille Limoges
University of Montreal
Montreal, Canada

Ernst Mayr
Harvard University
Cambridge, Massachusetts

William B. Provine
Cornell University
Ithaca, New York

Bernhard Rensch
Westfälische Wilhelms-Universität
Münster, Germany

Dudley Shapere
University of Chicago
Chicago, Illinois

G. Ledyard Stebbins
University of California
Davis

Alexander Weinstein
Harvard University
Cambridge, Massachusetts

* Deceased.

467

Conference Participants

In the workshop of May 23-25, 1974:

Ernst Mayr, chairman
Mark B. Adams
Ernest Boesiger*
Peter Buck
Richard W. Burkhardt, Jr.
Hampton L. Carson
William Coleman
C. D. Darlington
Theodosius Dobzhansky*
E. B. Ford
Bentley Glass
Stephen Jay Gould
Gerald Holton

David L. Hull
E. David Kohn
I. Michael Lerner*
Richard C. Lewontin
Camille Limoges
William B. Provine
Dudley Shapere
Otto Solbrig
G. Ledyard Stebbins
Frank Sulloway
Robert L. Trivers
Alexander Weinstein

In the workshop of October 11-12, 1974:

Ernst Mayr, chairman
Mark B. Adams
Garland E. Allen
Frederick B. Churchill
William Coleman
Irven DeVore
Michael T. Ghiselin
Stephen Jay Gould
John C. Greene
Wayne Gruner
Viktor Hamburger
E. David Kohn

Richard C. Lewontin
Camille Limoges
Everett C. Olson
Ronald Overmann
William B. Provine
Bobb Schaeffer
Dudley Shapere
Frank Sulloway
Daniel Todes
Robert L. Trivers
Alexander Weinstein

*Deceased.

Index